A
COMPLETE GCSE
MATHEMATICS

BASIC COURSE

Second Edition

By the same author

A Complete GCSE Mathematics — General Course
A Complete GCSE Mathematics — Higher Course

A Complete O-level Mathematics
New Comprehensive Mathematics for O-level
CSE Mathematics 1 The Core Course
CSE Mathematics 2 The Special Topics
A Concise CSE Mathematics

Arithmetic for Commerce
A First Course in Statistics

Revision Practice in Algebra
Revision Practice in Arithmetic
Revision Practice in Geometry and Trigonometry
Revision Practice in Multiple-Choice Maths Questions
Revision Practice in Short-Answer Questions in O-level Maths
Revision Practice in Statistics

with S. Llewellyn   Mathematics The Basic Skills

with G.W. Taylor   BTEC First Mathematics for Technicians
BTEC National NII Mathematics for Technicians
BTEC National NIII Mathematics for Technicians

A
COMPLETE GCSE
MATHEMATICS

BASIC COURSE

A. Greer

Formerly Senior Lecturer,
Gloucestershire College of Arts and Technology

SECOND EDITION

Stanley Thornes (Publishers) Ltd

First published in 1987 by
Stanley Thornes (Publishers) Ltd
Old Station Drive
Leckhampton
CHELTENHAM GL53 0DN

Reprinted 1987
Reprinted 1988
2nd Edition 1989

British Library Cataloguing in Publication Data

Greer, A (Alex)
    A complete GCSE mathematics: basic course.
    2nd ed.
    1. Mathematics
    I. Title
    510

    ISBN 0-7487-0076-5

Typeset by Tech-Set, Gateshead, Tyne & Wear in 10/12pt Century.
Printed and bound in Great Britain at The Bath Press, Avon.

Contents

Preface

This book is one of a series of three books intended to cover all the topics prescribed by the National Criteria in Mathematics at three levels. It is suitable for those students who expect to obtain a Grade E, F or G in the General Certificate of Secondary Education (GCSE), the target grade being F. Also I have tried to cover the additional topics introduced by the five regional Examining Groups.

Because this is a revision book the sections on Arithmetic, Algebra, Geometry, Graphical Work and Statistics have been dealt with separately. The emphasis is on a simple approach—a teacher and the class may work, if desired, through the book, chapter by chapter.

A large number of examples and exercises have been included which supplement and amplify the text. Many practical examples of mathematics in the real world have been used.

At the end of most chapters there is a 'miscellaneous exercise' which has been divided into two parts, Section A and Section B. This is because the examining boards have devised overlapping examination papers. The basic-level students will take papers 1 and 2, the general-level students will take papers 2 and 3, whilst the higher-level students will take papers 3 and 4. In this book the questions in Section A of the miscellaneous exercises are intended to cover paper 1, whilst those in Section B are intended to cover paper 2.

Also at the end of most chapters are sets of multi-choice questions and mental tests. It is hoped that these, together with the miscellaneous exercises, will give the student confidence in answering examination papers.

I would like to thank my son David for working out the answers.

A. Greer   Gloucester 1989

Coursework

We are aware of the demands made upon GCSE candidates and it is not suggested that they should attempt all the tasks which are offered in the chapter on coursework. The investigations and other studies are provided as a guide for those students who are not sure what is expected of them in this part of the examination.

The readers may like to choose a few of the suggestions for practice. We hope the topics will provide some inspiration when candidates attempt their own particular coursework projects.

I was pleased to accept the author's invitation to provide a section dealing with coursework and it is a source of personal satisfaction to be associated with this very successful series of books.

C.H. Hopkins   Gloucester 1989

Acknowledgements

The author and publishers are grateful to the following who provided material for inclusion:

British Gas for the bill, page 103.

British Rail for the timetables, pages 114 and 210.

City of Gloucester Buses for the network, page 211 and the timetable, page 214.

MEB for the bill, page 106.

The Post Office for the postal rates, page 216.

Whole Numbers

Place Value

As young children we start learning about numbers by counting objects 1, 2, 3, 4, 5, 6, 7, 8, 9, 10. These numbers are called **counting** or **natural numbers.**

Whole numbers are the numbers 0, 1, 2, 3, 4, 5, etc.

In our number system we use the symbols 0, 1, 2, 3, 4, 5, 6, 7, 8 and 9.

The positions of the figures in the number give the value of the number:

Hundreds	Tens	Units
5	7	6

576 is five hundred and seventy-six.

Each figure in the number has a **place value** because it is so many hundreds, tens or units.

In 817 the 7 means 7 units.

In 178 the 7 means 7 tens, i.e. 70.

In 781 the 7 means 7 hundreds, i.e. 700.

In these three numbers the 7 stands for a different value when it is in a different place.

It is important to remember that the figure with the smallest place value is always on the extreme right of the number. The figure with the greatest place value is always on the extreme left of the number.

Notice that in the number 4945, the first 4 has a value 4 thousand, i.e. 4000, whilst the other 4 has a value of 4 tens, i.e. 40.

Nought or zero plays an important part in our number system.

> 306 means 3 hundred and 6 units. The nought keeps the place for the missing tens.

> 360 means 3 hundred and sixty. The nought keeps the place for the missing units.

The Number Line

Whole numbers can be shown on a number line (Fig. 1.1). As we move to the right along the number line the numbers increase in value.

Fig. 1.1

Exercise 1.1 is a revision exercise which will tell you if you have understood the number system.

Write the following in figures:

1. forty-seven

2. four hundred and thirty-four

3. four hundred and three

1

4. two hundred and sixty

5. three thousand four hundred and sixty-three

6. seven thousand eight hundred and thirteen

7. seventy-two thousand

8. twenty-five thousand three hundred and thirty-five

9. seven hundred thousand

10. four hundred and thirty-three million

11. five hundred and sixty-three million three hundred and forty thousand five hundred and seven

12. nine hundred and eighty-nine million

Write the following numbers in words:

13. 354

14. 608

15. 202

16. 340

17. 8765

18. 72 000

19. 562 000

20. 905 000

21. 87 509

22. 204 037

23. 690 058

24. 6 893 024

25. 6 013 704

26. Put the following numbers in order of size, smallest first: 321, 123, 213, 312, 231, 132.

27. What is the place value of the figure 4 in each of the following numbers?
    (a) 408   (b) 114   (c) 341
    (d) 439 081   (e) 34 736 090

Adding Whole Numbers

(1)   To add a line of numbers

Example 1

Add 5, 4, 7 and 6.

$$5 + 4 + 7 + 6 = 22$$

Working in your head, add the first two numbers $(5 + 4 = 9)$. Then add the third number $(9 + 7 = 16)$. Finally add the fourth number $(16 + 6 = 22)$.

The order in which the numbers are added does not affect the answer.

Hence to check the answer the numbers may be added in reverse order.

Thus

$$6 + 7 + 4 + 5 = 22$$

Because the same answer has been produced by using two different methods we can say with confidence that the correct total is 22.

(2)   To add a column of figures

Example 2

By forming a column, add 8, 3, 5 and 4.

$$
\begin{array}{r}
8 \\
3 \\
5 \\
+\ \underline{4} \\
\underline{20}
\end{array}
$$

Working in your head, start at the bottom and add the first two numbers $(4 + 5 = 9)$. Next add the third number $(9 + 3 = 12)$. Finally add the fourth number $(12 + 8 = 20)$.

To check the answer, start at the top and add downwards.

There are many ways in which the process of addition may be specified.

Some of these are given in Example 3.

Example 3

(a) Add 7, 5 and 4.

$$7 + 5 + 4 = 16$$

(b) Find the sum of 5, 6 and 8.

$$5 + 6 + 8 = 19$$

Hence the sum of 5, 6, and 8 is 19.

(c) If 9 is added to 8, what is the total?

$$9 + 8 = 17$$

Hence when 9 is added to 8 the total is 17.

(d) Increase 7 by 5.

$$7 + 5 = 12$$

Hence 7 increased by 5 is 12.

Exercise 1.2

This exercise should be tackled mentally.

Work out:

1. $3 + 2 + 5 + 7$
2. $7 + 3 + 5 + 6$
3. $11 + 7 + 9 + 3$
4. $15 + 6 + 4 + 5$
5. $13 + 7 + 6 + 0 + 5$
6. Find the sum of 2, 7 and 8.
7. Add 3, 7, 5 and 8.
8. What is the total of 8, 7, 5 and 3?
9. Add 6 to 7.
10. What is the sum of 5, 8, 6 and 3?
11. Increase 3 by 8.
12. Calculate 5 plus 0 plus 3 plus 4.
13. Add 3, 7, 9, 8 and 11.
14. If 9 is added to 4, what is the total?
15. Work out the total of 8, 2, 7, 3 and 9.

Rounding Numbers

When numbers greater than 10 have to be added, subtracted, multiplied or divided a calculator is often used. A calculator in good condition does not make mistakes but human beings do. We need, therefore, a method for establishing that the answer produced by a calculator is sensible. The rounding of numbers allows us to do this.

It is usual to round off

a number between 10 and 100 to the nearest number of tens

a number between 100 and 1000 to the nearest number of hundreds

a number between 1000 and 10 000 to the nearest number of thousands

and so on.

Example 4

(a) 29 would be rounded up to 3 tens, i.e. 30.

(b) 643 would be rounded down to 6 hundreds, i.e. 600.

(c) 8748 would be rounded up to 9 thousands, i.e. 9000.

When a number is half-way between tens, hundreds, thousands, etc. the number is always rounded up.

Example 5

(a) 7500 would be rounded up to 8 thousands, i.e. 8000.

(b) 650 would be rounded up to 7 hundreds, i.e. 700.

Exercise 1.3

Round off the following numbers to the nearest number of tens:

1. 42   2. 8   3. 85   4. 92

5. 76

Round off the following numbers to the nearest number of hundreds:

6. 698   7. 703   8. 66   9. 350

10. 885

Round off the following numbers to the nearest number of thousands:

11. 2631   **12.** 791   **13.** 8179

14. 8500   **15.** 5765

The Electronic Calculator

There are very many different types of calculators on the market but one like that shown in Fig. 1.2 is good enough for most purposes.

The keyboard of this calculator has 10 number keys marked 0, 1, 2, 3, 4, 5, 6, 7, 8 and 9, and there are six function keys marked $+, -, \times, \div, \sqrt{}$ and %. In addition there are four memory keys M+ (memory plus), M− (memory minus), MR (memory recall) and MC (memory clear). There is an = key and two clear keys marked CE (correct error) and C (clear the calculator).

Fig. 1.2

Addition and Subtraction of Whole Numbers

Before or after attempting to add and subtract using a calculator, we should always make a rough estimate of the answer to make sure that the answer produced by the calculator is sensible. To make this rough

estimate we round off the numbers to the nearest ten, hundred, thousand, etc.

Example 6

Add 63, 72 and 348.

To make a rough estimate we round off the numbers to 60, 70 and 300.

Rough estimate,

$$60 + 70 + 300 = 430$$

The calculator is used as follows. (Take care to press the keys in the proper sequence and to check the display as you proceed.)

Input	Display
63	63.
+	63.
72	72.
+	135.
348	348.
=	483.

Hence

$$83 + 72 + 348 = 483$$

The rough estimate shows that the answer is sensible, but is it accurate?

Any answer produced by using a calculator should be carefully checked. One way of checking the answer to Example 6 is to input the numbers into the calculator in reverse order, i.e. 348 + 72 + 83. If the same answer is produced we can be confident that the answer is correct.

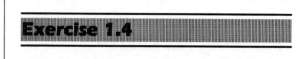

Exercise 1.4

Using a calculator, add the following numbers:

1. 147 and 2187

2. 4386 and 118

3. 156, 289 and 1385

4. 48, 276, 2359 and 89

5. 508, 72, 4883 and 18

6. 53, 281, 7469, 239 875, 2 368 754 and 93

7. 6 328 123, 15 768, 798 632, 54 698 and 900 599

8. Find the value of 287 + 3219 + 75.

9. Find the sum of 286, 73, 899 and 4387.

10. Calculate the total of 345 672, 18 769, 278 945 and 16 754 321.

Subtracting means taking away. — is the subtraction sign meaning minus. So 8 − 3 = 5 is read as eight minus three equals five or take three from eight.

Example 7

(a) Take 5 from 9.

$$9 - 5 = 4$$

(b) What is 8 minus 5?

$$8 - 5 = 3$$

(c) Subtract 6 from 9.

$$9 - 6 = 3$$

The **difference** between two numbers is the larger number minus the smaller number.

Example 8

Find the difference between 5 and 9.

$$\text{Difference} = 9 - 5 = 4$$

Exercise 1.5

This exercise should be done mentally.

1. Subtract 5 from 7.

2. Take 3 from 8.

3. Find the difference between 8 and 2.

4. How many less than 8 is 5?

5. Subtract 9 from 16.

6. Calculate 9 minus 4.

7. What is 7 minus 3.

8. Find the value of 15 − 8.

Example 9

Subtract 346 from 598.

$$
\begin{array}{r}
598 \\
- 346 \\
\hline
252 \\
\end{array}
$$

To check the subtraction, the bottom two numbers when added will equal the top number if the subtraction is correct. That is

$$
\begin{array}{r}
252 \\
+ 346 \\
\hline
598 \\
\end{array}
$$

So we can be confident that

$$598 - 346 = 252$$

Example 10

Use a calculator to subtract 763 from 2942.

Input	Display	
2942	2942.	
−	2942.	
763	763.	
=	2179.	(this is the answer)
+	2179.	
763	763.	
=	2942.	(this is the check)

Hence

$$2942 - 763 = 2179$$

If the answer is correct then

$$2179 + 763 = 2942$$

This check has been included in the program above.

Exercise 1.6

A calculator should be used for this exercise.

1. Take 497 from 983 and check the answer.

2. Calculate the difference between 79 and 3647.

3. Subtract 8657 from 20 243.

4. Take 18 736 from 248 987.

5. Work out $142\,879 - 98\,263$.

6. Work out 18 234 minus 4387.

7. What is the difference between 9874 and 11 468?

8. Find the value of $7498 - 6899$.

Positive and Negative Numbers

Positive numbers have a value greater than zero. They either have no sign in front of them or a $+$ sign. Some examples of positive numbers are $5, +9, 8, +21, +293$ and 678.

Negative numbers have a value less than zero. They are *always* written with a $-$ sign in front of them. Some examples of negative numbers are $-4, -15, -168$ and -783.

Directed numbers or **integers** are whole numbers, including zero and negative numbers. Some examples of integers are $-28, -15, 0, 2, +9, 24$ and -892. Directed numbers are used on thermometers to show Celsius temperatures above and below freezing point (Fig. 1.3). They are also used on graphs (Fig. 1.4). It is also possible to have negative time, for example -12 seconds to blast off.

An overdraft at a bank can be regarded as negative money. For instance, $-£200$ would represent an overdraft of £200 (i.e. the customer owes the bank £200).

In football leagues, goal difference is used. For instance if a team scores, in a season, 48 goals and concedes 25 goals, the goal difference is $48 - 25 = 23$. But if the team scores only 21 goals and concedes 28 goals, the goal difference is $21 - 28 = -7$.

Fig. 1.3

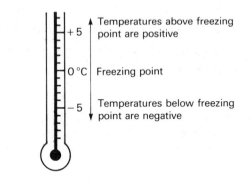

Fig. 1.4

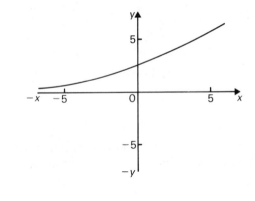

The Number Line

Directed numbers may be represented on a number line (Fig. 1.5). Numbers to the right of zero are positive whilst numbers to the left of zero are negative.

Fig. 1.5

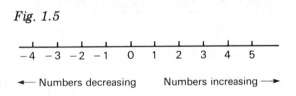

Negative numbers become smaller the further we move to the left of zero. For instance -8 is smaller in value than -3 and -57 is smaller in value than -24.

Positive numbers, however, become greater in value the further we move to the right of zero. For instance 12 has a greater value than 7 and $+84$ has a greater value than $+35$.

Example 11

Rearrange in order of size, smallest first, the integers $-3, 5, -7, -2, 8, -4$ and 3.

The order is

$$-7, -4, -3, -2, 3, 5 \text{ and } 8$$

Exercise 1.7

Copy the number line shown in Fig. 1.6 and use it to answer the following questions:

Fig. 1.6

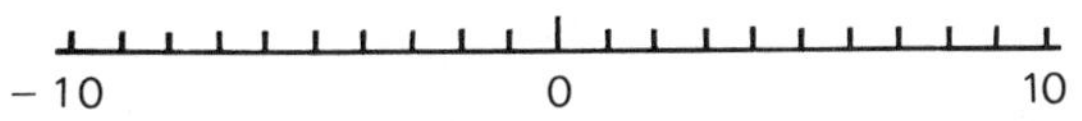

1. Mark the number line with the integers $-6, -3, +3$ and 7.

2. Put in order, lowest first, the integers $-1, 7, +5, -5, -7, +9, 6$ and -3.

3. Plants in a greenhouse will die if the temperature falls below $-2\,^\circ$C. Which of these temperatures means death for the plants: $5\,^\circ$C, $-3\,^\circ$C, $-5\,^\circ$C, $8\,^\circ$C, $0\,^\circ$C and $-1\,^\circ$C?

4. At 6 a.m. the temperature was $-4\,^\circ$C. At 11 a.m. the temperature was $7\,^\circ$C. How much has the temperature risen?

5. A thermometer reading is $-5\,^\circ$C. If the temperature falls by another $3\,^\circ$C, what is the new temperature?

6. Write down in order of size, largest first, the numbers $-3, 0, 5, 3$ and -5.

7. A team in the football league scored 58 goals during the course of the season but conceded 62. What was the goal difference?

8. The countdown for a rocket is -5 seconds to blast off. How long will it take for the rocket to have been in free flight for 7 seconds?

Combined Addition and Subtraction

It is the sign in front of a number which tells us what to do with it.

For example $9 - 7 + 4$ means nine take away seven and add on four. Since addition can be done in any order, we could add four to nine and take away seven. That is

$$9 - 7 + 4 = 9 + 4 - 7 = 13 - 7 = 6$$

When we have a string of positive and negative numbers we first add all the positive numbers together. Next we add all the negative numbers together. We are then left with two numbers, one positive and the other negative. The answer is then obtained by subtraction.

Example 12

Work out the value of
$9 + 7 - 6 - 5 + 3 + 4 - 8$.

$$9 + 7 - 6 - 5 + 3 + 4 - 8$$
$$= (9 + 7 + 3 + 4) - (6 + 5 + 8)$$
$$= 23 - 19$$
$$= 4$$

Sometimes the negative number will be larger than the positive number. In such cases the answer will be a negative number.

Example 13

Find the value of $3 - 4 + 7 + 5 - 7 - 9$.

$$3 - 4 + 7 + 5 - 7 - 9$$
$$= (3 + 7 + 5) - (4 + 7 + 9)$$
$$= 15 - 20$$
$$= -5$$

The number line (Fig. 1.7) shows how this answer has been obtained.

Fig. 1.7

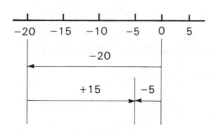

Exercise 1.8

Find the value of each of the following:

1. $8 + 7 + 5 - 3 - 4 - 6$

2. $8 + 9 - 7 - 6 - 5 + 4$

3. $-8 - 6 + 7 - 5 + 9 - 2$

4. $-2 - 6 + 3 - 9 + 5 - 11$

5. $18 - 9 - 7 + 6 - 3 - 8 + 11$

When using a calculator we can input positive and negative numbers as they occur. There is no need to add the positive and negative numbers separately.

Example 14

Using a calculator find the value of $23 - 27 - 32 + 75 + 29 - 19$.

Rough estimate,

$$= 20 - 30 - 30 + 80 + 30 - 20$$
$$= 50$$

Input	Display
23	23.
−	23.
27	27.
−	−4.
32	32.
+	−36.
75	75.
+	39.
29	29.
−	68.
19	19.
=	49.

This is the answer which is sensible according to the rough estimate.

To check the answer (it is 49) input the numbers in reverse order, that is

$$-19 + 29 + 75 - 32 - 27 + 23$$

If we obtain the same answer we can be confident that

$$23 - 27 - 32 + 75 + 29 - 19 = 49$$

A negative number can be displayed on a calculator by using the method shown in Example 15.

Example 15

Input the number -59 on a calculator.

Input	Display
−	0.
59	59.
=	−59.

Example 16

Find the value of $-298 - 53 + 76 - 81$.

Rough estimate,

$$-300 - 50 + 80 - 80 = -350$$

Input	Display
—	0.
298	298.
—	−298.
53	53.
+	−351.
76	76.
—	−275.
81	81.
=	−356.

To check the answer input the numbers in reverse order, i.e.

$$-81 + 76 - 53 - 298 = -356$$

Hence

$$-298 - 53 + 76 - 81 = -356$$

Note that the rough estimate shows that the answer is sensible.

Exercise 1.9

Use a calculator to work out the value of each of the following. In each case check your answer.

1. $20 - 4 + 15 - 35 + 28 - 7$

2. $-30 + 55 + 11 + 9 - 17 - 25 - 43$
   $- 12 + 27$

3. $-52 + 41 - 83 + 78 + 99 - 52 - 31$

4. $-98 - 71 + 183 + 63 + 85 - 95 + 17$

5. $-278 - 159 - 367 - 874 + 653$
   $+ 886 - 75 + 653$

6. $-4983 - 278 - 6985 + 8762 - 789$
   $+ 2347 - 8762 + 12\,678$

7. $15\,785 - 11\,654 - 7890 + 17\,564$
   $+ 8762 - 7534 - 2632$

Multiplication

$\times$ is the multiplication sign, meaning multiplied by, or times. Thus $6 \times 8 = 48$ which is read as six times eight equals forty-eight.

Multiplication is a quick way of adding equal numbers.

$$7 + 7 + 7 + 7 = 28$$

$4 \times 7 = 28$ (i.e. four sevens equal twenty-eight)

28 is called the **product** of 4 and 7.

The product of 4, 5 and 8 = $4 \times 5 \times 8 = 160$.

The order in which the numbers are multiplied does not affect the product.

This fact may be used:

(1) To simplify multiplication when it is done mentally:
$$4 \times 7 \times 25 = (4 \times 25) \times 7$$
$$= 100 \times 7$$
$$= 700$$

(2) To check a product by interchanging numbers:
$$56 \times 39 = 2184$$
$$39 \times 56 = 2184$$

The table below is useful when revising your multiplication tables, which are needed when doing mental arithmetic.

THE MULTIPLICATION TABLE

1	2	3	4	5	6	7	8	9	10
2	4	6	8	10	12	14	16	18	20
3	6	9	12	15	18	21	24	27	30
4	8	12	16	20	24	28	32	36	40
5	10	15	20	25	30	35	40	45	50
6	12	18	24	30	36	42	48	54	60
7	14	21	28	35	42	49	56	63	70
8	16	24	32	40	48	56	64	72	80
9	18	27	36	45	54	63	72	81	90
10	20	30	40	50	60	70	80	90	100

To find, for instance, the product of 6 and 8, look along the 6th row and down the 8th column. There you will find the number 48 which is the product of 6 and 8, i.e. $6 \times 8 = 48$.

Exercise 1.10

This exercise should be done mentally.

Find the values of

1. 3×5 2. 6×7

3. 8×9 4. $2 \times 3 \times 6$

5. $4 \times 5 \times 7$

Find the products of

6. 6 and 9   7. 7 and 8

8. 3 and 5   9. 8, 5 and 3

10. 2, 5 and 7

Multiplying using a Calculator

When the multiplication is too difficult to do mentally a calculator may be used.

Example 17

Find the value of 234×392.

For a rough estimate we will take

$$200 \times 400 = 80\,000$$

Input	Display
234	234.
×	234.
392	392.
=	91728.

Hence $234 \times 392 = 91\,728$

The rough estimate shows that the answer is sensible. To check its accuracy multiply the numbers in reverse order, i.e. 392×234.

Example 18

Calculate the product of 37, 18 and 43.

Rough estimate,

$$40 \times 20 \times 40 = 32\,000$$

Input	Display
37	37.
×	37.
18	18.
×	666.
43	43.
=	28638.

Hence

$$37 \times 18 \times 43 = 28\,638$$

The rough estimate shows that the product is sensible. To check the answer input the numbers in reverse order, i.e. $43 \times 18 \times 37$.

Exercise 1.11

A calculator should be used for this exercise.

1. Multiply 107 by 29.

2. Work out $436 \times 519 \times 15$.

3. Find the value of $634 \times 28 \times 122$.

4. What is the product of 347 and 192?

5. Find the product of 3872 and 4163.

6. Work out $987 \times 326 \times 21$.

7. Multiply 62 358 by 52.

8. What is 18 793 times 407?

Multiplication by 10, 100, 1000

In the table below notice how the number 587 differs from the number 5870.

Thousands	Hundreds	Tens	Units
	5	8	7
5	8	7	0

By writing a nought after the last figure of 587, each figure is moved one place to the left and takes a place value which is ten times the previous value. That is, 5870 is ten times larger than 587.

Therefore to multiply a whole number by 10, write a nought after the last figure of the number. Then move all the figures one place to the left.

	Hundreds	Tens	Units
		6	4
64 × 10	6	4	0

Similarly to multiply a whole number by 100, write two noughts after the last figure in the number. The figures will then be moved two places to the left as shown below.

Thousands	Hundreds	Tens	Units
		8	6

Thousands	Hundreds	Tens	Units
86 × 100	8	6	0	0

To multiply by 1000, three noughts are written after the last figure in the number. Thus

$$925 \times 1000 = 925\,000$$

Although we can multiply by 10, 100 and 1000 using a calculator, it is much quicker to multiply mentally. This will help you considerably when dealing with decimal currency and metric measurements.

Exercise 1.12

This exercise should be tackled mentally.

Multiply each of the following numbers by 10, 100 and 1000:

1.	8	**2.**	15	**3.**	80
4.	723	**5.**	817	**6.**	4321
7.	5060	**8.**	3000	**9.**	7050

10. 20 000

Division

Division is concerned with dividing into equal parts. ÷ is the division sign meaning divided by.

$35 \div 7$ means 'How many sevens are there in 35?' We can find this out by repeatedly taking 7 away from 35:

$$35 - 7 = 28$$
$$28 - 7 = 21$$
$$21 - 7 = 14$$
$$14 - 7 = 7$$
$$7 - 7 = 0$$

So there are 5 sevens in 35 and we say

$$35 \div 7 = 5$$

A quicker way is to use the multiplication tables.

We know that $35 = 5 \times 7$.

Hence $35 \div 7 = 5$.

There are several ways of writing division problems. The two commonest ways of writing, for instance, 18 divided by 3 are

$18 \div 3$ and $\dfrac{18}{3}$.

Most division problems will be solved by using a calculator.

Example 19

Divide 567 by 21.

Rough estimate,

$$600 \div 20 = 30$$

Input	Display
567	567.
÷	567.
21	21.
=	27.

Therefore $567 \div 21 = 27$.

The result of a division can always be checked by multiplying back. Thus in Example 19, if the answer produced by the calculator is correct then 27×21 should equal 567. The division and the check are shown below.

Input	Display	
567	567.	
÷	567.	
21	21.	
×	27.	(this is the answer)
21	21.	
=	567.	(this is the check)

Exercise 1.13

Work out each of the following and check your answers.

A calculator is essential for this exercise.

1. Divide 216 by 3.

2. $235 \div 5$

3. $255 \div 17$

4. Divide 912 by 16.

5. $\dfrac{1408}{44}$

6. $\dfrac{11\,736}{18}$

7. $8773 \div 283$

8. 754 944 divided by 768.

Remainders

Sometimes one number will not divide into another number a whole number of times. In such cases a remainder is left.

Example 20

Divide 22 by 5.

Since $4 \times 5 = 20$

$22 \div 5 = 4$ remainder 2

Remainders are often required when working with fractions.

Exercise 1.14

This exercise should be tackled mentally.

Work out the answers to the following and write down the remainders:

1. $15 \div 2$ 2. $23 \div 3$ 3. $19 \div 4$

4. $11 \div 6$ 5. $33 \div 5$ 6. $47 \div 8$

7. $29 \div 9$

The Order of Arithmetical Operations

In working out problems such as

$$3 \times (2 + 5)$$
$$16 - (8 - 3)$$
$$3 \times 5 + 2$$
$$3 + 6 \div 2$$

a certain order of operations must be observed.

(1) First work out the contents of any brackets:

$$3 \times (2 + 5) = 3 \times 7 = 21$$
$$16 - (8 - 3) = 16 - 5 = 11$$

We get the same answer (and sometimes make the working easier) if we expand the bracket in the way shown below:

$$3 \times (2 + 5) = 3 \times 2 + 3 \times 5$$
$$= 6 + 15$$
$$= 21$$

Sometimes the multiplication sign is not shown

$$4(9 - 3) \text{ means } 4 \times (9 - 3)$$
$$4(9 - 3) = 4 \times 6$$
$$= 24$$

or $$4(9 - 3) = 4 \times 9 - 4 \times 3$$
$$= 36 - 12$$
$$= 24$$

(2) Multiplication and/or division must be done before addition and subtraction:

$$11 - 12 \div 4 + 3 \times (6 - 2)$$
$$= 11 - 12 \div 4 + 3 \times 4$$
$$= 11 - 3 + 12$$
$$= 20$$

Exercise 1.15

Work out each of the following:

1. $8 + 5 \times 2$
2. $3 \times 7 - 2$
3. $6 \times 5 - 2 \times 3 + 4$
4. $7 \times 5 - 16 \div 4 + 7$
5. $3 + 4 \times (5 - 4)$
6. $8 - 3 \times (4 - 2)$
7. $11 - 12 \div 4 + 3(7 - 2)$
8. $15 \div (3 + 2) - 8 \times 3 + 7(5 + 3)$
9. $35 \div (20 - 15)$
10. $36 \div 3 - 36 \div 9$

Some Important Facts

(1) When zero is added to any number the sum is the number.

$$378 + 0 = 378 \quad \text{and} \quad 0 + 59 = 59$$

(2) When a number is multiplied by 1 the product is the number.

$$5932 \times 1 = 5932 \quad \text{and} \quad 1 \times 967 = 967$$

(3) When a number is multiplied by zero the product is zero.

$$178 \times 0 = 0 \quad \text{and} \quad 0 \times 5943 = 0$$

(4) It is impossible to divide a number by zero. The result is meaningless. Thus $53 \div 0$ is meaningless.

(5) If zero is divided by any number the result is zero. Thus $0 \div 89 = 0$ and $0 \div 9548 = 0$.

Exercise 1.16

This exercise should be tackled mentally.

Work out the value of each of the following:

1. $798 + 0$
2. 3562×1
3. 0×60
4. $0 \times 3 + 5$
5. $2 \times 7 \times 1$
6. $9 \times 0 \times 8$
7. $15 - 9 \times 0 + 3$
8. $14 - 18 \div 6 + 0 \times (7 - 3)$
9. $0 \div 42$
10. $(5 - 5) \times 5$
11. $7 \times (9 - 3) + 0 \div 5$
12. $18 \times 0 + 3(7 - 7) + 6$

Using the Memory Keys on a Calculator

By using the memory keys on a calculator, quite complicated arithmetic problems can be solved.

Example 21

Calculate $15 \times 19 + 32 \div 8 - 14 \times 7$.

Input	Display
15	15.
×	15.
19	19.
M+	285.
32	32.
÷	32.
8	8.
M+	4.
14	14.
×	14.
7	7.
M−	98.
MR	191.

$$15 \times 19 + 32 \div 8 - 14 \times 7 = 191$$

Exercise 1.17

Using a calculator work out the values of each of the following:

1. $19 \times 27 + 23 \times 8$

2. $32 \times 21 - 15 \times 17$

3. $48 \div 6 - 54 \div 9$

4. $42 \times 31 + 54 \times 23 + 47 \times 11$

5. $39 \times 48 - 63 \times 12 - 14 \times 18$

6. $64 \times 128 - 96 \div 12 - 58 \times 24 + 63 \times 15$

7. $32 \times (15 + 17) - 18 \times (21 - 9)$

8. $58 \times (62 - 32) - 24 \times (15 + 19)$

Sequences

A set of numbers connected by some definite law is called a **sequence of numbers**. Each of the numbers in the sequence is called a **term of the sequence**.

Example 22

(a) Find the next term of the sequence 1, 5, 9, 13, . . .

Each term in the sequence is formed by adding 4 to the previous term so the sequence is 1, 5, 9, 13, (13 + 4), . . .

That is 1, 5, 9, 13, 17, . . .

(b) Write down the next term of the sequence 18, 15, 12, 9, . . .

Since each term is formed by taking 3 away from the previous term the sequence is 18, 15, 12, 9, (9 − 3), . . .

That is 18, 15, 12, 9, 6, . . .

(c) Find the next term of the sequence 3, 15, 75, 375, . . .

Because each term is formed by multiplying the previous term by 5, the sequence is 3, 15, 75, 375, (375 × 5), . . .

That is 3, 15, 75, 375, 1875, . . .

(d) In the sequence 128, 64, 32, ?, 8, ?, . . . Write down the terms denoted by a question mark.

Each term is formed by dividing the previous term by 2, so the sequence is 128, 64, 32, (32 ÷ 2), 8, (8 ÷ 2), . . .

That is 128, 64, 32, 16, 8, 4, . . .

The two terms denoted by a question mark are 16 and 4.

Exercise 1.18

Write down the next two terms for each of the following sequences:

1. 2, 10, 50, . . .

2. 1, 5, 9, 13, . . .

3. 3, 9, 15, 21, . . .

4. 176, 88, 44, . . .

5. 1, 3, 5, 7, . . .

6. 29, 25, 21, 17, . . .

7. 2, 6, 18, 54, . . .

8. 1, 4, 16, 64, . . .

Write down the terms denoted by a question mark in the following sequences:

9. 0, 2, 4, ?, ?, 10, 12, 14, ?

10. 324, 108, 36, ?, ?

11. 3, 12, 48, ?, 768, ?

12. 5, 11, 17, ?, 29, ?

13. 162, 54, 18, ?, ?

14. 6, 12, 24, ?, 96, ?, 384

15. 1, 6, ?, 16, 21, ?, 31, 36

Practical Problems Involving Whole Numbers

The examples which follow are all of a practical nature. They all depend upon whole numbers for their solution.

Example 23

(a) Four resistors in an electrical system are connected in series. Their values are 746, 1289, 963 and 275 ohms. Their total resistance is found by adding these four values. What is it?

$$\text{Total resistance} = 746 + 1289 + 963$$
$$+ 275$$
$$= 3273 \text{ ohms}$$

(b) A farmer digs 11 608 kilograms of potatoes. He puts them in bags, each bag containing 55 kilograms. How many bags does he fill?

$$\text{Number of bags filled} = 11\,608 \div 55$$
$$= 211 \text{ remainder } 3$$

So he fills 211 bags.

(c) A typist finds that she types 14 words on a line and 26 lines of typing on a page.

    (i)   How many words can she type on a page?

   (ii)   How many pages will she need to type in order to complete a manuscript consisting of 114 660 words?

    (i)   Number of words per page
$$= 14 \times 26 = 364$$

   (ii)   Number of pages $= 114\,660 \div 364$
$$= 315$$

Exercise 1.19

1. Fig. 1.8 shows a shaft which is used in a diesel engine. Calculate the dimension marked x.

Fig. 1.8

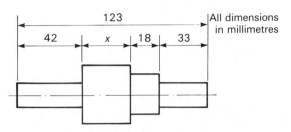

2. Two holes are drilled in a plate as shown in Fig. 1.9. What is the dimension marked x?

Fig. 1.9

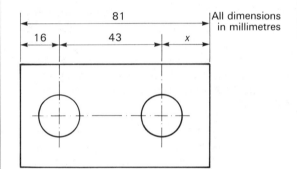

3. A small tank holds 875 litres of oil. If 59 litres are used, how much oil is left in the tank?

4. A housewife bought some furniture. She spent £132 on a wardrobe, £235 on a double bed and £98 on a table. What was the total cost of her purchases?

5. A screw has a weight of 9 grams. What is the weight of 895 such screws?

6. A reel of ribbon contains 78 metres initially. Lengths of 13 metres, 22 metres and 37 metres are cut off. How much ribbon remains?

7. A plank of wood 1836 millimetres long is cut into 102 equal parts. What is the length of each part?

8. A shopkeeper buys 1200 pens at a total cost of 7200 pence. How much does each pen cost?

9. A clothing shop on one day sells 3 suits costing £58 each, 5 coats costing £37 each and 8 pairs of trousers costing £13 each. How much are the total sales for the day?

10. To make a cake a housewife uses 400 grams of butter, 320 grams of caster sugar, 800 grams of self-raising flour, 200 grams of currants and 60 grams of mixed peel. What is the total weight of these ingredients?

11. A steel bar is 416 millimetres long. How many pieces each 16 millimetres long can be obtained from the bar?

12. In the first 2 hours of a shift an operator makes 29 soldered joints per hour. In the next 3 hours she makes 32 joints per hour. In the final 2 hours she makes 28 joints per hour. How many joints does the operator make in a 7-hour shift?

13. A typist can type 90 words per minute. She has to type a manuscript containing 630 000 words. How long will the typing take her?

14. A car travels 9 kilometres on 1 litre of petrol. How many litres are needed for a journey of 630 kilometres?

Miscellaneous Exercise 1

Section A

1. Consider the numbers 2, 6, 18, 54, 162.

   (a) These numbers form a pattern. What is the next number in the pattern?

   (b) If each number of the sequence can be used only once, which three numbers add up to 62?

2. On one winter's day the maximum temperature is 7 degrees. The minimum temperature is −3 degrees. By how many degrees has the temperature changed?

3. What is the remainder when 6372 is divided by 32?

4. A mail-order catalogue states that a suite of furniture can be purchased by 38 payments of £16. How much does a customer pay for the furniture?

5. The (cooking) instructions for cooking a duck are 'Cook for 20 minutes per pound with an additional 10 minutes'. What is the cooking time for a 5 pound duck?

6. Calculate the value of

   (a) $23 \times 0 + 5 \times (6 - 3)$

   (b) $23 \times 12 - 4 \times (9 - 8)$

7. Write down the terms denoted by ? in the sequence 1, 8, 15, ?, 29, 36, ?, ?.

Section B

1. The official attendance at a football match was 27 836. What was the attendance rounded off to the nearest thousand?

2. Knitting yarn is sold by weight. If 10 grams of wool has a length of 30 metres, what length of wool will have a weight of 60 grams?

3. Five resistors in an electrical system are connected in series. Their individual resistances are 732, 569, 1187, 945 and 278 ohms. Their total resistance is found by adding the five individual resistances. What is their total resistance?

4. Work out the values of

   (a) $(7 + 3) \times 4$ (b) $7 + 3 \times 4$

   (c) $7 \times 3 + 4$ (d) $7 \times 4 + 3 \times 4$

5. A woman on a diet is allowed 950 calories per day. For breakfast she has a poached egg, a slice of buttered toast and a cup of tea with milk. These items contain 58, 134 and 20 calories respectively. For lunch she has a ham salad which contains 173 calories and a cup

of coffee with milk which contains 25 calories. Work out:

(a) the number of calories in the breakfast

(b) the number of calories in the lunch

(c) the number of calories she can consume for her evening meal.

6. The numbers 1, 2, 4, 7, 11, 16 form a pattern.

(a) What is the next number in the pattern?

(b) If each number can be used only once, which three numbers add up to 25?

7. A ball bearing weighs 6 grams. A consignment weighs 1932 grams. How many ball bearings are there in the consignment?

Multi-Choice Questions 1

1. Which is the smallest of these four numbers?

A 20 002   B 22 000
C 20 200   D 20 020

2. Consider the sequence of numbers 1, 3, 6, 10, 15, 21, ... The sum of the next two numbers in the sequence is

A 51   B 53   C 55   D 64

3. What is the difference between the two numbers 90 and 144?

A 46   B 54   C 66   D 234

4. Consider the sequence of numbers 2, 5, 8, 11, ... The difference between the next two numbers in the sequence is

A 31   B 17   C 14   D 3

5. Which is the largest of these four numbers?

A 40 003   B 43 000
C 40 300   D 40 030

6. The sum of eleven thousand and eleven hundred is

A 11 100   B 11 110
C 12 100   D 111 000

7. $8 + 5 \times (4 - 2)$ is equal to

A 2   B 18   C 26   D 50

8. If the temperature rises from $-5°$C to $15°$C, what is the increase in temperature?

A 52°   B 20°   C 10°   D $-10°$

Mental Test 1

Try to write down the answers to the following questions without writing anything else:

1. Work out 20 times 8.

2. Divide 72 by 9.

3. Write down the number forty-one thousand and thirty-five.

4. Subtract 17 from 38.

5. What is the difference between 8 and 15?

6. Calculate the product of 3 and 12.

7. What is the sum of 8 and 11?

8. Multiply 8 by 5 and add 9.

9. Add twenty-four and twelve.

10. Add 11 and 15 and then subtract 9 from their sum.

Factors and Multiples

Odd and Even Numbers

If we count in twos starting with 2, the sequence is 2, 4, 6, 8, 10, 12, 14, ... If we continue the sequence we discover that each number in it ends in either 0, 2, 4, 6 or 8. These numbers are called **even numbers**. Thus 80, 192, 234, 9756 and 7398 are all even numbers.

If we now start at 1 and count in twos we get the sequence 1, 3, 5, 7, 9, 11, 13, 15, 17, 19, ... If we carry on the sequence, we discover that each number in it ends in either 1, 3, 5, 7 or 9. These numbers are called **odd numbers**. Thus 61, 83, 425, 687 and 4869 are all odd numbers. Note that any number which when divided by 2 has a remainder of 1 is an odd number.

Exercise 2.1

State whether each of the numbers below is an odd number or an even number:

1. 75
2. 824
3. 752
4. 623
5. 4829
6. 7778
7. 9386
8. 59 751
9. 230 000
10. 8267

11. From the set of numbers 15, 34, 55, 63, 88, 91, 126, 139, write down
    (a) all the even numbers
    (b) all the odd numbers.

12. Write down all the even numbers between 15 and 27.

13. Consider the numbers 14, 17, 31, 48, 79, 127 and 134. Write down all the odd numbers and find their sum.

14. Add up all the odd numbers between 10 and 18.

Powers of Numbers

The quantity $5 \times 5 \times 5$ is usually written as 5^3, and 5^3 is called the third **power** of 5. The number 3 which indicates the number of fives to be multiplied together is called the **index** (plural: indices).

$$2^4 = 2 \times 2 \times 2 \times 2 = 16$$
$$9^2 = 9 \times 9 = 81$$

Squares of Numbers

When a number is multiplied by itself the result is called the **square** of the number. For example

$$7 \text{ squared} = 7 \times 7 = 7^2 = 49$$
$$12 \text{ squared} = 12 \times 12 = 12^2 = 144$$

49 and 144 are also called **square numbers**.

The square of any number can be found by using a calculator in the way shown in Example 1.

Example 1

Find the square of 31 (i.e. find the value of 31×31).

Input	Display
31	31.
$\times$	31.
$=$	961.

Hence 31 squared $= 31^2 = 31 \times 31$
$= 961$.

Cubes of Numbers

The **cube** of a number is the number raised to the power of 3. Thus

$$\text{Cube of } 4 = 4^3 = 4 \times 4 \times 4$$
$$= 64$$
$$\text{Cube of } 85 = 85^3 = 85 \times 85 \times 85$$
$$= 614\,125$$

The cube of any number can be found by using a calculator in the way shown in Example 2.

Example 2

Find the cube of 19 (i.e. find the value of $19 \times 19 \times 19$).

Input	Display
19	19.
$\times \times$	19.
$=$	361.
$=$	6859.

Note that the number of equals signs is one less than the power to which the number has been raised (in this case there are only two equals signs).

Hence cube of $19 = 19^3 = 19 \times 19 \times 19$
$= 6859$.

Square Roots of Numbers

The **square root** of a number is the number whose square equals the given number.

The square of 5 is $5 \times 5 = 25$.

The square root of 25 is 5.

$\sqrt{}$ is the square-root sign and means 'the square root of'.

Similarly, the square of 7 is $7 \times 7 = 49$

$$\sqrt{49} = 7$$

To find the square root of numbers which are multiplied together we proceed as follows:

(1) Find the square root of each of the individual numbers.

(2) Multiply each of these individual square roots together to give the required answer. Thus

$$\sqrt{4 \times 9} = \sqrt{4} \times \sqrt{9} = 2 \times 3 = 6$$

Example 3

Find the square root of $16 \times 25 \times 49$.

$$\sqrt{16 \times 25 \times 49} = \sqrt{16} \times \sqrt{25} \times \sqrt{49}$$
$$= 4 \times 5 \times 7$$
$$= 140$$

Exercise 2.2

Find the squares of the following numbers:

1. 7	2. 8	3. 13
4. 27	5. 44	6. 35
7. 89	8. 123	9. 543
10. 356		

Find the cubes of the following numbers:

11. 3	12. 9	13. 17
14. 38	15. 59	16. 132

17. 234   **18.** 384   **19.** 120

20. 200

Find the square roots of the following numbers:

21. 16   **22.** 36   **23.** 64

24. 81   **25.** 144   **26.** 961

27. 3844   **28.** 7921   **29.** 17 956

30. 207 936

Find the square roots of the following:

31. 9×36 **32.** 64×64

33. 16×25 **34.** $9 \times 25 \times 36$

35. $36 \times 49 \times 64$

Number Patterns

A **square number** can be shown as a pattern of dots in the shape of a square.

$$9 = 3^2 = 3 \times 3 \quad 16 = 4^2 = 4 \times 4$$

$$25 = 5^2 = 5 \times 5$$

The sequence of square numbers is $1^2, 2^2, 3^3, 4^2, 5^2, 6^2, \ldots$ or $1, 4, 9, 16, 25, 36, \ldots$

A **rectangular number** can be shown as a pattern of dots in the shape of a rectangle.

$$6 = 2 \times 3 \qquad 24 = 6 \times 4$$

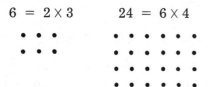

$$24 = 8 \times 3$$

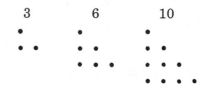

Note that 1 is not regarded as being a rectangular number. An example of a sequence of rectangular numbers is $4, 6, 8, 10, \ldots$ Another might be $6, 9, 12, \ldots$

A **triangular number** can be shown as a pattern of dots in the shape of a triangle.

3   6   10

The sequence of triangular numbers is $1, 3, 6, 10, 15, 21, \ldots$

Note that

$$3 = 2 + 1;$$
$$6 = 3 + 2 + 1;$$
$$10 = 4 + 3 + 2 + 1;$$
$$15 = 5 + 4 + 3 + 2 + 1$$

Exercise 2.3

1. Write down the next two terms of the sequence of square numbers starting from $100, 121, \ldots$

2. Write down the next two terms of the sequence of rectangular numbers starting from $22, 24, \ldots$

3. Write down the next three terms of the sequence of triangular numbers starting from $28, 36, \ldots$

From the set of numbers $15, 34, 57, 63, 67$ and 81, write down one that is

4. a square number

5. a rectangular number

6. a triangular number.

The Digit Sum of a Number

The **digits** of a number are the single figures which make up the number. For instance, the digits of the number 243 are 2, 4 and 3.

The **digit sum** of a number is found by adding the digits together. Thus the digit sum of the number 243 is $2 + 4 + 3 = 9$.

Sometimes the digit sum is 10 or more. When this happens we add the digits of the sum together until a number less than 10 is obtained.

Example 4

Find the digit sum of the number 3642.

Adding the digits 3, 6, 4 and 2

$$3 + 6 + 4 + 2 = 15$$

Adding the digits 1 and 5

$$1 + 5 = 6$$

The digit sum of the number 3642 is 6.

Exercise 2.4

Find the digit sum for each of the following numbers:

1. 215	2. 306	3. 423
4. 982	5. 721	6. 750
7. 6284	8. 9321	9. 8762
10. 8105		

Tests for Divisibility

If a number is even then 2 will divide into it without leaving a remainder, that is, the number is divisible by 2. Numbers divisible by 2 are called **multiples** of 2.

Exercise 2.5

Which of the following numbers are divisible by 2?

1. 8	2. 7	3. 11
4. 29	5. 36	6. 47
7. 53	8. 62	9. 1786
10. 5302	11. 18 735	12. 19 006

3 will divide into a number without leaving a remainder if it divides exactly into the digit sum of the number.

Example 5

Is the number 3582 divisible by 3?

Adding the digits of 3582:

$$3 + 5 + 8 + 2 = 18$$

Adding the digits of 18:

$$1 + 8 = 9$$

Digit sum of 3582 is 9.

3 will divide into 9 exactly so 3582 is divisible by 3.

Numbers which are divisible by 3 are called multiples of 3.

Example 6

Write down all the multiples of 3 between 13 and 25.

The multiples are 15, 18, 21 and 24

because each of these numbers is divisible by 3.

Exercise 2.6

Which of the following numbers are divisible by 3?

1. 315	2. 426	3. 287

4. 6135   **5.** 8272   **6.** 65

7. 15 640   **8.** 29 739

State which of the following numbers are multiples of 3:

9. 72   **10.** 143   **11.** 132

12. 4872   **13.** 902   **14.** 5631

15. 16 247

5 will divide into a number without leaving a remainder only if the number ends in 0 or 5.

20, 35, 85, 9000 and 16 345 are all divisible by 5. 21, 73, 198, 4372 and 32 453 are not divisible by 5 because they do not end in 0 or 5.

Numbers which are divisible by 5 are called multiples of 5.

Exercise 2.7

State which of the following numbers are divisible by 5:

1. 95   **2.** 384   **3.** 987

4. 2640   **5.** 9358   **6.** 8000

7. 45 654   **8.** 19 740

State which of the following numbers are multiples of 5:

9. 31   **10.** 90   **11.** 765

12. 928   **13.** 7865   **14.** 75 654

15. 87 945

Numbers which ends in 0 are divisible by 10. Thus 8730, 90, 780 and 12 640 are all divisible by 10. However 38, 439, 1872 and 16 737 are not divisible by 10 because they do not end in 0.

Exercise 2.8

Which of the following numbers are divisible by 10?

1. 892   **2.** 1100   **3.** 1560

4. 983   **5.** 8940   **6.** 12 764

7. 8960   **8.** 67 893

State which of the following numbers are multiples of 10:

9. 7035   **10.** 9870

11. 18 630   **12.** 5240

13. 6 000 000   **14.** 894 555

15. 786 740

Factors

A number is a **factor** of another number if it divides into that number without leaving a remainder.

6 is a factor of 42 because it divides into 42 without leaving a remainder.

42 is also divisible by 7, and so 7 is also a factor of 42.

42 has other factors, namely, 1, 2, 3, 14, 21 and 42 as well as 6 and 7.

3 is not a factor of 8.

5 is not a factor of 22.

Exercise 2.9

1. Is 6 a factor of
   (a) 18   **(b)** 21   **(c)** 24
   (d) 36   **(e)** 56?

2. Is 5 a factor of
   (a) 4   **(b)** 12   **(c)** 15
   (d) 25   **(e)** 32?

3. Is 9 a factor of
    (a) 20   (b) 30   (c) 72
    (d) 108   (e) 126?

4. Is 10 a factor of
    (a) 25   (b) 30   (c) 45
    (d) 70   (e) 120?

5. Which of the following numbers are factors of 63?
    (a) 2   (b) 3   (c) 4
    (d) 5   (e) 6   (f) 9
    (g) 21

6. Which of the following numbers are factors of 72?
    (a) 1   (b) 2   (c) 3
    (d) 4   (e) 5   (f) 6
    (g) 7   (h) 8   (i) 9
    (j) 10   (k) 11   (l) 12
    (m) 13   (n) 14

7. Write down all the positive whole numbers which are factors of 9.

Prime Numbers

Every number has itself and 1 as factors. If a number has no other factors it is said to be a **prime number**. It is useful to learn all the prime numbers up to 100. They are

2, 3, 5, 7, 11, 13, 17, 19, 23, 29, 31, 37, 41, 43, 47, 53, 59, 61, 67, 71, 73, 79, 83, 89, 97

Notice that, with the exception of 2, all the prime numbers are odd. Also note that although 1 is a factor of all other numbers it is not regarded as being a prime number.

Prime Factors

A factor which is a prime number is called a **prime factor**. In the statement $42 = 7 \times 6$, 7 is a prime factor but 6 is not (since it equals 2×3).

Example 7

Find all the factors of 80.

The factors are:

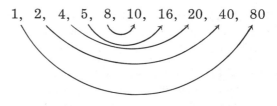

1, 2, 4, 5, 8, 10, 16, 20, 40, 80

Note that 1 and 80 are regarded as factors.

If we pair off the factors from each end, as shown above, and multiply them together we get

$1 \times 80 = 80$ $2 \times 40 = 80$

$4 \times 20 = 80$ $5 \times 16 = 80$

$8 \times 10 = 80$

We see that each pair gives a product of 80 and we can check that we have found all the factors by using this method of pairing.

Exercise 2.10

Write down all the factors of the following numbers and check by pairing:

1. 36   2. 56   3. 100
4. 84   5. 60   6. 105
7. 70

Lowest Common Multiple (LCM)

The multiples of 3 are 3, 6, 9, 12, 15, 18, 21, 24, 27, 30, 33, 36, etc.

The multiples of 4 are 4, 8, 12, 16, 20, 24, 28, 32, 36, 40, etc.

We see that 12, 24 and 36 are multiples of both 3 and 4. The lowest of these common multiples is 12 and we say that the **LCM** of 3 and 4 is 12. Notice that 12 is the smallest number into which 3 and 4 will divide exactly.

We see then that the LCM of a set of numbers is the smallest number into which each of the numbers of the set will divide exactly.

The LCM of many sets of numbers can sometimes be found by just looking at them but sometimes the method of Example 8 (c) has to be used.

Example 8

(a) The LCM of 2, 3 and 4 is 12 because $12 \div 2 = 6$, $12 \div 3 = 4$ and $12 \div 4 = 3$.

(b) The LCM of 5, 10 and 20 is 20 because $20 \div 5 = 4$, $20 \div 10 = 2$ and $20 \div 20 = 1$.

(c) To find the LCM of 4, 5 and 6 we can use this method:

The multiples of 4 are 8, 12, 16, 20, 24, 28, 32, 36, 40, 44, 48, 52, 56, 60, 64, 68, ...

The multiples of 5 are 10, 15, 20, 25, 30, 35, 40, 45, 50, 55, 60, 65, 70, ...

The multiples of 6 are 12, 18, 24, 30, 36, 42, 48, 54, 60, 66, 72, ...

We see that the lowest common multiple is 60 and this is the smallest whole number into which 4, 5 and 6 will divide exactly.

Lowest common multiples are needed when adding fractions (see Chapter 3).

Exercise 2.11

Find the LCM of the following sets of numbers:

1. 2 and 3
2. 4 and 8
3. 4 and 5
4. 6 and 9
5. 2, 3 and 5
6. 6 and 15
7. 3, 5 and 6
8. 12 and 16
9. 5 and 15
10. 3, 5 and 10
11. 2, 3, 4 and 9
12. 3, 4 and 5

Miscellaneous Exercise 2

Section A

1. Given the numbers 2, 3, 4, 5, 6, 7, 8 and 9, write down
   (a) all the even numbers
   (b) the multiples of 3
   (c) all the prime numbers.

2. Given the numbers 47, 55, 60, 63, 81 and 122
   (a) Write down the multiples of 5.
   (b) Find the total of the multiples of 3.
   (c) One of the numbers has a square root which is a whole number. Write down the square root of this number.

3. Three prime numbers are 11, 13 and 37:
   (a) Add them up. Is the sum a prime number?
   (b) Multiply them together. Is the answer a prime number?

4. Consider the numbers 13, 24, 31, 65, 75 and 125:
   (a) Two of these numbers are prime. Which are they?
   (b) Three of these numbers have a common factor. What is it?

5. From the numbers 16, 32, 64, 81 and 132, write down
   (a) the odd numbers
   (b) the numbers which have square roots which are whole numbers
   (c) the numbers which are multiples of 2.

Section B

1. Write down two prime numbers whose sum is 16 and whose difference is 6.

2. List the following numbers:

    (a) multiples of 8 which are less than 48

    (b) the next three prime numbers greater than 23.

3   Find the smallest number that is divisible by 3, 4, 6 and 9.

4. Evaluate $13^2 - 12^2$.

5. Work out the value of $\sqrt{16 \times 49}$.

6. Add up all the prime numbers between 10 and 20.

7. Consider the number sequence 1, 3, 6, 10, 15, 21, ... What is the next number in the sequence which has a square root that is a whole number?

8. A baker wants to bake a batch of cakes that can be packed in all three of the following ways: only in threes, only in fives or only in tens. What is the smallest number of cakes he needs to bake?

Multi-Choice Questions 2

1. What is the value of 2^3?

    A 8   B 6   C 4   D 3

2. The sum of all the prime numbers between 10 and 20 is

    A 90   B 75   C 70   D 60

3. The smallest whole number that can be divided exactly by 3, 4, 6 and 9 is

    A 36   B 24   C 18   D 12

4. The difference between the cube of 2 and the square of 3 is

    A 17   B 12   C 1   D 0

5. The number of prime numbers between 10 and 30 is

    A 2   B 5   C 6   D 8

6. The lowest common multiple of 3, 4 and 6 is

    A 12   B 13   C 24   D 72

7. What is the value of $4^2 + 7^2$?

    A 11   B 14   C 28   D 65

8. Consider the five prime numbers 53, 59, 67, 71, 73 and 79. What is the other prime number between 50 and 80?

    A 57   B 61   C 74   D 77

9. Consider the numbers 13, 24, 31, 65, 75 and 125. Three of these numbers have a common factor. What is it?

    A 3   B 5   C 13   D 25

10. What is the value of 61^2?

    A 122 B 261 C 3601 D 3721

Mental Test 2

Try to answer the following questions without writing down anything except the answer:

1. Write down the next two even numbers after 35.

2. Is 596 759 an odd number?

3. Write down all the even numbers between 7 and 15.

4. What is the square root of 9?

5. Write down the value of 8 squared.

6. Find the square root of 25×49.

7. Is 345 divisible by 3?

8. Write down all the multiples of 3 between 8 and 20.

9. Write down all the factors of 10.

10. Is 387 divisible by 5?

11. Is 7 a factor of 347?

12. From the set of numbers 13, 18, 35, 38, 48, 53 and 85, write down all the even numbers.

13. Which of the following numbers are factors of 72?

    1, 2, 3, 4, 5, 6, 7, 8 and 9

14. Given the set of numbers 7, 9, 15, 19, 27, 31 and 53, write down all the prime numbers.

15. What are the prime factors of 10?

16. Find the LCM of 4, 8 and 16.

Fractions

Introduction

The circle shown in Fig. 3.1 has been divided into eight equal parts. Each of these parts is called one-eighth of the circle and is written $\frac{1}{8}$. The number below the line shows how many equal parts there are and it is called the **denominator**. The number above the line shows how many of these equal parts are taken and it is called the **numerator**. If five of the eight equal parts are taken then we have taken $\frac{5}{8}$ of the circle.

Fig. 3.1

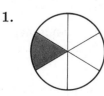

$\frac{1}{8}$ of circle

Exercise 3.1

Fig. 3.2 shows some drawings of circles. Write down the fractions represented by the shaded portions:

Fig. 3.2

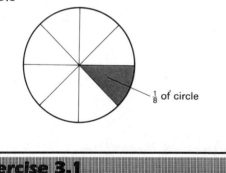

1.   2.

3.   4.

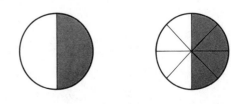

5.

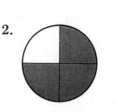

In a similar way to that shown in Fig. 3.2, sketch five circles and shade:

6. $\frac{7}{8}$ 7. $\frac{2}{5}$ 8. $\frac{5}{12}$

9. $\frac{3}{10}$ 10. $\frac{4}{7}$

Equivalent Fractions

In Fig. 3.3, we have shaded $\frac{1}{2}$ of a circle and $\frac{4}{8}$ of a circle. From the diagrams you can see that the same area has been shaded in both circles.

Hence

$$\frac{1}{2} = \frac{4}{8}$$

Fig. 3.3

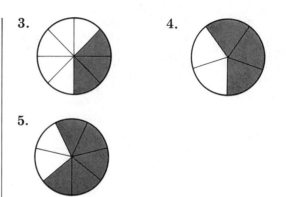

We can see that to change $\frac{1}{2}$ into $\frac{4}{8}$ we have multiplied top and bottom by 4.

By drawing more sketches of fractions (Fig. 3.4) we find that if we multiply or divide the top and bottom of a fraction by the same number (provided it is not zero) we do not alter its value.

Fig. 3.4

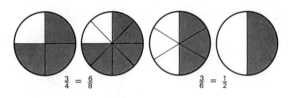

$$\frac{3}{4} = \frac{6}{8} \qquad\qquad \frac{3}{6} = \frac{1}{2}$$

$\frac{2}{3}, \frac{4}{6}, \frac{6}{9}, \frac{8}{12}$ and $\frac{10}{15}$ all have the same value and so they are called **equivalent fractions**.

Example 1

(a) $\frac{3}{8}$ is equivalent to $\dfrac{3 \times 2}{8 \times 2} = \dfrac{6}{16}$

(b) $\frac{21}{30}$ is equivalent to $\dfrac{21 \div 3}{30 \div 3} = \dfrac{7}{10}$

Exercise 3.2

Draw suitable sketches to show that

1. $\frac{2}{3} = \frac{4}{6}$ 2. $\frac{3}{5} = \frac{6}{10}$ 3. $\frac{1}{4} = \frac{2}{8}$

4. $\frac{3}{4} = \frac{12}{16}$ 5. $\frac{1}{6} = \frac{2}{12}$

Copy the following fractions and fill in the missing numbers:

6. $\frac{3}{4} = \frac{}{20}$ 7. $\frac{2}{3} = \frac{}{12}$ 8. $\frac{5}{7} = \frac{}{14}$

9. $\frac{3}{8} = \frac{}{32}$ 10. $\frac{5}{6} = \frac{}{30}$ 11. $\frac{9}{12} = \frac{}{4}$

12. $\frac{8}{16} = \frac{}{2}$ 13. $\frac{6}{15} = \frac{}{5}$ 14. $\frac{36}{42} = \frac{}{7}$

15. $\frac{72}{88} = \frac{}{11}$ 16. $\frac{5}{12} = \frac{}{60}$ 17. $\frac{21}{49} = \frac{}{7}$

18. $\frac{49}{56} = \frac{}{8}$ 19. $\frac{3}{8} = \frac{}{64}$ 20. $\frac{18}{27} = \frac{}{3}$

Reducing a Fraction to its Lowest Terms

When the numerator and denominator of a fraction possess no common factors the fraction is said to be in its **lowest terms**. Thus $\frac{7}{16}$ and $\frac{9}{64}$ are fractions in their lowest terms.

However, the fraction $\frac{6}{8}$ is not in its lowest terms because the numerator and denominator can be divided by 2 to give the fraction $\frac{3}{4}$.

Example 2

Reduce $\frac{42}{48}$ to its lowest terms.

$$\frac{42}{48} \text{ is equivalent to } \frac{42 \div 6}{48 \div 6} = \frac{7}{8}$$

Exercise 3.3

Reduce each of the following fractions to its lowest terms:

1. $\frac{9}{18}$ 2. $\frac{25}{35}$ 3. $\frac{3}{9}$ 4. $\frac{18}{24}$

5. $\frac{6}{15}$ 6. $\frac{11}{33}$ 7. $\frac{15}{20}$ 8. $\frac{9}{27}$

Types of Fraction

Fractions in which the numerator is less than the denominator are called **proper fractions**. $\frac{3}{4}, \frac{7}{8}$ and $\frac{17}{20}$ are all proper fractions.

A fraction whose numerator is greater than its denominator is called an **improper fraction** or a **top-heavy** fraction. So $\frac{9}{4}, \frac{20}{3}$ and $\frac{127}{100}$ are all top-heavy fractions.

A number like $1\frac{3}{4}$ is called a **mixed number** because it consists of a whole number and a fraction.

Every improper fraction can be written as a mixed number.

Example 3

(a) Change $\frac{5}{3}$ into a mixed number.

$$\frac{5}{3} \text{ may be written } \frac{3+2}{3} = \tfrac{3}{3} + \tfrac{2}{3}$$

$$= 1 + \tfrac{2}{3}$$

$$= 1\tfrac{2}{3}$$

(b) Change $\frac{19}{8}$ into a mixed number.

$$\frac{19}{8} \text{ may be written } \frac{16+3}{8} = \tfrac{16}{8} + \tfrac{3}{8}$$

$$= 2 + \tfrac{3}{8}$$

$$= 2\tfrac{3}{8}$$

To change a top-heavy fraction into a mixed number the rule is:

Divide the top number by the number underneath. The answer goes as the whole number part of the mixed number. The remainder goes on top of the fractional part and the number underneath stays the same.

Example 4

Change $\frac{27}{4}$ into a mixed number.

$$27 \div 4 = 6 \text{ remainder } 3$$

$$\text{Therefore } \tfrac{27}{4} = 6\tfrac{3}{4}.$$

All mixed numbers can be changed into top-heavy fractions.

To change a mixed number into a top-heavy fraction the rule is:

Multiply the whole number by the number on the bottom of the fraction. To this number add the number on top of the fractional part. The result of this addition becomes the top number of the top-heavy fraction and its bottom number stays the same.

Example 5

Change $7\frac{3}{4}$ into an improper fraction.

$$7\tfrac{3}{4} = \frac{(7 \times 4) + 3}{4}$$

$$= \frac{28 + 3}{4}$$

$$= \tfrac{31}{4}$$

Exercise 3.4

Change the following top-heavy fractions into mixed numbers:

1. $\frac{7}{2}$ 2. $\frac{17}{8}$ 3. $\frac{26}{5}$ 4. $\frac{18}{7}$

5. $\frac{23}{9}$ 6. $\frac{13}{4}$ 7. $\frac{29}{6}$ 8. $\frac{37}{5}$

Change the following mixed numbers into top-heavy fractions:

9. $3\frac{1}{5}$ 10. $1\frac{3}{4}$ 11. $5\frac{6}{7}$ 12. $2\frac{3}{8}$

13. $8\frac{5}{7}$ 14. $7\frac{2}{9}$ 15. $4\frac{5}{6}$ 16. $6\frac{7}{10}$

Comparing the Values of Fractions

When putting fractions in order of size, first express them as fractions with the same denominator. This common denominator should be the LCM (see page 23) of the denominators of the fractions to be compared.

Example 6

Arrange the fractions $\frac{2}{3}, \frac{3}{4}, \frac{5}{6}$ and $\frac{4}{5}$ in order of size, smallest first.

The LCM of the denominators 3, 4, 5 and 6 is 60. We now change each of the fractions so that it has a denominator of 60.

$$\frac{2}{3} = \frac{2 \times 20}{3 \times 20} = \frac{40}{60}$$

$$\frac{3}{4} = \frac{3 \times 15}{4 \times 15} = \frac{45}{60}$$

$$\frac{5}{6} = \frac{5 \times 10}{6 \times 10} = \frac{50}{60}$$

$$\frac{4}{5} = \frac{4 \times 12}{5 \times 12} = \frac{48}{60}$$

After the fractions have been expressed with the same denominator, all we have to do is compare their numerators.

Therefore the order is

$$\frac{40}{60}, \frac{45}{60}, \frac{48}{60} \text{ and } \frac{50}{60} \text{ or } \frac{2}{3}, \frac{3}{4}, \frac{4}{5} \text{ and } \frac{5}{6}$$

Exercise 3.5

Put the following sets of fractions in order of size, smallest first:

1. $\frac{3}{5}, \frac{7}{10}, \frac{11}{20}$ 2. $\frac{7}{10}, \frac{4}{5}, \frac{2}{3}$

3. $\frac{1}{2}, \frac{5}{6}, \frac{2}{3}$ 4. $\frac{3}{4}, \frac{5}{6}, \frac{7}{12}$

5. $\frac{3}{4}, \frac{5}{8}, \frac{9}{16}, \frac{17}{32}$ 6. $\frac{3}{8}, \frac{4}{7}, \frac{5}{9}, \frac{3}{5}$

7. $\frac{1}{3}, \frac{1}{4}, \frac{1}{2}, \frac{1}{5}$ 8. $\frac{3}{5}, \frac{3}{4}, \frac{5}{8}, \frac{4}{7}$

Addition and Subtraction of Fractions

When adding or subtracting fractions the procedure is as follows:

(1) Find the LCM of the denominators of the fractions to be added or subtracted.

(2) Express each of the fractions with this common denominator.

(3) Add or subtract the numerators of these equivalent fractions to give the numerator of the answer. The denominator of the answer is the LCM found in (1).

Example 7

(a) Work out $\frac{2}{3} + \frac{1}{4}$.

The LCM of the denominators 3 and 4 is 12. Expressing each fraction with a denominator of 12 gives:

$\frac{2}{3}$ is equivalent to $\frac{2 \times 4}{3 \times 4} = \frac{8}{12}$

$\frac{1}{4}$ is equivalent to $\frac{1 \times 3}{4 \times 3} = \frac{3}{12}$

$$\frac{2}{3} + \frac{1}{4} = \frac{8}{12} + \frac{3}{12}$$
$$= \frac{8 + 3}{12}$$
$$= \frac{11}{12}$$

The work may be set out as follows:

$$\frac{2}{3} + \frac{1}{4} = \frac{(2 \times 4) + (1 \times 3)}{12}$$
$$= \frac{8 + 3}{12}$$
$$= \frac{11}{12}$$

(b) Find the sum of $\frac{2}{5}$ and $\frac{3}{4}$.

The LCM of the denominators 5 and 4 is 20.

$$\frac{2}{5} + \frac{3}{4} = \frac{(2 \times 4) + (3 \times 5)}{20}$$
$$= \frac{8 + 15}{20}$$
$$= \frac{23}{20}$$
$$= 1\frac{3}{20}$$

If mixed numbers are present then we add the whole numbers first.

Example 8

Add $5\frac{2}{3}$ and $3\frac{5}{8}$.

The LCM of the denominators 3 and 8 is 24.

$$5\frac{2}{3} + 3\frac{5}{8} = 5 + 3 + \frac{2}{3} + \frac{5}{8}$$

$$= 8 + \frac{(2 \times 8) + (5 \times 3)}{24}$$

$$= 8 + \frac{16 + 15}{24}$$

$$= 8 + \frac{31}{24}$$

$$= 8 + 1\frac{7}{24}$$

$$= 9\frac{7}{24}$$

Example 9

(a) Work out $\frac{2}{3} - \frac{1}{2}$.

The LCM of 3 and 2 is 6.

$$\frac{2}{3} - \frac{1}{2} = \frac{(2 \times 2) - (1 \times 3)}{6}$$

$$= \frac{4 - 3}{6}$$

$$= \frac{1}{6}$$

(b) Subtract $3\frac{7}{10}$ from $6\frac{4}{5}$.

The LCM of 5 and 10 is 10.

$$6\frac{4}{5} - 3\frac{7}{10} = 6 - 3 + \frac{4}{5} - \frac{7}{10}$$

$$= 3 + \frac{4}{5} - \frac{7}{10}$$

$$= 3 + \frac{(4 \times 2) - (7 \times 1)}{10}$$

$$= 3 + \frac{8 - 7}{10}$$

$$= 3 + \frac{1}{10}$$

$$= 3\frac{1}{10}$$

Sometimes we have to 'borrow' from the whole number as shown in Example 10.

Example 10

Work out $4\frac{1}{3} - 2\frac{11}{12}$.

The LCM of 3 and 12 is 12.

$$4\frac{1}{3} - 2\frac{11}{12} = 4 - 2 + \frac{1}{3} - \frac{11}{12}$$

$$= 2 + \frac{1}{3} - \frac{11}{12}$$

$$= 2 + \frac{4}{12} - \frac{11}{12}$$

We cannot take $\frac{11}{12}$ from $\frac{4}{12}$ so we 'borrow' 1 from the 2 and say $2 = 1 + \frac{12}{12}$. We now have

$$4\frac{1}{3} - 2\frac{11}{12} = 1 + \frac{12}{12} + \frac{4}{12} - \frac{11}{12}$$

$$= 1 + \frac{12 + 4 - 11}{12}$$

$$= 1 + \frac{5}{12}$$

$$= 1\frac{5}{12}$$

'Borrowing' can be avoided by changing the mixed numbers into top-heavy fractions before attempting to subtract.

Thus

$$4\frac{1}{3} - 2\frac{11}{12} = \frac{13}{3} - \frac{35}{12}$$

$$= \frac{(13 \times 4) - (35 \times 1)}{12}$$

$$= \frac{52 - 35}{12}$$

$$= \frac{17}{12}$$

$$= 1\frac{5}{12}$$

Exercise 3.6

Work out the following:

1. $\frac{1}{3} + \frac{1}{5}$ 2. $\frac{3}{4} + \frac{1}{8}$ 3. $\frac{3}{4} + \frac{1}{2}$

4. $\frac{1}{2} + \frac{1}{3}$ 5. $\frac{3}{5} + \frac{7}{10}$ 6. $\frac{5}{9} + \frac{2}{3}$

7. $\frac{4}{5} + \frac{2}{3}$ 8. $\frac{2}{7} + \frac{3}{4}$ 9. $3\frac{1}{2} + 4\frac{1}{3}$

10. $2\frac{3}{4} + 3\frac{5}{8}$ 11. $5\frac{2}{3} + 3\frac{1}{8}$ 12. $4\frac{2}{5} + 5\frac{3}{4}$

13. $2\frac{3}{5} + \frac{7}{10} + \frac{1}{2}$ 14. $5\frac{1}{3} + \frac{5}{6} + \frac{3}{4}$

15. $7\frac{5}{6} + \frac{7}{12} + 2\frac{1}{4}$ 16. $\frac{7}{8} - \frac{1}{3}$

17. $\frac{2}{3} - \frac{3}{5}$ 18. $\frac{5}{6} - \frac{1}{3}$ 19. $\frac{4}{5} - \frac{3}{10}$

20. $\frac{3}{4} - \frac{2}{5}$ 21. $3\frac{3}{4} - 2\frac{2}{3}$ 22. $5 - 2\frac{15}{16}$

23. $1\frac{1}{4} - \frac{7}{8}$ 24. $2\frac{2}{3} - 1\frac{10}{11}$ 25. $3\frac{1}{4} - 2\frac{3}{10}$

Multiplication of Fractions

When multiplying fractions first multiply the top numbers to obtain the top number of the product. Then multiply the bottom numbers to obtain the bottom number of the product. Mixed numbers must be changed into top-heavy fractions before attempting to multiply.

Example 11

(a) Work out $\frac{2}{3} \times \frac{4}{5}$.

$$\frac{2}{3} \times \frac{4}{5} = \frac{2 \times 4}{3 \times 5}$$

$$= \frac{8}{15}$$

(b) Find the product of $\frac{3}{5}$ and $2\frac{3}{4}$.

$$\frac{3}{5} \times 2\frac{3}{4} = \frac{3}{5} \times \frac{11}{4}$$

$$= \frac{3 \times 11}{5 \times 4}$$

$$= \frac{33}{20}$$

$$= 1\frac{13}{20}$$

Exercise 3.7

Work out each of the following:

1. $\frac{2}{3} \times \frac{4}{5}$ 2. $\frac{1}{2} \times \frac{1}{3}$ 3. $\frac{3}{4} \times \frac{3}{5}$

4. $\frac{2}{5} \times \frac{3}{7}$ 5. $1\frac{1}{2} \times \frac{1}{5}$ 6. $\frac{1}{3} \times 2\frac{2}{3}$

7. $2 \times 1\frac{1}{5}$ 8. $1\frac{3}{4} \times 1\frac{1}{2}$ 9. $1\frac{2}{5} \times 3\frac{1}{2}$

10. $2\frac{1}{2} \times 2\frac{1}{3}$

Cancelling

Example 12

Find the value of $\frac{2}{3} \times 1\frac{7}{8}$.

$$\frac{2}{3} \times 1\frac{7}{8} = \frac{2}{3} \times \frac{15}{8}$$

$$= \frac{2 \times 15}{3 \times 8}$$

$$= \frac{30}{24}$$

$$= \frac{5}{4}$$

$$= 1\frac{1}{4}$$

The step of reducing $\frac{30}{24}$ to its lowest terms has been done by dividing the top and bottom numbers by 6. This step could have been avoided by cancelling before multiplication, as shown below.

$$\frac{\overset{1}{\cancel{2}}}{\underset{1}{\cancel{3}}} \times \frac{\overset{5}{\cancel{15}}}{\underset{4}{\cancel{8}}} = \frac{1 \times 5}{1 \times 4}$$

$$= \frac{5}{4}$$

$$= 1\frac{1}{4}$$

We have divided 2 into 2 (a top number) and 8 (a bottom number). Also 3 into 15 (a top number) and 3 (a bottom number). We have, in fact, divided top and bottom by the same amount. Notice carefully that we can cancel between only a top number and a bottom number.

A whole number can be written as a fraction with a denominator of 1. For instance $8 = \frac{8}{1}$.

Example 13

$$\frac{2}{3} \times 6 = \frac{2}{\underset{1}{\cancel{3}}} \times \frac{\overset{2}{\cancel{6}}}{1}$$

$$= \frac{2 \times 2}{1 \times 1}$$

$$= \frac{4}{1}$$

$$= 4$$

Sometimes in calculations with fractions the word 'of' appears. It should always be taken as meaning multiply.

Example 14

What is $\frac{4}{5}$ of 20?

$$\frac{4}{5} \text{ of } 20 = \frac{4}{\cancel{5}_1} \times \frac{\cancel{20}^4}{1}$$

$$= \frac{4 \times 4}{1 \times 1}$$

$$= \frac{16}{1}$$

$$= 16$$

Work out the following:

1. $\frac{3}{4} \times \frac{8}{9}$ 2. $\frac{2}{3} \times \frac{3}{8}$ 3. $\frac{5}{8} \times \frac{4}{15}$

4. $\frac{7}{12} \times \frac{8}{21}$ 5. $\frac{3}{4} \times 1\frac{7}{9}$ 6. $5\frac{1}{5} \times \frac{10}{13}$

7. $1\frac{5}{8} \times \frac{7}{26}$ 8. $20 \times \frac{2}{5}$ 9. $\frac{5}{8} \times \frac{7}{10}$

10. $3\frac{3}{4} \times 16$ 11. $\frac{3}{4}$ of 16 12. $\frac{5}{7}$ of 140

13. $\frac{2}{3}$ of 60 14. $\frac{4}{5}$ of 40 15. $1\frac{1}{2} \times 20$

Division of Fractions

To divide by a fraction we invert it and then proceed as in multiplication.

Example 15

(a) Divide $\frac{3}{5}$ by $\frac{2}{7}$.

$$\frac{3}{5} \div \frac{2}{7} = \frac{3}{5} \times \frac{7}{2}$$

$$= \frac{3 \times 7}{5 \times 2}$$

$$= \frac{21}{10}$$

$$= 2\frac{1}{10}$$

(b) Work out $10 \div \frac{5}{8}$.

$$10 \div \frac{5}{8} = \frac{\cancel{10}^2}{1} \times \frac{8}{\cancel{5}_1}$$

$$= \frac{2 \times 8}{1 \times 1}$$

$$= \frac{16}{1}$$

$$= 16$$

Work out the following:

1. $\frac{2}{3} \div \frac{3}{4}$ 2. $\frac{3}{5} \div \frac{5}{8}$ 3. $\frac{3}{8} \div \frac{9}{16}$

4. $\frac{5}{8} \div \frac{15}{32}$ 5. $\frac{4}{5} \div 1\frac{1}{3}$ 6. $2 \div \frac{1}{4}$

7. $3\frac{3}{4} \div 2\frac{1}{2}$ 8. $20 \div \frac{2}{5}$ 9. $35 \div \frac{7}{8}$

10. $5 \div 5\frac{1}{5}$ 11. $\frac{2}{3} \div 4$ 12. $\frac{3}{4} \div 12$

Sequences

It is possible to have sequences consisting of whole numbers, fractions and a mixture of fractions and whole numbers.

Example 16

Find the next two terms of the sequence $\frac{1}{3}, \frac{2}{3}, 1\frac{1}{3}, \ldots$

By converting the mixed number into a top-heavy fraction we get the sequence $\frac{1}{3}, \frac{2}{3}, \frac{4}{3}, \ldots$ We see that each term is obtained by multiplying the previous term by 2. Therefore

$$4\text{th term} = \frac{4}{3} \times \frac{2}{1}$$

$$= \frac{4 \times 2}{3 \times 1}$$

$$= \frac{8}{3}$$

$$= 2\frac{2}{3}$$

$$\text{5th term} = \frac{8}{3} \times \frac{2}{1}$$

$$= \frac{8 \times 2}{3 \times 1}$$

$$= \frac{16}{3}$$

$$= 5\frac{1}{3}$$

Exercise 3.10

Write down the next two terms of the following sequences:

1. $\frac{1}{2}, \frac{3}{4}, 1, \ldots$

2. $\frac{2}{5}, \frac{4}{5}, \frac{6}{5}, \ldots$

3. $\frac{1}{8}, \frac{1}{4}, \frac{1}{2}, \ldots$

4. $\frac{2}{3}, 2, 6, \ldots$

5. $4, \frac{4}{3}, \frac{4}{9}, \ldots$

6. $\frac{1}{2}, \frac{1}{4}, \frac{1}{8}, \ldots$

7. $\frac{1}{5}, \frac{3}{5}, 1, \ldots$

8. $7, 5\frac{1}{2}, 4, \ldots$

The Square Root of a Fraction

To find the square root of a fraction we find the square roots of the numerator (top number) and the denominator (bottom number) separately as shown in Example 17.

Example 17

Find the square root of $\frac{25}{64}$.

$$\sqrt{\frac{25}{64}} = \frac{\sqrt{25}}{\sqrt{64}}$$

$$= \frac{5}{8}$$

If the numbers under a square-root sign are connected by a plus or minus sign we must add or subtract before attempting to find the square root.

Example 18

(a) Find the square root of $9 + 16$.

$$\sqrt{9 + 16} = \sqrt{25}$$

$$= 5$$

(b) Find the square root of $25 - 16$.

$$\sqrt{25 - 16} = \sqrt{9}$$

$$= 3$$

(c) Find the square root of $\frac{5 + 11}{25 - 16}$.

$$\sqrt{\frac{5 + 11}{25 - 16}} = \sqrt{\frac{16}{9}}$$

$$= \frac{\sqrt{16}}{\sqrt{9}}$$

$$= \frac{4}{3}$$

Exercise 3.11

Find the square roots of the following:

1. $\frac{4}{9}$

2. $\frac{9}{25}$

3. $\frac{25}{49}$

4. $\frac{36}{81}$

5. $25 + 144$

6. $65 - 29$

7. $25 - 9$

8. $\frac{9 + 16}{25 - 9}$

9. $\frac{25 + 144}{25 - 16}$

10. $\frac{35 - 19}{86 - 37}$

Problems

Example 19

In a class of schoolchildren, $\frac{1}{4}$ come to school by bus, $\frac{1}{3}$ ride bicycles to school and the remainder walk to school. If there are 24 children in the class, find

(a) the number who use the bus

(b) the number who walk to school.

(a) The number who use the bus

$$= \tfrac{1}{4} \text{ of } 24$$

$$= \frac{1}{\cancel{4}_1} \times \frac{\cancel{24}^6}{1}$$

$$= \frac{1 \times 6}{1 \times 1}$$

$$= 6$$

(b) Fraction who use either the bus or bicycle

$$= \tfrac{1}{4} + \tfrac{1}{3}$$

$$= \frac{3 + 4}{12}$$

$$= \tfrac{7}{12}$$

The complete class is a whole unit, i.e. 1.

Fraction who walk to school $= \tfrac{1}{1} - \tfrac{7}{12}$

$$= \frac{12 - 7}{12}$$

$$= \tfrac{5}{12}$$

Number who walk to school $= \tfrac{5}{12}$ of 24

$$= \frac{5}{\cancel{12}_1} \times \frac{\cancel{24}^2}{1}$$

$$= \frac{5 \times 2}{1 \times 1}$$

$$= 10$$

Exercise 3.12

1. Margaret spends $\tfrac{1}{3}$ of her pocket money and has 90 pence left. How much did she have to start with?

2. A woman in a clothing factory takes $\tfrac{3}{4}$ of a minute to sew a buttonhole. How long will it take her to sew 60 buttonholes?

3. A boy spends $\tfrac{1}{4}$ of his pocket money on a toy and $\tfrac{2}{3}$ on a record. What fraction of his pocket money has he spent and what fraction has he left?

4. In a youth club, $\tfrac{1}{3}$ of those present are playing table tennis whilst $\tfrac{1}{5}$ are playing pool. What fraction are not playing either game?

5. A snail crawls 1 metre in $1\tfrac{1}{4}$ minutes. How long will it take to crawl 32 metres?

6. A watering can holds $5\tfrac{3}{4}$ litres of water. It is filled 12 times from a tank containing 250 litres originally. How much water is left in the tank?

7. At a school, $\tfrac{1}{8}$ of the time is spent in mathematics classes, $\tfrac{1}{4}$ of the time in language classes and $\tfrac{1}{16}$ of the time playing games. The total time spent at school is 32 hours per week. Calculate

    (a) the amount of time spent in mathematics classes

    (b) the total time spent in mathematics and language classes

    (c) the amount of time spent in other subjects.

8. In a class of 30 children, 8 live in houses with gardens, 12 live in houses without gardens and the remainder live in flats. What fraction of the children

    (a) live in houses without gardens

    (b) live in houses?

9. Mary can iron a shirt in $4\tfrac{3}{4}$ minutes. How long will it take her to iron 12 shirts?

10. A cook adds $2\tfrac{1}{2}$ cups of milk to a pudding. If the cup holds $\tfrac{1}{10}$ of a litre, how much milk is added?

11. A bricklayer takes $\tfrac{3}{8}$ of a minute to lay one brick. How many bricks will he lay in 2 hours?

12. A comprehensive school has 1200 pupils. $\frac{2}{5}$ are in the upper school. $\frac{3}{8}$ in the middle school and the remainder in the lower school. Calculate the number of pupils

   (a) in the upper school

   (b) in the lower school.

Miscellaneous Exercise 3

Section A

1. Which of the fractions $\frac{1}{3}, \frac{1}{4}, \frac{1}{5}$ and $\frac{1}{6}$ is the largest?

2. What fraction of the shape shown in Fig. 3.5 has been shaded?

Fig. 3.5

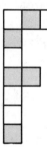

3. By how much is $\frac{8}{5}$ greater than $\frac{5}{8}$?

4. Divide $\frac{9}{16}$ by $\frac{5}{8}$.

5. Calculate $\frac{5}{6} + \frac{7}{8}$.

6. Calculate

$$\frac{7 \times (23 - 15)}{4 \times (12 + 9)}$$

leaving the answer as a fraction in its lowest terms.

7. What is $\frac{2}{3}$ of $1\frac{1}{2}$ hours?

8. Using only the numbers 2, 3, 4, 5 and 6, make up a fraction equivalent to $\frac{2}{3}$.

Section B

1. Express $\frac{3}{10} \times \frac{5}{6}$ as a single fraction.

2. Using only the numbers 2, 3, 4, 5, 6 make up a fraction between $\frac{3}{5}$ and $\frac{4}{5}$ in value.

3. Calculate $3\frac{2}{5} + 2\frac{1}{4}$.

4. Find the value of $1\frac{1}{4} + 2\frac{1}{8} - \frac{5}{8}$.

5. Calculate $\frac{1}{2} + \frac{1}{3} + \frac{1}{4}$.

6. A brand of tea is sold in packets containing $\frac{1}{4}$ kilogram. How many packets can be obtained from a tea-chest containing 103 kilograms?

7. Write down the next two terms of the sequence

$$1, \frac{1}{2}, \frac{1}{4}, \ldots$$

8. Consider the fractions $\frac{1}{2}, \frac{1}{3}, \frac{1}{4}, \frac{1}{5}$:

   (a) Which fraction is the largest?

   (b) What is the value of $\frac{1}{2} - \frac{1}{3}$?

   (c) What is the value of $\frac{1}{3} \times \frac{1}{5}$?

Multi-Choice Questions 3

1. The fraction $\frac{12}{16}$ is the same as
   A $\frac{1}{4}$ B $\frac{2}{6}$ C $\frac{3}{4}$ D $\frac{4}{3}$

2. What is the value of $\frac{3}{4} + \frac{1}{3}$?
   A $1\frac{1}{12}$ B $1\frac{1}{7}$ C $\frac{4}{7}$ D $\frac{3}{12}$

3. What is $\frac{1}{2} + (\frac{1}{2} \times \frac{1}{2})$?
   A $\frac{1}{8}$ B $\frac{1}{2}$ C $\frac{3}{4}$ D 1

4. Work out the value of $3 \times \frac{3}{4}$.
   A 1   B $1\frac{3}{4}$ C $2\frac{1}{4}$ D $3\frac{3}{4}$

5. Look at the fractions $\frac{2}{3}, \frac{1}{2}, \frac{7}{10}, \frac{5}{6}$. The largest fraction is
   A $\frac{2}{3}$ B $\frac{1}{2}$ C $\frac{7}{10}$ D $\frac{5}{6}$

6. The value of $2\frac{1}{4} \times 3\frac{1}{2}$ is

   A $6\frac{1}{8}$ B $6\frac{7}{8}$ C $7\frac{1}{8}$ D $7\frac{7}{8}$

7. The value of $\frac{3}{5} - \frac{5}{12}$ is

   A $\frac{2}{60}$ B $\frac{2}{17}$ C $\frac{11}{60}$ D $\frac{2}{7}$

8. What is the value of $3\frac{1}{5} \div \frac{4}{5}$?

   A $\frac{3}{4}$ B $3\frac{1}{4}$ C $3\frac{3}{5}$ D 4

Mental Test 3

Try to write down the answers to the following without writing anything else.

1. Fill in the missing number in $\frac{2}{3} = \frac{*}{9}$.

2. Reduce $\frac{3}{12}$ to its lowest terms.

3. Change $\frac{5}{4}$ to a mixed number.

4. Change $3\frac{1}{2}$ into a top-heavy fraction.

5. Is $\frac{2}{3}$ bigger than $\frac{3}{4}$?

6. Work out the value of $\frac{1}{4} + \frac{1}{2}$.

7. What is the value of $1 - \frac{2}{3}$?

8. What is $\frac{1}{2}$ of 20?

9. What is $6 \times \frac{1}{3}$?

10. What is $3 \div \frac{1}{2}$?

The Decimal System

Place Value

The number 4945 means four thousand nine hundred and forty-five. Each figure in the number has a place value because it is so many thousands, hundreds, tens or units according to its place in the number.

Notice that in the number 4945, the first four has a value of 4 thousand, i.e. 4000, whilst the other four has a value of 4 tens, i.e. 40.

The number 4945 means

$$4 \times 1000 + 9 \times 100 + 4 \times 10 + 5 \times 1$$

Money is usually written in decimal form, for instance £39.03, meaning thirty-nine pounds and three pence. The dot, called the **decimal point**, separates the pounds from the pence.

Now £39.03 is an amount of money between £39 and £40.

The 3p is part of a whole pound, i.e. it is a fraction of a whole pound. Since there are one hundred pence in a pound then three pence is three hundredths of a pound.

Pounds	Pence
39	03

Now consider the number 39.03. By comparing it with £39.03 we see that it means thirty nine and three hundredths.

Tens	Units	Tenths	Hundredths
3	9	0	3

We see that the figures to the left of the decimal point are whole numbers whilst the figures to the right of the decimal point are a fraction of a whole number. The number 39.03 means

$$3 \times 10 + 9 \times 1 + \frac{3}{100}$$

Notice that the nought in the first column after the decimal point keeps the place for the missing tenths.

Next consider the number 7.38.

Units	Tenths	Hundredths
7 ·	3	8

It means

$$7 \times 1 + \frac{3}{10} + \frac{8}{100}$$

We see then that:

The first place after the decimal point represents tenths.

The second place after the decimal point represents hundredths.

This can be extended to show:

The third place after the decimal point represents thousandths.

The fourth place after the decimal point represents ten-thousandths and so on.

Note carefully that one-tenth is ten times greater than one-hundredth and one-hundredth is ten times greater than one-thousandth and so on. Each column in the table opposite represents a number ten times larger than the number after it.

Units	Tenths	Hundredths	Thousandths
3 .	3	3	3

The number 3.333 means

$$\frac{3}{1} + \frac{3}{10} + \frac{3}{100} + \frac{3}{1000}$$

We can write three-tenths as .3 or $\frac{3}{10}$.

When there are no whole numbers it is usual to put a nought in front of the decimal point so that .3 becomes 0.3.

$$\frac{47}{100} \text{ can be written as } 0.47$$

Numbers such as 0.3, 8.59 and 736.28 are called **decimal numbers**.

Example 1

(a) Write down the place value of the figure 3 in the number 42.43.

Since the 3 occurs in the second column after the decimal point its value is 3 hundredths.

(b) Find the difference between the actual value of the two sevens in the number 7074.

The first seven has a value of 7000.
The second seven has a value of 70.
The difference between the actual values of the two sevens is
7000 − 70 = 6930.

Writing Fractions as Decimals

Example 2

Write the following fractions as decimals:

(a) $\frac{4}{10}$ **(b)** $\frac{7}{100}$ **(c)** $\frac{19}{100}$

(d) $\frac{8}{1000}$ **(e)** $\frac{29}{1000}$ **(f)** $\frac{287}{1000}$

(g) $5\frac{9}{10}$ **(h)** $8\frac{1}{100}$ **(i)** $9\frac{36}{1000}$

		Decimal number		
Fractional number	Units	Tenths $\frac{1}{10}$	Hundredths $\frac{1}{100}$	Thousandths $\frac{1}{1000}$
(a) $\frac{4}{10}$	0 .	4		
(b) $\frac{7}{100}$	0 .	0	7	
(c) $\frac{19}{100}$	0 .	1	9	
(d) $\frac{8}{1000}$	0 .	0	0	8
(e) $\frac{29}{1000}$	0 .	0	2	9
(f) $\frac{287}{1000}$	0 .	2	8	7
(g) $5\frac{9}{10}$	5 .	9		
(h) $8\frac{1}{100}$	8 .	0	1	
(i) $9\frac{36}{1000}$	9 .	0	3	6

In (c) $\frac{19}{100} = \frac{1}{10} + \frac{9}{100}$, so the 1 is written in the tenths column and the 9 in the hundredths column.

Similarly, in (e) $\frac{29}{1000} = \frac{2}{100} + \frac{9}{1000}$, so the 2 is written in the hundredths column and the 9 in the thousandths column.

In (f) $\frac{287}{1000} = \frac{2}{10} + \frac{8}{100} + \frac{7}{1000}$, so the 2 is written in the tenths column, the 8 in the hundredths column and the 7 in the thousandths column.

Exercise 4.1

Write down the place value of

1. The figure 7 in
   (a) 21.67   **(b)** 2.567
   (c) 274.34

2. The figure 5 in
   (a) 569.08   **(b)** 0.058
   (c) 19.765

3. The figure 8 in
   (a) 0.378   **(b)** 2.874
   (c) 2834.5

Write the following numbers as fractions with denominators (bottom numbers) of 10, 100 or 1000:

4. 0.3   **5.** 3.7   **6.** 20.08

7. 25.27   **8.** 0.308   **9.** 60.009

10. 0.004

11. What is the difference between the actual values of the two eights in the number 8058?

12. Write down the figures which is **(a)** in the tenths place of 4.523, **(b)** in the thousandths place of 189.2365, **(c)** in the hundredths place of 36.782.

13. Write down the following fractions in decimal form:

     (a) $\frac{9}{10}$ **(b)** $\frac{273}{1000}$ **(c)** $\frac{45}{100}$

     (d) $\frac{58}{1000}$ **(e)** $\frac{3}{100}$

14. Write down the following decimal numbers as fractions with denominators of 10, 100 or 1000:

     (a) 0.8   **(b)** 0.57   **(c)** 0.603

     (d) 0.009   **(e)** 0.053

15. **(a)** What does the 5 in 12.58 mean?

     (b) What does the 8 in 96.583 mean?

     (c) What does the 7 in 875.3 mean?

The Number Line for Decimals

Positive and negative numbers do not have to be whole numbers. They can also be fractions or decimal numbers, for example,

$\frac{3}{4}, -\frac{5}{8}, +\frac{1}{3}, 0.35, -0.687, 5.624,$

$-8.16, +9.2.$

Fig. 4.1 shows a number line for decimal numbers. The numbers -1.3, 2.4 and 5.8 have been labelled.

Fig. 4.1

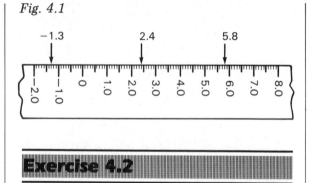

Exercise 4.2

Make a copy of the scale shown in Fig. 4.1. Draw arrows and label the following points:

1. 7.2   **2.** 2.9   **3.** 4.6

4. 0.8   **5.** -0.7 **6.** 1.2

7. 3.9   **8.** -1.9

By looking at the number line (Fig. 4.1) decide which number is the higher:

9. 0.4 or 0.9   **10.** 3.3 or 4.8

11. -2.0 or 2.0   **12.** -0.8 or 0.3

13. 3.6 or 6.3   **14.** 0.2 or 2.0

15. -1.5 or 1.5   **16.** -0.5 or -1.8

17. -1.8 or -1.2 **18.** 0.1 or -1.3

Necessary and Unnecessary Zeros

A possible source of confusion is deciding which zeros are necessary and which are not:

(1) Zeros are not needed after the last figure of a number (but they are necessary to show how accurate a number is. For instance, 8.300 metres indicates that the measurement is correct to the nearest millimetre, i.e. to the nearest 0.001 metre)

     $5.400 = 5.4$ $8.00 = 8$

     $9.70 = 9.7$ $16.0 = 16$

(2) Zeros are not needed in front of whole numbers.

     $07.3 = 7.3$

(3) Zeros are needed to keep the place for missing hundreds, tens, units, tenths, hundredths, thousandths, etc. The noughts in these numbers are needed:

$$70, 306, 800, 7.005$$

(4) One zero is usually left before the decimal point if there are no whole numbers but this is not essential:

.837 is usually written 0.837

.058 is usually written 0.058

(5) Sometimes noughts are placed in front of whole numbers. For instance James Bond is known as 007, and a bearing is sometimes stated as a three figure number, for instance 036°.

Exercise 4.3

1. (a) Is 4.3 a positive decimal number?
   (b) Is $-\frac{3}{8}$ a positive fraction?
   (c) Is +28 a positive whole number?

2. Does 5 equal
   (a) 5.0? (b) 50? (c) 5.00?
   (d) 500?

3. Write the following numbers without a decimal point:
   (a) 7.00   (b) 60.0
   (c) 500.00   (d) 400.000

4. Write the following numbers in order of size, smallest first:
   (a) 103, 1.03, 10.3, 0.103
   (b) 50.4, 504, 5.04, 0.504
   (c) 0.025, −2.5, 25, −0.25, 0.25
   (d) 0.151, 0.0151, 0.15, 0.115
   (e) 5.205, −52.05, 5.502, 5.025, −5.052

5. Write the following numbers without unnecessary zeros (assuming no indication of accuracy is intended):
   (a) 48.90   (b) 4.000
   (c) 0.5000   (d) 500.00
   (e) 108 070   (f) 00.065
   (g) 0.90   (h) 14 000 000.00

Adding and Subtracting Decimals

When adding or subtracting decimals manually, it is important to line up the decimal points. If whole numbers are present then put a decimal point at the end of the whole number and add zeros as required. This will help in lining up the number correctly.

Example 3

Add 127.35, 28, 0.37 and 19.6.

28.00 is the same as 28 and 19.60 is the same as 19.6

$$\begin{array}{r} 127.35 \\ 28.00 \\ 0.37 \\ \underline{19.60\,+} \\ \underline{175.32} \end{array}$$

To do the addition, add upwards. To check the answer add downwards.

Note that the decimal point in the answer is placed underneath all the other decimal points.

When subtracting it is often helpful to add zeros so that there is the same number of figures following the decimal point in both numbers.

Example 4

Subtract 38.192 from 93.2.

$$93.200$$
$$\underline{38.192} -$$
$$\underline{55.008}$$

Therefore

$$93.2 - 38.192 = 55.008$$

The answer can be checked by adding the bottom two lines (38.192 + 55.008). If the calculation is correct their sum will equal the top line (i.e. 93.2).

Most problems involving addition and subtraction of decimal numbers will be done on a calculator.

Example 5

Using a calculator, find the sum of 58.36, 0.2, 8 and 37.532.

Input	Display
58.36	58.36
+	58.36
0.2	0.2
+	58.56
8	8.
+	66.56
37.532	37.532
=	104.092

Hence

$$58.36 + 0.2 + 8 + 37.532 = 104.092$$

The addition may be checked by inputting the numbers in reverse order, i.e. 37.532 + 8 + 0.2 + 58.36.

Example 6

Find the value of $37.46 - 9.2$.

Input	Display
37.46	37.46
−	37.46
9.2	9.2
=	28.26

Hence

$$37.46 - 9.2 = 28.26$$

This subtraction can be checked by adding 9.2 to 28.26. If this equals 37.46 the answer is correct.

Example 7

Work out $8.23 - 25.7 - 48.31 + 5.63 + 64.2$.

Input	Display
8.23	8.23
−	8.23
25.7	25.7
−	−17.47
48.31	48.31
+	−65.78
5.63	5.63
+	−60.15
64.2	64.2
=	4.05

Hence

$$8.23 - 25.7 - 48.31 + 5.63 + 64.2 = 4.05$$

The work may be checked by inputing the numbers in reverse order, i.e. $64.2 + 5.63 - 48.31 - 25.7 + 8.23$.

Exercise 4.4

Using a calculator find the values of

1. $2.389 + 0.325$

2. $5.16 + 12.2$

3. $5.189 + 17.23 + 873.2$

4. $0.362 + 0.085 + 0.009 + 0.169$

5. Add 87.26 to 1.075

6. Find the total of 18.37, 5.162 and 398.03.

7. Find the sum of 0.065, 0.198, 0.003 and 0.798.

8. Increase 15.635 by 0.078.

9. Find the value of 13.18 minus 12.06.

10. Work out $28.153 - 9.6$.

11. Find the difference between 27.076 and 84.6.

12. Subtract 298.075 from 387.1.

13. Add together 65.3, 39, 0.675 and 18.23.

Work out each of the following:

14. $87.3 - 14.62 + 2.83 - 54.61$

15. $16.34 - 27.8 - 39 - 8.26 + 87.34 + 28.67$

16. $7.21 - 3.48 - 11.57 - 53.2 + 36.91 + 85.76$

17. $72.1 - 11.81 - 14.29 + 5.62 + 28.32 - 2.89$

18. $76.25 + 87.634 - 76.5 - 98.237 + 234.2 - 223.1$

19. $13.6 + 98.2 - 12.3 - 0.7 - 6.08$

20. $23.55 + 34.98 - 24.5 - 23.8 + 11.32$

Multiplying by 10, 100 and 1000

Decimal numbers can be easily multiplied by 10, 100, 1000, etc. These operations are needed when working with metric quantities.

Example 8

(a)
$$5.3 \times 10 = 5\frac{3}{10} \times \frac{10}{1}$$
$$= \frac{53}{10} \times \frac{10}{1}$$
$$= \frac{53 \times 1}{1 \times 1}$$
$$= 53$$

(b)
$$5.3 \times 100 = 5\frac{3}{10} \times \frac{100}{1}$$
$$= \frac{53}{10} \times \frac{100}{1}$$
$$= \frac{53 \times 10}{1 \times 1}$$
$$= 530$$

From Example 8 we see that:

(1) To multiply by 10, move all the figures one place to the left:
$$9.632 \times 10 = 96.32$$
$$0.078 \times 10 = 0.78$$

(2) To multiply by 100, move all the figures two places to the left:
$$87.34 \times 100 = 8734$$
$$9.635 \times 100 = 963.5$$

(3) Using the same method we can show that to multiply by 1000, move all the figures three places to the left:
$$875.632 \times 1000 = 875\,632$$
$$0.005\,68 \times 1000 = 5.68$$
$$6.32 \times 1000 = 6320$$

Dividing by 10, 100 and 1000

Decimal numbers can also be easily divided by 10, 100, 1000, etc.

Example 9
$$2.7 \div 10 = 2\frac{7}{10} \div \frac{10}{1}$$
$$= \frac{27}{10} \times \frac{1}{10}$$
$$= \frac{27 \times 1}{10 \times 10}$$
$$= \frac{27}{100}$$
$$= 0.27$$

From Example 9 we see that to divide a number by 10, move all the figures one place to the right:

$$47.36 \div 10 = 4.736$$

$$0.358 \div 10 = 0.0358$$

Using a similar method we can show that to divide a number by 100, move all the figures two places to the right:

$$498.75 \div 100 = 4.9875$$

$$3.685 \div 100 = 0.03685$$

We can also show that to divide a number by 1000, move all the figures three places to the right:

$$8765 \div 1000 = 8.765$$

$$25.362 \div 1000 = 0.025362$$

Exercise 4.5

(This exercise should be tackled mentally.) Multiply each of the following numbers by **(a)** 10, **(b)** 100 and **(c)** 1000:

1. 0.35

2. 5.983

3. 0.038

4. 98.2345

5. 8.1624

6. 0.046

7. 0.00058

8. 0.009

Divide each of the following numbers by **(a)** 10, **(b)** 100 and **(c)** 1000:

9. 189

10. 18.13

11. 527.31

12. 0.03

13. 0.325

14. 0.0028

15. 5.62

Multiplication of Decimals

Generally the operation of multiplying decimal numbers will be performed on a calculator. However, very simple numbers should be multiplied mentally.

The main problem when multiplying decimal

numbers is that of placing the decimal point in the answer. The following rules will help in this respect:

(1) First multiply the numbers as if there were no decimal point.

(2) Add together the number of figures (including noughts) after the decimal points in the two original numbers. This gives the number of figures following the decimal point in the answer.

The number of figures following the decimal point is called the **number of decimal places.** Thus

0.3 has one decimal place

3.65 has two decimal places

Example 10

(a) Multiply 7.2 by 3.

$$7.2 \times 3 = 21.6 \quad 72 \times 3 = 216$$
(1 place) (no places) (1 place)

(b) Multiply 0.2×0.03

$$0.2 \times 0.03 = 0.006 \quad 2 \times 3 = 6$$
(1 place) (2 places) (3 places)

(c) Multiply 20 by 0.4.

$$20 \times 0.4 = 8.0 \quad 20 \times 4 = 80$$
(no places) (1 place) (1 place)

Exercise 4.6

Without using a calculator, work out the value of each of the following:

1. 0.4×5

2. 0.3×6

3. 0.5×3

4. 0.04×0.3

5. 0.1×0.1

6. 0.003×0.2

7. 0.001×0.03

8. 0.4×0.03

9. 4.1×0.3

10. 6.2×0.2

11. 5.3×0.002

12. 20×0.03

13. 8.3×0.001 **14.** 30×0.02

15. 200×0.3 **16.** 300×0.001

17. 50×0.002 **18.** 5000×0.003

Example 11

Using a calculator multiply 16.25 by 0.35.

Input	Display
16.25	16.25
$\times$	16.25
0.35	0.35
$=$	5.6875

Hence

$$16.25 \times 0.35 = 5.6875$$

This multiplication can be checked by multiplying in the reverse order, i.e. 0.35×16.25.

Exercise 4.7

Using a calculator work out the following:

1. 6.9×5 **2.** 4.3×7

3. 16.9×6 **4.** 43.5×3

5. 27.19×15 **6.** 432.6×27

7. 1.6×0.2 **8.** 5.9×1.3

9. 27.12×0.47 **10.** 4.03×2.73

11. 83.6×4.9 **12.** 0.07×1.8

13. 4.25×2.72 **14.** 2.06×1.9

15. 2.251×0.93

16. Multiply 15.12 by 0.95.

17. Find the product of 0.78 and 0.03.

18. Multiply 0.075×3.26.

19. Find the value of 20.23×0.062.

20. Calculate the product of 3.175 and 1.23.

Division of Decimals

Simple numbers may be divided without using a calculator.

Example 12

Divide 4.8 by 2.

$$\begin{array}{r} 2\overline{)4.8} \\ 2.4 \end{array}$$ keep the figures and the decimal points in line

$$4.8 \div 2 = 2.4$$

Example 13

Divide 0.012 by 0.4.

$0.012 \div 0.4$ can be written $\dfrac{0.012}{0.4}$

From Example 12 we know how to divide by a whole number so we need an equivalent fraction with a denominator of 4 instead of 0.4. Therefore we multiply the numerator and denominator by 10.

$$\frac{0.012}{0.4} = \frac{0.012 \times 10}{0.4 \times 10} = \frac{0.12}{4} = 0.03$$

$$\begin{array}{r} 4\overline{)0.12} \\ 0.03 \end{array}$$

A division can always be checked by multiplying back.

In Example 12 the check is $2 \times 2.4 = 4.8$.

In Example 13 the check is $0.4 \times 0.03 = 0.012$.

Exercise 4.8

This exercise should be tackled without using a calculator.

Work out the value of each of the following:

1. $3.6 \div 4$ **2.** $2.5 \div 5$

3. $4.6 \div 2$ **4.** $9.6 \div 3$

5. $7.2 \div 9$ 6. $0.4 \div 0.2$

7. $0.8 \div 0.02$ 8. $0.09 \div 0.3$

9. $90 \div 0.03$ 10. $0.14 \div 0.07$

11. $0.56 \div 0.008$ 12. $8.4 \div 0.7$

13. $0.0015 \div 0.03$ 14. $4.8 \div 0.08$

15. $6.5 \div 0.005$

Division will generally be done by using a calculator.

Example 14

Use a calculator to divide 3.62 by 0.25.

Input	Display	
3.62	3.62	
÷	3.62	
0.25	0.25	
×	14.48	(this is the answer)
0.25	0.25	
=	3.62	(this is the check)

The check depends upon the fact that if

$$3.62 \div 0.25 = 14.48$$

then $14.48 \times 0.25 = 3.62$

Exercise 4.9

A calculator may be used for this exercise.

Work out the value of each of the following:

1. $52.5 \div 3$ 2. $48.5 \div 5$

3. $45.32 \div 4$ 4. $715.6 \div 4$

5. $0.819 \div 9$ 6. $4.24 \div 0.4$

7. $35.7 \div 0.07$ 8. $0.128 \div 0.8$

9. $8.8 \div 0.11$ 10. $0.3372 \div 0.003$

11. $1.428 \div 0.35$

12. $0.007\,263 \div 0.807$

13. Divide 0.0196 by 0.14.

14. Divide 0.002 55 by 0.015.

15. Divide 10.8 by 1.25.

Changing Fractions into Decimals

In Chapter 3 we discovered that $\frac{3}{4}$ was the same as $3 \div 4$. The line separating the numerator and the denominator acts as a division sign.

Example 15

(a) $\frac{2}{5} = 2 \div 5 = 0.4$

(b) $2\frac{3}{8} = 2 + \frac{3}{8} = 2 + (3 \div 8) = 2 + 0.375$
$$= 2.375$$

Exercise 4.10

Use a calculator to change the following fractions into decimal numbers:

1. $\frac{3}{4}$ 2. $\frac{1}{5}$ 3. $\frac{7}{8}$

4. $\frac{13}{16}$ 5. $\frac{1}{20}$ 6. $\frac{3}{25}$

7. $\frac{9}{20}$ 8. $\frac{4}{5}$ 9. $3\frac{2}{5}$

10. $1\frac{5}{8}$ 11. $7\frac{1}{2}$ 12. $6\frac{1}{4}$

13. $4\frac{1}{8}$ 14. $6\frac{1}{5}$ 15. $9\frac{5}{16}$

Converting Decimals into Fractions

To convert a decimal into a fraction we use the fact that decimal numbers are fractions with denominators of 10, 100, 1000, etc.

Example 16

To change a decimal number into a fraction we must remember the headings in the table below.

Units	Tenths	Hundredths	Thousandths
0 ·	7		

We see that

$$0.7 = \tfrac{7}{10} \text{ (seven tenths)}$$

Units	Tenths	Hundredths	Thousandths
0 .	5	3	
0 .	8	7	9

$$0.53 = \tfrac{5}{10} + \tfrac{3}{100} = \tfrac{50}{100} + \tfrac{3}{100} = \tfrac{53}{100}$$

$$0.879 = \tfrac{8}{10} + \tfrac{7}{100} + \tfrac{9}{1000}$$

$$= \tfrac{800}{1000} + \tfrac{70}{1000} + \tfrac{9}{1000}$$

$$= \tfrac{879}{1000}$$

Sometimes we need to reduce these fractions to their lowest terms by dividing top and bottom by the same number:

$$0.8 = \tfrac{8}{10} = \tfrac{4}{5}$$

$0.07 = \tfrac{7}{100}$ which is already in its lowest terms

$$0.875 = \tfrac{875}{1000} = \tfrac{175}{200} = \tfrac{35}{40} = \tfrac{7}{8}$$

(dividing each time by 5)

Example 17

Calculate the value of $2\tfrac{5}{16} - 2.3014$.

$$2\tfrac{5}{16} = 2 + (5 \div 16) = 2 + 0.3125$$

$$= 2.3125$$

$$2.3125 - 2.3014 = 0.0111$$

Exercise 4.11

Convert the following decimal numbers into fractions in their lowest terms:

1. 0.3   2. 0.55   3. 0.38

4. 0.52   5. 0.16   6. 0.175

7. 0.125   8. 0.625   9. 7.36

10. 2.06

11. Find the difference between $\tfrac{2}{5}$ and 0.58, giving the answer as a decimal.

12. Calculate the value of $2\tfrac{3}{8} + 0.763$, stating the answer as a decimal.

13. What is the difference between $5\tfrac{3}{4}$ and 5.91?

14. Work out the sum of $6\tfrac{1}{5}$ and 0.36 stating the answer as a decimal.

15. Calculate the value of $3\tfrac{7}{8}$ minus 2.29 stating the answer as a decimal.

Problems Involving Decimals

The examples which follow are of a practical nature. They all depend upon decimals for their solution.

Example 18

(a) A train consists of 35 trucks. 18 of them carry 18.4 tonnes each and the remainder carry 17.6 tonnes each. What is the total weight carried by the train?

Weight carried by 18 trucks

$$= 18 \times 18.4 \text{ tonnes}$$

$$= 331.2 \text{ tonnes}$$

Weight carried by 17 trucks

$$= 17 \times 17.6 \text{ tonnes}$$

$$= 299.2 \text{ tonnes}$$

Total weight carried

$$= (331.2 + 299.2) \text{ tonnes}$$

$$= 630.4 \text{ tonnes}$$

(b) Fig. 4.2 shows a shaft for an electrical machine. Calculate the dimension marked A in the diagram.

Fig. 4.2

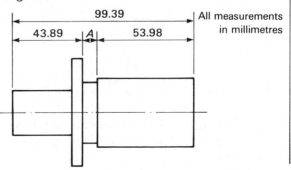

Dimension $A = 99.39 - (43.89 + 53.98)$

$$= 99.39 - 97.87$$

$$= 1.52 \text{ millimetres}$$

This calculation is best done using a calculator as follows:

Input	Display
99.39	99.39
—	99.39
43.89	43.89
—	55.5
53.98	53.98
=	1.52

Therefore

Dimension $A = 99.39 - 43.89 - 53.98$

$$= 1.52 \text{ millimetres}$$

Exercise 4.12

1. A lorry weighs 2.7 tonnes when empty. It is loaded with 52 steel bars each weighing 0.038 tonnes. What is the total weight of the loaded lorry?

2. A herd of cows consists of 34 cows. If the herd yields 622.2 litres of milk in a morning, how much milk, on average, does each cow yield?

3. Fig. 4.3 shows a plate with two holes drilled in it. Calculate the dimension marked A in the diagram.

Fig. 4.3

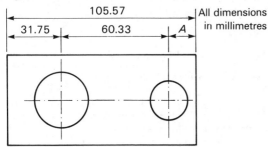

All dimensions in millimetres

4. A car travels 76.5 kilometres on 6 litres of petrol. How far does it travel per litre?

5. 16 holes spaced 47.3 millimetres apart are to be marked off on a sheet metal detail. 17.6 millimetres is to be allowed between the centres of the end holes and the edges of the plate. Calculate the total length of plate required.

6. A lady goes to her bank on four occasions. On the first three occasions she deposits £21.36, £87.53 and £39.84 whilst on the fourth occasion she withdraws £109.35. How much of the money she deposited has she left in her account?

7. It takes 37 pieces of wallpaper to paper a room. If each piece is 2.35 metres long, calculate the total length of wallpaper required. If a roll of wallpaper contains 7 metres, calculate the number of rolls needed to paper the room.

8. The resistance of a copper wire is 53.8 ohms per metre length. Find the resistance of a wire 19.3 metres long.

Estimation

Suppose that you want to put an extension on to your house. You would explain to the builder what you require and ask him to prepare an **estimate**. This estimate is written to give you some idea how much the extension is likely to cost. It may be a little more or a little less than the actual cost.

To make his estimate the builder would have to work out roughly what he would need to charge for building materials, electrical work, plumbing work and labour. His estimate might be as follows:

Building materials	£2500
Electrical work	£300
Plumbing work	£500
Labour	£3600
Final estimate	£6900

His costs would only be approximate (i.e. not exactly correct but nearly correct). The costs given opposite are correct to the nearest £100 and we say that the figures have been **rounded** to the nearest £100.

The rounding of whole numbers has been dealt with on page 3. We now deal with the rounding of decimal numbers.

Decimal Places

Decimal numbers can be rounded off to the nearest number of tenths, hundredths, etc. For instance

5.63 is 5.6 rounded off to the nearest tenth

17.378 is 17.38 rounded off to the nearest hundredth

0.35 rounded off to the nearest tenth is 0.4.

However, with decimal numbers it is more usual to state the number correct to so many **decimal places** (often shortened to d.p.). The number of decimal places means the number of figures after the decimal point:

0.36 has two decimal places

7.93 has two decimal places

52.038 has three decimal places

The rule for decimal places is:

If the first figure to be discarded is 5 or greater, then the previous figure is increased by 1.

Example 19

93.7257 is 93.726 correct to 3 d.p.
is 93.73 correct to 2 d.p.
is 93.7 correct to 1 d.p.

0.007 682 is 0.007 68 correct to 5 d.p.
is 0.0077 correct to 4 d.p.
is 0.008 correct to 3 d.p.

8.703 is 8.70 correct to 2 d.p.
is 8.7 correct to 1 d.p.

Exercise 4.13

Write down the following numbers correct to the number of decimal places stated:

1. 19.372   (a) 2 d.p.   (b) 1 d.p.

2. 0.007 519   (a) 5 d.p.   (b) 3 d.p.
(c) 2 d.p.

3. 4.9703   (a) 3 d.p.   (b) 2 d.p.

4. 153.2617   (a) 3 d.p.   (b) 2 d.p.
(c) 1 d.p.

5. 34.1572   (a) 3 d.p.   (b) 2 d.p.
(c) 1 d.p.

The Accuracy of Arithmetic Calculations

Most calculators have an eight figure display, but for most calculations this is far too many figures. Frequently the figures used in a calculation have themselves been rounded. Thus if we have to calculate the product of 7.36 and 8.62 it is likely that both numbers have been stated correct to 2 d.p. Therefore the product should not be stated to an accuracy greater than this. Thus

$$7.36 \times 8.62 = 63.4432$$
$$= 63.44 \text{ correct to 2 d.p.}$$

If the numbers in a calculation have been rounded, the answer should not contain more decimal places than the least number of decimal places used amongst the given numbers.

Example 20

$$1.3 \times 7.231 \times 1.24 = 11.656 372$$
$$= 11.7 \text{ correct to 1 d.p.}$$

Note that the least number of decimal places in the numbers is 1 (for the number 1.3) so the answer should not be stated to an accuracy greater than 1 d.p.

Exercise 4.14

Each number in this exercise has been rounded off to the accuracy given.

Find the value for each calculation, correctly rounded:

1. 15.64×19.75

2. $14.6 \times 8.73 \times 9.156$

3. $13.96 \div 0.425$

4. $15.78 \div 13.279$

5. $9.153 \div 3.2$

Rough Estimates for Calculations

Before or after doing a calculation you should always make a rough estimate to make sure that the answer you produce is sensible. You can often see if there is a mistake by looking at the size of the answer. In doing a rough estimate try to select numbers which are easy to add, subtract, multiply or divide.

Example 21

(a) Multiply 32.4 by 0.259.

Since $0.25 = \frac{25}{100} = \frac{1}{4}$, for a rough estimate we will take

$$32 \times 0.25 = 32 \times \tfrac{1}{4} = 8$$

Accurate calculation:

$32.7 \times 0.259 = 8.4$ (correct to 1 d.p.)

The rough estimate shows that the answer is sensible.

(b) Add $5.32, 0.925$ and 17.81.

For a rough estimate we will take

$$5 + 1 + 18 = 24$$

Accurate calculation:

$5.32 + 0.925 + 17.81 = 24.055$

The rough estimate shows that the answer is sensible.

(c) Work out $\dfrac{47.5 \times 36.52}{11.3 \times 2.75}$.

For the rough estimate we should use numbers which will cancel, i.e.

$$\frac{\overset{5}{\cancel{50}} \times \overset{12}{\cancel{36}}}{\underset{1}{\cancel{10}} \times \underset{1}{\cancel{3}}} = \frac{5 \times 12}{1 \times 1} = 60$$

Accurate calculation:

$\dfrac{47.5 \times 36.52}{11.3 \times 2.75} = 55.8$ (correct to 1 d.p.)

The rough estimate shows that the answer is sensible.

Exercise 4.15

Do a rough estimate for each of the following and then, using a calculator, work out the accurate answers to the correct number of decimal places:

1. 0.23×29.62

2. $4.81 + 0.824 + 27.32$

3. $77.915 + 38.23 - 9.45 - 27.62$

4. $0.832 \times 0.097 \times 2.22$

5. $91.56 \div 31.35$

6. $0.092 \div 0.034$

7. $\dfrac{3.28 \times 8.19}{4.16}$

8. $\dfrac{29.92 \times 31.32}{10.89 \times 2.95}$

Miscellaneous Exercise 4

Section A

1. Calculate:
   (a) 2.4×6 (b) $2.4 \div 6$
   (c) 2.4×0.6 (d) $2.4 \div 0.6$

2. Calculate:
   (a) 36.9×9 (b) 2.4×1.1
   (c) $32.24 \div 20$

3. Write as decimal numbers:
   (a) $\frac{5}{8}$ (b) $\frac{7}{16}$

4. Express $81 \div 108$ as
   (a) a fraction in its lowest terms
   (b) a decimal number.

5. The cost of making a telephone call is 4.3 pence per call. Work out the cost of making 309 calls.

6. Calculate:
   (a) 73.8×1000 (b) $73.8 \div 100$

7. Which is the larger, 23 or 5.6, and by how much?

8. Work out $3 + 0.36 - 0.47 - 2.86 + 1.4$.

9. In the number 43.046
   (a) write down the actual value of the digit 6
   (b) calculate the difference between the actual values of the two digits 4.

10. A recent cricket match between Lancashire and Yorkshire attracted a crowd of 9478. Round off this attendance to the nearest 100.

11. Rearrange the numbers 0.3, 0.03 and 0.11 in order of size, starting with the number with the greatest value.

12. Find the total of 2.8, 174.08 and 0.314.

13. Express as decimal numbers:
    (a) $\frac{7}{8}$ (b) $\frac{5}{32}$

14. Express as common fractions in their lowest terms:
    (a) 0.7   (b) 0.12

Section B

1. Use a calculator to work out
   (a) 15.8×80.7 correct to the nearest whole number
   (b) the square of 14.7
   (c) the square root of 44.5 correct to 1 d.p.
   (d) $498 \div 3.14$ to the nearest tenth.

2. By rounding off each number to the nearest number of tens, find the best estimate for
$$\frac{27.5 \times 60.52}{11.3 \times 20.51}$$

3. The resistance of a copper wire 55.2 metres long is 3030.48 ohms. Work out the resistance per metre.

4. Find the value of
   (a) 7.54×9.7 to the nearest hundredth
   (b) $\dfrac{2.6 + 2.94}{4.54 + 3.77}$ correct to 2 d.p.

5. A roll of wallpaper contains 10.05 metres. To paper a room, lengths 2.35 metres long are cut from the roll:
   (a) How many full lengths can be cut from a roll?
   (b) What length is left? Give the answer to the nearest tenth of a metre.

6. A consignment of steel nuts weighs 5271 grams. If each nut weighs 7.53 grams, calculate the number of nuts in the consignment.

7. Which is the greater, $\frac{7}{16}$ or 0.4573, and by how much?

8. A steel angle bar has a weight of 19.05 kg per metre length. Calculate the weight of an angle bar which is 18.2 metres long.

Multi-Choice Questions 4

1. In the number 9.8392 the value of the digit 3 is

    A $\frac{3}{1000}$ B $\frac{3}{100}$ C $\frac{3}{10}$ D 300

2. 304.847 correct to 2 decimal places is

    A 30   B 31
    C 304.84   D 304.85

3. $2\frac{1}{5}$ written as a decimal number is

    A 2.01 B 2.02 C 2.1 D 2.2

4. The number 0.030 905 correct to 3 decimal places is

    A 0.030   B 0.0309
    C 0.031   D 0.039

5. 0.72 expressed as a fraction in its lowest terms is

    A $\frac{72}{1000}$ B $\frac{18}{25}$ C $\frac{7}{10}$ D $7\frac{1}{5}$

6. The largest of the numbers 0.22, 2.20, 0.022 and 2.02 is

    A 2.02   B 0.22
    C 2.20   D 0.022

7. $\frac{5}{8}$ written as a decimal number is

    A 0.13 B 0.58 C 0.625 D 1.6

Mental Test 4

Try to write down the answers to the following questions without writing anything else.

1. Write down the place value of the digit 5 in the number 43.85.

2. What is the difference between the actual values of the two threes in the number 13.32?

3. Write the fraction $\frac{31}{100}$ as a decimal number.

4. Write the fraction $\frac{9}{1000}$ as a decimal number.

5. Add $5 + \frac{1}{100} + \frac{7}{1000}$ giving the answer as a decimal number.

6. Write down the figure which is in the hundredths place of the number 342.578.

7. Write down which number, 5.6 or 6.5, is the larger.

8. Write 23 as a decimal number with two zeros after the decimal point.

9. Write the following numbers in order of size, smallest first: 203, 2.03, 20.3.

10. Add 1.2, 1.3 and 2.5.

11. Find the sum of 0.21 and 2.1.

12. Subtract 0.5 from 5.0.

13. Find the difference between 4.0 and 0.4.

14. Multiply 0.6 by 4.

15. Divide 0.68 by 2.

16. Change $\frac{4}{5}$ into a decimal number.

17. Round off £273 to the nearest £10.

18. Round off 883 to the nearest 100.

19. Round off 7.36 to the nearest tenth.

20. Round off 0.769 to the nearest hundredth.

Measurement

Measurement of Length

In the metric system the standard unit of length is the metre (abbreviation m). For some purposes the metre is too large a unit and so it is split up into smaller units as follows:

$$1 \text{ metre (m)} = 10 \text{ decimetres (dm)}$$
$$= 100 \text{ centimetres (cm)}$$
$$= 1000 \text{ millimetres (mm)}$$

In dealing with long distances the kilometre is used.

$$1 \text{ kilometre (km)} = 1000 \text{ metres (m)}$$

Because the metric system is essentially a decimal system, it is easy to convert from one unit to another by multiplying or dividing by 10, 100 and 1000. The decimetre is rarely used but the other units are frequently used.

Example 1

(a) Convert 15 m into millimetres.
$$15 \text{ m} = 15 \times 1000 \text{ mm}$$
$$= 15\,000 \text{ mm}$$

(b) Change 900 cm into metres.
$$900 \text{ cm} = 900 \div 100$$
$$= 9 \text{ m}$$

(c) Change 850 mm into centimetres.
$$850 \text{ mm} = 850 \div 10$$
$$= 85 \text{ cm}$$

(d) Change 72 000 m into kilometres.
$$72\,000 \text{ m} = 72\,000 \div 1000 = 72 \text{ km}$$

Exercise 5.1

This exercise should be done mentally.

1. Change to millimetres:
   (a) 3 cm   (b) 28 cm
   (c) 134 cm   (d) 5 m
   (e) 63 m   (f) 4.6 m

2. Change to centimetres:
   (a) 60 mm   (b) 240 mm
   (c) 68 mm   (d) 4 m
   (e) 56 m   (f) 3.74 m

3. Change to metres:
   (a) 5000 mm   (b) 700 cm
   (c) 890 cm   (d) 5643 mm
   (e) 5 km   (f) 6.42 km

4. Change to kilometres:
   (a) 7000 m   (b) 6340 m
   (c) 8325 m

In the imperial system small lengths are measured in inches whilst longer lengths are measured in feet and yards. Very long lengths are measured in miles.

$$12 \text{ inches (in)} = 1 \text{ foot (ft)}$$
$$3 \text{ feet (ft)} = 1 \text{ yard (yd)}$$
$$1760 \text{ yards (yd)} = 1 \text{ mile}$$

Example 2

(a) Change 48 in into feet.

$$48 \text{ in} = 48 \div 12 \text{ ft}$$
$$= 4 \text{ ft}$$

(b) Change 24 ft into yards.

$$24 \text{ ft} = 24 \div 3 \text{ yd}$$
$$= 8 \text{ yd}$$

(c) Change 12 320 yd into miles.

$$12\,320 \text{ yd} = 12\,320 \div 1760 \text{ miles}$$
$$= 7 \text{ miles}$$

Exercise 5.2

Use a calculator where necessary for this exercise.

1. Change to feet:
   (a) 24 in   (b) 60 in
   (c) 144 in   (d) 5 yd
   (e) 80 yd

2. Change to yards:
   (a) 6 ft   (b) 72 ft
   (c) 900 ft   (d) 2 miles
   (e) 28 miles

3. Change to inches:
   (a) 5 ft   (b) 20 ft
   (c) 34 ft   (d) 3 yd
   (e) 15 yd

4. Change to miles:
   (a) 5280 yd   (b) 15 840 yd
   (c) 35 200 yd

Weight

In the metric system, light objects are weighed in milligrams or grams. Heavier objects are weighed in kilograms but very heavy objects are weighed in tonnes.

$$1 \text{ gram (g)} = 1000 \text{ milligrams (mg)}$$
$$1 \text{ kilogram (kg)} = 1000 \text{ grams (g)}$$
$$1 \text{ tonne (t)} = 1000 \text{ kilograms (kg)}$$

Example 3

(a) Change 8000 g into kilograms.

$$8000 \text{ g} = 8000 \div 1000 \text{ kg}$$
$$= 8 \text{ kg}$$

(b) Change 5 kg into grams.

$$5 \text{ kg} = 5 \times 1000 \text{ g}$$
$$= 5000 \text{ g}$$

(c) Change 18 000 mg into grams.

$$18\,000 \text{ mg} = 18\,000 \div 1000 \text{ g}$$
$$= 18 \text{ g}$$

Exercise 5.3

This exercise should be done mentally.

1. Change to grams:
   (a) 8 kg   (b) 19 kg
   (c) 15 kg   (d) 12 000 mg
   (e) 27 000 mg

2. Change to kilograms:
   (a) 5000 g   (b) 18 000 g
   (c) 3 t   (d) 18 t

3. Change to milligrams:
   (a) 7 g (b) 24 g (c) 0.5 g

4. Change to tonnes:
   (a) 8000 kg (b) 427 000 kg
   (c) 600 kg

In the imperial system light objects are weighed in ounces. Heavier objects are measured in pounds whilst very heavy objects are measured in hundredweights or tons.

16 ounces (oz) = 1 pound (lb)

112 pounds (lb) = 1 hundredweight (cwt)

1 ton = 20 hundredweight (cwt)

= 2240 lb

Example 4

(a) How many ounces are there in 2 lb?

$$2\,lb = 2 \times 16\,oz$$
$$= 32\,oz$$

(b) Change 368 oz to pounds.

$$368\,oz = 368 \div 16\,lb$$
$$= 23\,lb$$

(c) Change 672 lb into hundredweight.

$$672\,lb = 672 \div 112\,cwt$$
$$= 6\,cwt$$

(d) How many pounds are there in 4 tons?

$$4\,tons = 4 \times 2240\,lb$$
$$= 8960\,lb$$

Exercise 5.4

A calculator may be used for this exercise.

1. Change to ounces:
   (a) $\frac{1}{4}$ lb (b) $\frac{1}{2}$ lb (c) $4\frac{1}{2}$ lb
   (d) 5 lb (e) 8 lb

2. Change to pounds:
   (a) 12 oz (b) 48 oz
   (c) 8 oz (d) 96 oz
   (e) 5 cwt (f) 3 tons

3. Change to hundredweight:
   (a) 5 tons (b) 8 tons
   (c) 560 lb

4. Change to tons:
   (a) 60 cwt (b) 6720 lb
   (c) 17 920 lb

Capacity

The amount of fluid that a container will hold is called its **capacity**.

In the metric system capacities are measured in millilitres, centilitres and litres.

1 litre (ℓ) = 100 centilitres (cℓ)

= 1000 millilitres (mℓ)

Example 5

(a) How many centilitres are there in 4 litres?

$$4\,\ell = 4 \times 100\,c\ell$$
$$= 400\,c\ell$$

(b) Change 23 000 mℓ into litres.

$$23\,000\,m\ell = 23\,000 \div 1000\,\ell$$
$$= 23\,\ell$$

In the imperial system capacities are measured in fluid ounces, pints and gallons.

20 fluid ounces (fl oz) = 1 pint (pt)

8 pints (pt) = 1 gallon (gal)

Example 6

(a) How many pints are there in 5 gal?

$$5 \text{ gal} = 5 \times 8 \text{ pt}$$
$$= 40 \text{ pt}$$

(b) How many fluid ounces are there in 3 pt?

$$3 \text{ pt} = 3 \times 20 \text{ fl oz}$$
$$= 60 \text{ fl oz}$$

Exercise 5.5

This exercise should be done mentally.

1. Change to centilitres:
    (a) 70 mℓ   (b) 560 mℓ
    (c) 6 ℓ

2. Change to millilitres:
    (a) 5 cℓ   (b) 26 cℓ
    (c) 8 ℓ

3. Change to litres:
    (a) 600 cℓ   (b) 50 cℓ
    (c) 3000 mℓ

4. Change to fluid ounces:
    (a) 3 pt   (b) $\frac{1}{2}$ pt   (c) 2 gal

5. Change to pints:
    (a) 40 fl oz   (b) 120 fl oz
    (c) 6 gal

6. Change to gallons:
    (a) 16 pt   (b) 64 pt

Conversion of Metric and Imperial Units

Approximate imperial/metric conversions are:

1 inch is slightly more than $2\frac{1}{2}$ cm
(i.e. 2 in is about 5 cm)

1 foot is about 30 cm

1 yard is slightly less than 1 metre

1 kilometre is about $\frac{5}{8}$ mile
(i.e. 5 miles is about 8 km)

1 kilogram is about $2\frac{1}{4}$ lb
(i.e. 4 kg is about 9 lb)

1 fluid ounce is about 30 mℓ

1 litre is about $1\frac{3}{4}$ pt
(i.e. 4 litres is about 7 pints)

1 gallon is about $4\frac{1}{2}$ litres
(i.e. 2 gallons is about 9 litres)

Example 7

(a) Find in pounds the approximate weight of 20 kg of potatoes.

$$20 \text{ kg} = 20 \times 2\frac{1}{4}$$
$$= 20 \times \frac{9}{4}$$
$$= 45 \text{ lb}$$

(b) The distance between Calais and Lille is 96 km. How many miles is this?

$$96 \text{ km} = 96 \times \frac{5}{8}$$
$$= 60 \text{ miles}$$

Exercise 5.6

Use the approximate conversions given above to answer the following questions.

1. Convert 64 kg into pounds.

2. Convert 5 ft into centimetres.

3. Change 90 cm into feet.

4. How many yards are approximately equivalent to 200 m?

5. Change 18 lb into kilograms.

6. In Great Britain the speed limit on motorways is 70 miles per hour. How many kilometres per hour is this?

7. How many millilitres are there in 5 fluid ounces?

8. Convert 21 pints into litres.

9. Change 12 litres into pints.

10. How many gallons are equivalent to 27 litres?

The Arithmetic of Metric Quantities

Metric quantities are added, subtracted, multiplied and divided in the same way as decimal numbers. It is important that all quantities used are in the same units. For instance we cannot add millimetres and centimetres directly but if we change the millimetres to centimetres then we can add the two quantities.

Example 8

(a) Add 15.2 m, 39 cm and 140 mm giving the answer in metres.

$$\begin{array}{r} 15.2\,\text{m} \\ 39\,\text{cm} = 0.39\,\text{m} \\ 140\,\text{mm} = \underline{0.14\,\text{m}} + \\ \underline{15.73\,\text{m}} \end{array}$$

(b) A bottle contains 3 litres of water. 90 centilitres are poured out of the bottle. How much water remains?

$$\begin{array}{r} 3.00\,\ell \\ 90\,\text{c}\ell = \underline{0.90\,\ell} - \\ \underline{2.10\,\ell} \end{array}$$

2.10 litres of water remain in the bottle.

(c) 57 lengths of wood each 95 cm long are needed by a builder. Assuming no waste in cutting the wood, calculate, in metres, the total length of wood needed.

$$\begin{aligned} \text{Total length needed} &= 57 \times 95\,\text{cm} \\ &= 57 \times 0.95\,\text{m} \\ &= 54.15\,\text{m} \end{aligned}$$

(d) Frozen peas are packed in bags containing 450 grams. How many full bags can be filled from 2 tonnes of peas?

$$\begin{aligned} 2\,\text{t} &= 2000\,\text{kg} \\ &= 2\,000\,000\,\text{g} \\ \text{Number of bags filled} &= 2\,000\,000 \div 450 \\ &= 4444 \end{aligned}$$

Exercise 5.7

A calculator may be used for this exercise.

1. Add 50 cm, 5.8 m and 400 mm giving your answer in metres.

2. Subtract 80 mm from 27 cm giving the answer in centimetres.

3. A reel contains 24 metres of ribbon. Lengths of 30 cm and 700 mm are cut from it. What length of ribbon remains?

4. A bottle contains 75 centilitres of medicine. How many 5-millilitre doses can be obtained from it?

5. A bottle of lemonade contains 2 litres. 15 centilitres are poured from the bottle. How much lemonade, in litres, remains?

6. 95 lengths of steel bar each 127 mm long are required by a toy manufacturer. Work out, in metres, the total length of steel bar needed.

7. 209 lengths of cloth each 135 cm long are required by a clothing manufacturer. What total length of cloth is required?

8. Garlic salt is packed in jars containing 32 g. How many jars can be filled from 1 kg of garlic salt?

9. How many lengths of string 50 cm long can be cut from a ball containing 3.5 m?

10. A tablet has a weight of 40 milligrams. How many of these tablets weigh 60 grams?

Money

The British system of currency uses the pound sterling as the basic unit. The only sub-unit used is pence.

$$100 \text{ pence (p)} = £1$$

A decimal point (or sometimes a dash) is used to separate the pounds and the pence. For instance £3.58 or £3-58 means three pounds and 58 pence.

There are two ways of writing amounts less than £1. For instance 74 pence can be written 74p or £0.74. Five pence may be written as 5p or as £0.05.

The addition, subtraction, multiplication and division of sums of money are performed in exactly the same way as for decimal numbers.

Example 9

(a) Add £8.94, £3.28 and 95p.

$$£8.94 + £3.28 + 95p = £8.94 + £3.28$$
$$+ £0.95$$
$$= £13.17$$

(b) A five-pound note is tendered for purchases of 84p. How much change should be obtained?

$$£5.00 - 84p = £5.00 - £0.84$$
$$= £4.16$$

(c) An article costs 82p. How much do 35 of these articles cost?

$$\text{Total cost} = 35 \times 82p$$
$$= 35 \times £0.82$$
$$= £28.70$$

(d) 37 similar articles cost £34.78. How much does each article cost?

$$\text{Cost of each article} = £34.78 \div 37$$
$$= £0.94 \text{ or } 94p$$

Exercise 5.8

1. Add £2.15, £7.28 and £6.54.

2. Find the total of 27p, 82p, 97p and £2.36.

3. Add £15.36, £10.42, 75p, 86p and £73.25.

4. Subtract 87p from £3.26.

5. A lady goes shopping in a departmental store where she has a credit rating of £200. She spends £43.64, £59.76 and £87.49. How much more can she spend on credit in the store?

6. A man deposited £540 in his bank account but filled out cheques for £138.26, £57.49 and £78.56 before he deposited a further £436. How much money is in his account?

7. An article costs £12.63. How much do 87 of these articles cost?

8. Bread rolls cost 7p each. 140 are needed for a dinner party. How much is the total cost of the rolls?

9. An electricity bill for a quarter (13 weeks) costs £74.36. What is the average cost of electricity per week?

10. 41 similar articles cost £102.09. How much does each article cost?

Miscellaneous Exercise 5

Section A

1. (a) Change 155 mm into metres.
    (b) Change 1360 mℓ into litres.
    (c) Change 3500 g into kilograms.

2. How much change will I get from £10 when I buy 2 lb of butter at 92p per pound.

3. If $1 \text{ kg} = 2\frac{1}{4}$ lb, work out the weight, in pounds, of a packet of salt weighing 8 kg.

4. (a) How many millimetres are there in 3 metres?

   (b) How many millilitres are there in 3.4 ℓ.

5. A housewife buys 1.5 kg of bacon. It is packed in packs weighing 250 g. How many packs does she buy?

6. If 1 in = 25 mm, find in centimetres the length of a 12 in rule.

7. Sugar is sold in bags containing 2 kg. Taking 1 kg = $2\frac{1}{4}$ lb, find the weight in pounds of 4 bags of sugar.

8. The rent of a house is £38.54 per week. How much rent is paid in six weeks?

Section B

1. What is the total cost of renting a television set for six months at £16.32 per month?

2. Calculate the total cost of three pairs of jeans at £18.83 per pair and four pairs of socks at 84p per pair.

3. How much change would I get from £5 when I buy 4 packets of coffee at 92p per packet.

4. (a) Change 12.34 kg into grams.

   (b) Change 123.4 cm into metres.

   (c) Change 1234 mℓ to litres.

5. How many pieces of string, each 30 cm long, can be cut from a piece of string 4.2 m long.

6. If 4 kg = 9 lb approximately, work out the weight of a bag of flour weighing 3 kg.

7. A large piece of cheese weighs $2\frac{1}{4}$ kg. Smaller pieces weighing 525 g, 485 g and 370 g are cut from it. What weight of cheese, in grams, is left?

8. A bottle contains 20 cℓ of medicine. How many 5 mℓ doses can be obtained from it?

9. 1 km = $\frac{5}{8}$ of a mile approximately. How far, in miles, is a distance of 160 km.

10. In a town the speed limit for cars is 30 miles per hour. What is the speed limit in kilometres per hour?

Multi-Choice Questions 5

1. Before baking a clay pot weighed 2 kg. After baking it weighed 1.8 kg. What was the loss of weight, in grams?

   A 0.2   B 2   C 20   D 200

2. A table is 1.22 m wide. How many millimetres is this?

   A 12.2   B 122
   C 1220   D 12 200

3. A piece of ribbon 4 m long is cut into 16 equal pieces. What is the length of each piece?

   A 12.5 cm   B 16 cm
   C 24 cm   D 25 cm

4. If one packet of a chemical weighs 23 grams, what is the total weight, in kilograms, of 100 such packets?

   A 2300   B 230   C 2.3   D 0.23

5. If 1 mile = 1.6 km approximately, how far, in kilometres, is 30 miles?

   A 24   B 30   C 48   D 60

6. 650 mm written in metres is

   A 0.65   B 6.5   C 65   D 6500

7. Add 2.5 m, 105 cm and 6250 mm. The total, in metres, is

   A 9.8   B 10.25
   C 102.5   D 113.75

8. What is the cost of 1 kg of tea if a packet weighing 125 g costs 63p?

   A £2.52   B £5.00
   C £5.04   D £5.12

9. How many packets of tea containing 120 g can be made up from a tea-chest containing 60 kg of tea?

   A 5   B 50   C 500   D 5000

10. Calculate the cost of 400 g of bacon if 1 kg costs £3.05. The answer is

    A 30p   B 61p
    C £1.22   D £12.20

11. The cost of four articles at 73p each is

    A £2.91   B £2.92
    C £2.96   D £29.20

12. The value of £6.92 − 75p + £0.96 is

    A £5.21   B £5.23
    C £7.13   D £7.15

13. If £154.80 is shared equally between 36 people then each person will receive

    A £4.03   B £4.30
    C £40.30   D £43.00

14. The total cost of 2 metres of material at £1.20 per metre and 3 metres of better material at £1.50 per metre is

    A £2.70   B £6.60
    C £6.90   D £7.10

Mental Test 5

1. Convert 8 m into millimetres.

2. Change 5 cm into millimetres.

3. Convert 830 cm into metres.

4. Change 3400 mm into metres.

5. Convert 80 000 m into kilometres.

6. Change 93 km into metres.

7. Change 72 inches into feet.

8. How many yards are there in 108 inches?

9. Change 8 yards to feet.

10. If 1 mile = 1.5 km, how many kilometres is equivalent to 4 miles?

11. If 1 in = $2\frac{1}{2}$ cm, how many inches is 10 cm?

12. Change 1800 g to kilograms.

13. Change 9 kg to grams.

14. Convert 23 000 mg into grams.

15. Convert 6 tonnes into kilograms.

16. How many ounces are there in 3 lb?

17. How many hundredweight are there in 6 tons?

18. 1 ounce is approximately equal to 30 grams. How many ounces are equivalent to 210 grams?

19. Change 3 litres to centilitres.

20. How many millilitres are equivalent to 9 litres?

21. Change 800 cℓ into litres.

22. Convert 5000 mℓ into litres.

23. How many pints are there in 7 gallons?

24. How many fluid ounces are there in 4 pints?

25. How many pints are equivalent to 60 fluid ounces?

26. Add £1.36 and £2.64.

27. Find the cost of 10 articles if they cost £3.25 each.

28. Find the cost, in pounds, of 100 articles if they cost 12p each.

29. Find the cost of one article, in pence, if 10 cost £1.80.

30. 100 similar articles cost £1.50. Find the cost, in pence of one of these articles.

31. Find the cost, in pounds, of 20 articles costing 15p each.

32. 25 articles cost £5. How much does each article cost?

33. Find the cost of 5 articles at 50p each.

34. An article is bought for £3.50. How much change should be received from a £5 note?

35. Find the total cost, in pounds, of 50 articles, if each costs 60p.

Ratio and Proportion

Ratio

Mortar is made by mixing sand and cement in the **ratio** 3:1 (three to one) and then adding water. You could put three sacks of sand to one sack of cement, or three bucketfuls of sand to one bucketful of cement or three cups of sand to one cup of cement. These ratios are all 3:1.

As long as sand and cement are kept in the same proportion, when mixed with water they will make mortar. The actual amounts of sand and cement used affect only the amount of mortar made.

Example 1

A suitable batter for pancakes uses 120 grams of flour, 1 egg and 250 millilitres of milk. This is sufficient for 12 pancakes.

(a) Work out the amount of each ingredient if 24 pancakes are to be made.

(b) If 360 grams of flour are used, how many eggs are needed?

(c) How many grams of flour are needed if 1 litre of milk is used?

 (a) Since 24 pancakes are to be made we need twice the amount of each ingredient. Hence

$$\text{Amount of flour} = 2 \times 120\,\text{g}$$
$$= 240\,\text{g}$$
$$\text{Number of eggs} = 2 \times 1$$
$$= 2$$
$$\text{Amount of milk} = 2 \times 250\,\text{m}\ell$$
$$= 500\,\text{m}\ell$$

(b) Because three times the amount of flour is used we need three times the number of eggs. Therefore 3 eggs are required.

(c) 1 litre of milk is four times the amount used in the original recipe. We therefore need four times the amount of flour. Hence we need 4×120 grams $= 480$ grams of flour.

Model trains are usually made to a scale of 1:72. Every measurement on the model is $\frac{1}{72}$th of the real measurement. The model measurements and the real measurements are in the ratio of 1:72.

Example 2

A model ship is made to a scale of 1:120. If the actual ship is 60 metres long, what is the length of the model?

 Every measurement on the model is $\frac{1}{120}$th of the real measurement and therefore

$$\text{Length of model} = 60 \div 120\,\text{m}$$
$$= 0.5\,\text{m}$$

Exercise 6.1

1. A recipe for a cake to serve 4 people is as follows:

  400 g butter

  320 g of caster sugar

  1 egg separated

  800 g self-raising flour

(a) Work out the amount of each ingredient if a cake to serve 8 people is to be made.

(b) If 960 g of caster sugar is used, how much butter is needed?

(c) If 2.4 kg of self-raising flour is used, how many eggs are needed?

2. A mortar mixture is made of cement, sand and water in the ratio $1:2:4$. If 80 kg of cement is used, how much sand and water is required?

3. A metal alloy is made by mixing copper and zinc in the ratio $5:1$. If 8 kg of zinc is used, how much copper is required?

4. Solder is made by mixing tin and lead in the ratio $3:1$. If 5 kg of lead is used work out

(a) the weight of tin required

(b) the total weight of solder made.

5. A model of an aeroplane is made to a scale of $1:80$. If the real wing span is 20 metres, what is the wing span on the model, in centimetres?

6. A curried stew is made by using the following ingredients:

   1 onion

   2 carrots

   250 g beef

   340 g can of oxtail soup

   salt and pepper

   2 teaspoons of curry powder

   50 g raisins

This is sufficient for 4 people.

(a) How many carrots are needed if 500 g of beef is used?

(b) What amount of raisins is required for a curry containing 125 g of beef?

(c) Write down the amounts of all the ingredients needed for a curry to serve 8 people.

Simplifying Ratios

Problems are often made easier by putting the ratios into simpler terms.

The ratios $1:5$, $3:7$ and $9:4$ are all in their **simplest terms** because they are all whole numbers and there is no number which will divide exactly into both sides of the ratio.

The ratio $8:6$ is not in its simplest terms because 2 will divide exactly into 8 and 6 to give $4:3$. $4:3$ is equivalent to $8:6$ and it is in its simplest terms.

Brass is made by alloying copper and zinc in the ratio $4:3$. We would get exactly the same brass if we mixed copper and zinc in the ratio $8:6$.

Example 3

Put the ratio $72:84$ in its simplest terms.

12 divides exactly into 72 and 84, therefore $72:84$ is the same as $72 \div 12 : 84 \div 12$, that is, $6:7$ which is in its simplest terms.

Exercise 6.2

Put each of the following ratios in its simplest terms:

1. $5:20$	2. $8:16$	3. $4:12$
4. $42:49$	5. $25:30$	6. $35:42$
7. $16:24$	8. $25:20$	9. $36:24$
10. $56:64$		11. $5000:7500$
12. $70:60$		

Simplifying Ratios with Units

We can simplify the ratio $30\,cm:80\,cm$ by

(1) removing the units because they are the same

(2) dividing both sides by 10.

Therefore

$$30 \text{ cm} : 80 \text{ cm} \text{ is the same as } 3 : 8$$

The ratio 5 cm : 2 m can be simplified by making the units the same and removing them. Thus

$$5 \text{ cm} : 2 \text{ m} \text{ is the same as } 5 \text{ cm} : 200 \text{ cm}$$

Removing the units (because they are now the same) and dividing each side by 5

$$5 \text{ cm} : 2 \text{ m} \text{ is equivalent to } 1 : 40$$

Exercise 6.3

Simplify each of the following ratios by removing the units (if they are the same) and putting the ratio in its lowest terms:

1. 400 cm : 1 cm   2. 20 kg : 80 kg

3. 5 mℓ : 25 mℓ   4. 500 cm : 1 m

5. 60 g : 2 kg   6. £1.00 : 50 p

7. 3 cm : 9 mm   8. 3 kg : 90 g

9. £8.00 : 25 p   10. 2 km : 400 m

Changing Ratios into Fractions

Exercises 6.2 and 6.3 should bring out the similarities between ratios and fractions. Compare, for instance,

$$42 : 49 = 6 : 7 \quad \text{with} \quad \frac{42}{49} = \frac{6}{7}$$

and

$$5 : 20 = 1 : 4 \quad \text{with} \quad \frac{5}{20} = \frac{1}{4}$$

Note that the first quantity in the ratio becomes the numerator (top number) of the fraction and the second number becomes its denominator (bottom number).

Example 4

(a) Change the ratio 5 : 9 into a fraction.

$$5 : 9 \text{ is equivalent to } \frac{5}{9}$$

(b) Change 15 : 20 into a fraction in its lowest terms.

$$15 : 20 \text{ is equivalent to } \frac{15}{20} = \frac{3}{4}$$

Exercise 6.4

Change the following ratios into fractions in their lowest terms:

1. 7 : 20   2. 5 : 10   3. 7 : 14

4. 12 : 15   5. 16 : 24   6. 80 : 100

7. 3 : 9   8. 21 : 24

Proportional Parts

The line AB (Fig. 6.1) whose length is 15 cm, has been divided into two parts in the ratio 2 : 3. The line has been divided into its **proportional parts** and, as can be seen from the diagram, the line has been divided into a total of five equal parts. The length AC contains 2 of these equal parts and the length BC contains 3 of them. Each part is 15 cm ÷ 5 = 3 cm long. Hence

$$AC = 2 \times 3 \text{ cm}$$
$$= 6 \text{ cm}$$

and

$$BC = 3 \times 3 \text{ cm}$$
$$= 9 \text{ cm}$$

Fig. 6.1

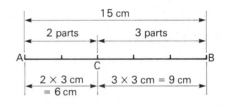

The problem of dividing the line AB into two parts in the ratio 2 : 3 could be tackled in the following way:

Total number of parts $= 2 + 3$

$= 5$

Length of each part $= 15\,\text{cm} \div 5$

$= 3\,\text{cm}$

Length of AC $= 2 \times 3\,\text{cm}$

$= 6\,\text{cm}$

Length of BC $= 3 \times 3\,\text{cm}$

$= 9\,\text{cm}$

Example 5

Divide £480 in the ratio $7:3:2$.

Total number of parts $= 7 + 3 + 2$

$= 12$

Amount of each part $= £480 \div 12$

$= £40$

Largest part $= 7 \times £40$

$= £280$

Middle part $= 3 \times £40$

$= £120$

Smallest part $= 2 \times £40$

$= £80$

Example 6

Two lengths are in the ratio of $6:5$. The longest length is 48 cm. What is

(a) the second length

(b) the total length?

The longest length is represented by 6 parts.

Hence 6 parts $= 48\,\text{cm}$

1 part $= 48\,\text{cm} \div 6$

$= 8\,\text{cm}$

(a) The second length $= 5 \times 8\,\text{cm}$
$= 40\,\text{cm}$.

(b) Total length $= 48\,\text{cm} + 40\,\text{cm}$
$= 88\,\text{cm}$.

Exercise 6.5

1. Divide £400 in the ratio $5:3$.

2. Divide 320 kg in the ratio $7:3$.

3. Divide a length of 240 m into two parts in the ratio $4:1$.

4. A line 3.36 m long is to be divided into three parts in the ratio $2:7:11$. Find, in millimetres, the length of each part.

5. A sum of money is divided into two parts in the ratio $5:7$. If the smaller amount is £200, what is the larger amount?

6. An alloy consists of copper, zinc and tin in the ratio $2:3:5$. If 45 kg of zinc are used, how much alloy will be made?

7. A sum of money is divided into three parts in the ratio $2:4:5$.

   If the largest share is £80, calculate the sum of money divided.

Simple Direct Proportion

If we know the cost of one article it is easy to find the cost of seven articles. Again, if we know the amount a man is paid for one hour's work we can find how much he will be paid for four hours' work.

Example 7

The cost of 1 kg of sugar is 80 p. What is the cost of

(a) 3 kg of sugar

(b) $\frac{1}{2}$ kg of sugar?

(a) 1 kg costs 80 p
3 kg costs $3 \times 80\,\text{p} = 240\,\text{p}$
$= £2.40$

(b) $\frac{1}{2}$ kg costs $\frac{1}{2} \times \frac{80}{1} = 40\,\text{p}$

If we know the cost of 5 articles, by dividing by 5 we can find the cost of 1 article.

Example 8

A car goes 120 km on 8 litres of petrol. At the same rate how far will it travel on

(a) 1 litre?   (b) 5 litres?

Using 8 litres the car travels 120 km

(a) Using 1 litre the car travels
$120 \div 8$ km = 15 km.

(b) Using 5 litres the car travels
5×15 km = 75 km

Exercise 6.6

1. In 1 hour an electric fire uses 2 units of electricity. How many units will it use in 5 hours?

2. A car travels 14 km on 1 litre of petrol. How far will it travel on 6 litres?

3. The cost of 1 kg of butter is 90p. What is the cost of

    (a) 3 kg of butter

    (b) $\frac{1}{2}$ kg of butter?

4. 6 pencils cost 42p. What is the cost of 1 pencil?

5. A car travels 50 miles in 2 hours. How far does it travel in

    (a) 1 hour   (b) 3 hours

    if it keeps up the same speed?

6. Dress material costs £20 for 4 metres. How much does

    (a) 1 metre   (b) 6 metres

    cost?

7. The cost of running a deep freezer for 5 hours is 30p. Calculate the cost of running the freezer for

    (a) 1 hour   (b) 2 hours.

8. Curtain material costs £15 for 6 metres. Find the cost of

    (a) 1 metre   (b) 9 metres.

Direct Proportion

Two quantities are in **direct proportion** if an increase (or decrease) in one quantity is matched by an increase (or decrease) in the same ratio in the second quantity. If 8 pencils cost 80p then 16 pencils will cost 160p. The number of pencils has been doubled and the cost has also been doubled.

Four pencils cost 40p. The number of pencils has been halved and the cost has also been halved. The number of pencils and their cost are in direct proportion.

Example 9

If 7 pens cost 56p, how much will 5 pens cost?

7 pens cost 56p

1 pen costs $56 \div 7 = 8$p

5 pens cost 5×8p = 40p

This method is known as the **unitary method** and it can be used for all problems in direct proportion.

Exercise 6.7

1. 3 oranges cost 60p. How much will 7 oranges cost?

2. 9 rubbers cost 54p. How much will 12 rubbers cost?

3. 3 grapefruit cost 72p. How much will 7 cost?

4. 8 plates cost £35.20. How much will 5 plates cost?

5. A car travels 400 km on 40 litres of petrol. How many litres will be needed for a journey of 520 km?

6. 18 metres of stair carpet costs £168. How much will 12 metres cost?

7. A machine makes 60 articles in 2 hours. How many articles will it make in

(a) 4 hours   (b) $\frac{1}{2}$ hour?

8. A man earns £12 in 4 hours. How much will he earn in

(a) 3 hours   (b) 30 minutes?

Best Buys

I can buy 100 mℓ of shampoo for £1.40, 150 mℓ for £1.68 and 200 mℓ for £2.80. Which is the best buy?

The **best buy** is the one which costs the least per millilitre. By using a calculator it is easy to work this out.

100 mℓ for £1.40 is equivalent to $140 \div 100 = 1.40$ p per millilitre

150 mℓ for £1.68 is equivalent to $168 \div 150 = 1.12$ p per millilitre

200 mℓ for £2.80 is equivalent to $280 \div 200 = 1.40$ p per millilitre.

The best buy is therefore 150 mℓ for £1.68.

Exercise 6.8

Which are the best buys?

1. 20 g for 60p or 50 g for £1.00.

2. 850 g for £2.04 or 2500 g for £7.00.

3. 250 g of cornflour for 28p or 300 g for 50p.

4. 250 g packet of cornflakes for 55p or a 750 g packet for £1.12.

5. 100 g for £1.50, 200 g for £2.50 or 250 g for £4.28.

Inverse Proportion

Some quantities are not directly proportional to one another.

Suppose that I have a bag of sweets to share between 10 children and that each child receives 6 sweets. If I now share the same bag of sweets amongst 20 children then each child will get only 3 sweets.

By doubling the number of children we halve the number of sweets that each child gets.

Now suppose we share the sweets amongst 5 children. Each child will get 12 sweets. By halving the number of children we double the number of sweets that each child receives.

In mathematics we say that $\frac{1}{2}$ is the **inverse** of 2 and that 2 is the **inverse** of $\frac{1}{2}$.

In the same way $\frac{1}{5}$ is the inverse of 5 and $\frac{1}{9}$ is the inverse of 9.

We see that the number of children is **inversely proportional** to the number of sweets that each child receives.

In dealing with problems on inverse proportion it is often easier to use the following method:

Total number of sweets $= 10 \times 6 = 60$

Shared amongst 5 children:

Number each child gets $= 60 \div 5 = 12$

Shared amongst 20 children:

Number each child gets $= 60 \div 20 = 3$

If an increase (or decrease) in one quantity produces a decrease (or increase) in a second quantity in the same ratio, the two quantities are in **inverse proportion**.

Example 10

Five men building a wall take 20 days to complete it. How long will it take

(a) 4 men   (b) 10 men

to complete it?

5 men take 20 days so 1 man will take 5 × 20 days = 100 days (one man takes longer than 5 men, so multiply).

(a) 4 men take 100 ÷ 4 days = 25 days (4 men take less time than one man so divide).

(b) 10 men take 100 ÷ 10 = 10 days.

Exercise 6.9

1. If six men take 12 days to dig a ditch, how long would it take two men?

2. A farmer employed 15 people to pick his fruit. It takes them 3 days. How long would it have taken 9 people to pick the fruit?

3. A bag of sweets is shared between 5 children and each child gets 3 sweets. How many sweets would each get if a similar bag were shared between 3 children?

4. Seven men make 21 toys in 30 minutes. How long would it take 21 men to make the toys?

5. Five men take 20 hours to pick a field of beans. How long would it take 20 men?

6. Twelve men take 2 hours to dig a hole in the road. How long would it have taken three men?

7. Eight women can do a piece of work in 60 hours. How many women would be needed to complete the work in 20 hours?

8. Ten men building a wall take 30 days to complete it. How long would it have taken six men to build the wall?

Exercise 6.10

1. A car goes 8 km on 1 litre of petrol. How far will it travel on $5\frac{1}{2}$ litres?

2. 7 men take 1 hour to make 21 model cars. How long would it take 21 men?

3. 10 men dig up a row of potatoes in 40 minutes. How long would it take 5 men?

4. A car travels 80 miles in 2 hours. If it maintains the same speed, how far will it travel in

    (a) 1 hour?   **(b)** 5 hours?

5. If 100 grams of sweets cost 44 p, how much will 300 grams cost?

6. 8 women take 5 hours to pick a row of raspberries. How long would it take 4 women to do the work?

7. It took 4 men 2 hours to dig a hole in a road. How long would 12 men have taken?

8. 15 kg of potatoes cost £3.00. Calculate the cost of

    (a) 5 kg   **(b)** 40 kg

9. By working for 8 hours a man earned £32. At the same rate of pay, how much would he earn for a 40-hour week?

10. A builder estimates that he can build a garage in 3 weeks using 5 men. If he employs 6 men, how long would the job take? Assume that all the men work at the same rate.

Measures of Rate

If a car travels 8 km on 1 litre of petrol we say that its fuel consumption is 8 km per litre of petrol. This is the **rate** at which the car consumes fuel.

The flow of water from a pipe or tap is usually measured in gallons per minute.

This is the rate of flow of water.

When a car has a speed of 40 miles per hour this is the rate at which it travels, i.e. 40 miles in 1 hour.

Example 11

(a) A car has a fuel consumption of 7 km per litre. How much fuel will be needed for a journey of 105 km?

$$\text{Fuel required} = \frac{\text{Length of journey}}{\text{Fuel consumption}}$$

$$= \frac{105}{7} \text{ litres}$$

$$= 15 \text{ litres}$$

(b) The flow of water from a tap is 10 litres per minute. How long will it take to fill a container with a capacity of 30 litres.

$$\text{Time taken} = \frac{\text{Capacity of container}}{\text{Rate of flow}}$$

$$= \frac{30}{10} \text{ minutes}$$

$$= 3 \text{ minutes}$$

We see from this example that if we know the rate of flow we can work out the time needed to fill a container of known capacity. We can also work out the amount of water delivered in a given time.

Example 12

The flow of water from a pipe is 20 litres per minute. How much water will be delivered in 8 minutes?

Amount of water delivered

$$= \text{Rate of flow} \times \text{Time taken}$$

$$= 20 \times 8 \text{ litres}$$

$$= 160 \text{ litres}$$

Exercise 6.11

1. The flow of water from a tap is measured to be 60 litres in 5 minutes:

   (a) Work out the rate of flow in litres per minute.

   (b) Calculate the time needed to fill a container with a capacity of 120 litres.

   (c) How much water will be delivered in 8 minutes if this rate of flow is maintained?

2. A car travels 80 miles on 2 gallons of petrol:

   (a) Calculate the fuel consumption of the car in miles per gallon.

   (b) How far would the car be expected to travel on 5 gallons of fuel?

   (c) How many gallons of fuel are needed for a journey of 240 miles?

3. The density of aluminium is 2700 kg per cubic metre. What is the weight of 3 cubic metres of aluminium?

4. A schoolboy found that he walked 100 metres in 50 seconds:

   (a) What is his speed in metres per second?

   (b) If he maintains this speed, how far will he walk in 150 seconds?

   (c) How long will it take him to walk a further 80 metres?

5. A sheet of metal weighs 5 lb per square foot:

   (a) How much will 12 square feet of the sheet weigh?

   (b) A piece of this metal weighs 20 lb. What is its area?

Rate of Exchange

Every country has its own monetary system. If there is to be trade and travel between any two countries there must be a rate at which the money of one country can be converted into money of the other country. This rate is called the **rate of exchange**.

Foreign Exchange Rates at June 1985

Country	Rate of Exchange
Austria	29.90 schillings = £1
Belgium	77.75 francs = £1
Canada	1.71 dollars = £1
Eire	1.23 punt = £1
France	11.73 francs = £1
West Germany	3.85 marks = £1
Holland	4.34 guilders = £1
Italy	2450 lire = £1
Spain	225 pesetas = £1
Switzerland	3.23 francs = £1
United States	1.43 dollars = £1

Example 13

(a) If 225 pesetas = £1, find to the nearest penny in British money the value of 30 000 pesetas.

$$30\ 000 \text{ pesetas} = £\frac{30\ 000}{225} = £133.33$$

(b) A traveller changes traveller's cheques for £50 into US dollars. If the exchange rate is $1.26 = £1, how many dollars does he get?

$$£50 = \$1.26 \times 50 = \$63.00$$

Exercise 6.12

Where necessary round off the answers to the nearest hundredth.

Using the exchange rates given above, find

1. How many French francs you would get for £20?

2. How many German marks you would get for £50?

3. How many Italian lire is equivalent to £15?

4. How many Spanish pesetas would you get for £10?

5. How much is 20 Swiss francs worth in pounds?

6. How much is 40 punts worth in pounds?

7. A traveller to Holland changes a traveller's cheque for £50 into guilders. How many guilders would she receive?

8. If the exchange rate in Austria is 27.00 schillings to the £1, how much is £100 worth in schillings?

9. Find the number of pounds equivalent to 7000 Belgian francs.

10. How many Canadian dollars is equivalent to £80?

Scales on Maps and Drawings

When drawings of large objects, such as houses and ships, are to be made they are always drawn to a scale. If, for example, 1 cm represents 10 m, this means that 10 m would be represented by 1 cm on the drawing.

On maps and drawings, scales are often stated as a ratio, for instance 1:1000. This means that a distance of 1000 m on the ground would be represented by a distance of 1 m on the map or drawing.

Example 14

(a) The scale on a plan is 1:100. What would

  (i) 5 cm on the plan represent
  (ii) 0.5 cm on the plan represent?

  Since the scale is 1:100, 1 cm on the plan represents 100 cm = 1 m on the ground.

  (i) 5 cm on the plan represents
  5 × 1 m = 5 m.

  (ii) 0.5 cm on the plan represents
  0.5 × 1 m = 0.5 m.

(b) A map is drawn to a scale of $1\,\mathrm{cm} = 50\,\mathrm{km}$. Express this scale as a ratio.

$$1\,\mathrm{cm} = 50\,\mathrm{km} = 50\,000\,\mathrm{m}$$
$$= 50\,000 \times 100\,\mathrm{cm}$$
$$= 5\,000\,000\,\mathrm{cm}$$

Hence the scale is $1 : 5\,000\,000$.

Exercise 6.13

1. A map has a scale of $1\,\mathrm{cm} = 1\,\mathrm{km}$. Find the distance that
   (a) $2\,\mathrm{cm}$ on the map represents
   (b) $3.5\,\mathrm{cm}$ on the map represents.

2. The scale of a map is $1 : 20\,000$. What distance, in kilometres, does $8\,\mathrm{cm}$ on the map represent?

3. The scale of a map is $1 : 100\,000$. The actual distance between two towns is $25\,\mathrm{km}$. What is the distance, on the map, in centimetres, between the two towns?

4. The scale of a map is $1\,\mathrm{cm} = 5\,\mathrm{km}$.
   (a) What distance on the ground is represented by
      (i) $0.3\,\mathrm{cm}$
      (ii) $4\,\mathrm{cm}$
      (iii) $15\,\mathrm{cm}$
      on the map?
   (b) What lengths on the map represent
      (i) $10\,\mathrm{km}$
      (ii) $25\,\mathrm{km}$
      (iii) $2.5\,\mathrm{km}$
      on the ground?

5. The drawing of a house is made to a scale of $1\,\mathrm{cm} = 2\,\mathrm{m}$:
   (a) What is this scale as a ratio?
   (b) On the drawing the length of the kitchen measures $3\,\mathrm{cm}$. What is the actual length of the kitchen?

6. A map is drawn to a scale of $2\,\mathrm{cm} = 1\,\mathrm{km}$:
   (a) Express this scale as a ratio.
   (b) The distance between two points on the map is $10.8\,\mathrm{cm}$. How far apart, in kilometres, are the two points on the ground?

7. The distance between two islands in the West Indies is measured as $8.1\,\mathrm{cm}$ on a map drawn to a scale of $1 : 5\,000\,000$. What is the actual distance between the two islands? Give the answer in kilometres.

Miscellaneous Exercise 6

Section A

1. Solder is made by mixing tin and lead in the ratio of $5 : 2$. If $10\,\mathrm{kg}$ of tin is used, how much solder will be made altogether?

2. The model of a car is made to a scale of $1 : 50$. If the width of the car is $6\,\mathrm{ft}$, calculate, in inches, the width of the model.

3. Write the ratio $24 : 36$ as a fraction in its lowest terms.

4. Simplify the ratio $100\,\mathrm{g} : 2\,\mathrm{kg}$ by removing the units (after equalising) and putting the ratio in its lowest terms.

5. Divide £500 into two parts in the ratio $7 : 3$.

6. 3 pears cost 36p. How much will 8 pears cost?

7. I can buy $200\,\mathrm{m}\ell$ of washing-up liquid for 30p, $500\,\mathrm{m}\ell$ for 48p or $1\,\ell$ for 99p. Which is the best buy?

8. 4 men digging a ditch take 25 days to complete the job. How long would it take 5 men to dig the ditch?

9. The scale on a map is $1 : 10\,000$. What distance, in metres, does $6\,\mathrm{cm}$ on the map represent?

10. The drawing of a house uses a scale of 1 cm = 2 m. The length of the kitchen is 8 m. Work out the length of the kitchen on the drawing.

Section B

1. If 5 pencils cost 45p, work out the cost of 12 pencils.

2. A car does 7 km per litre of petrol. How far will it go on $4\frac{1}{2}$ litres?

3. If a man walks 3 km in 40 minutes, how many kilometres will he have walked in 60 minutes if he keeps going at the same rate?

4. At a wedding reception for 120 people it was estimated that one bottle of wine would be sufficient for 8 people.

   (a) How many bottles would supply the 120 guests?

   (b) At a cost of £2.90 per bottle, what would be the total bill for wine?

5. Simplify the ratio 175 : 200.

6. Simplify the ratio 2 km : 500 m.

7. Express the ratio 8 : 64 as a fraction in its lowest terms.

8. Is 500 g of cornflour at 56p a better buy than 1 kg at £1.06?

9. A mortar mixture is made of cement, sand and water in the ratio 1 : 2 : 4. If 80 kg of cement is used, what weight of sand and water is needed?

10. A small business has three partners who share the profits in the ratio 4 : 3 : 2. In one month the profits were £900. How much does each partner get?

11. The scale on a map is 1 : 20 000. Two towns are 15 km apart. Calculate the distance between the towns on the map.

12. The wing span of an aeroplane is 50 m. A model is made to a scale of 1 : 75. What is the wing span on the model?

Multi-Choice Questions 6

1. A mortar mixture is made of cement, sand and water mixed by weight in the ratio 1 : 2 : 4 respectively. What weight of sand is contained in the mixture if 8 kg of cement is used?

   A 8 kg   B 16 kg   C 28 kg   D 32 kg

2. Which one of the following is not the same as 45 : 72?

   A 5 : 9   B 10 : 16
   C 25 : 40   D 65 : 104

3. Flour and salt are mixed in the ratio 49 : 1 when making bread. How many grams of salt are there in 5000 g of the mixture?

   A 5   B 10   C 50   D 100

4. Express 15 : 45 as a fraction in its lowest terms. The answer is

   A $\frac{15}{45}$ B $\frac{1}{3}$ C $\frac{45}{15}$ D $\frac{3}{1}$

5. Express 8 : 50 as a fraction in its lowest terms. It is

   A $\frac{8}{50}$ B $\frac{4}{25}$ C $\frac{50}{8}$ D $\frac{25}{4}$

6. Simplify the ratio 20 cm : 5 m. It is

   A 20 : 5   B 5 : 20
   C 1 : 25   D 25 : 1

7. A model is made to a scale of 1 : 50. The length of the model is 6 cm. Hence the length of the object is

   A 6 cm   B 50 cm
   C 300 cm   D 350 cm

8. Shampoo is sold in four sizes. The best buy is

   A 100 g for £1.60   B 150 g for £2.55
   C 200 g for £2.80   D 250 g for £3.75

Mental Test 6

Try to answer the following questions without writing anything down except the answer.

1. Express $3:6$ as a fraction in its lowest terms.

2. Express $£4:50\text{p}$ as a fraction in its lowest terms.

3. If 5 kg of apples cost £2, how much do 7 kg cost?

4. Express $1:2$ as a fraction.

5. If 20 articles cost £4, how much do 10 articles cost?

6. A train travelled a distance of 200 km in 4 hours. If it goes at the same speed how far will it travel in 5 hours?

7. 12 men take 10 days to complete a job. How long would it have taken 6 men?

8. Express the ratio $50:200$ as a fraction in its lowest terms.

9. A car travels 10 km on 1 litre of fuel. How much fuel is needed for a journey of 200 km?

10. A car travels 60 miles and uses 2 gallons of petrol. Work out its fuel consumption in miles per gallon.

11. If 200 pesetas = £1, find the number of pesetas equivalent to £10.

12. If 4 German marks = £1, how much in British money are 60 marks worth?

Percentages

Introduction

Suppose that in an arithmetic test, Mary scored 57 marks out of 100. A short way of writing out of 100 is by using the words **per cent**. Thus we say that Mary scored 57 per cent of the possible marks. The symbol % is usually written instead of the words per cent and we write that Mary scored 57% of the marks.

From Chapters 3 and 4 it will be remembered that 57 out of 100 could also be written as $\frac{57}{100}$ or 0.57.

Fractions written with a denominator of 100 are called **percentages** and we see that percentages are a third way of writing fractions.

$$\tfrac{1}{2} = \tfrac{50}{100} = 50\% = 0.5$$

$$\tfrac{3}{4} = \tfrac{75}{100} = 75\% = 0.75$$

In solving many problems it is useful to express a fraction or a decimal as a percentage.

Changing Fractions and Decimals into Percentages

To change a fraction into a percentage express it with a denominator (bottom number) of 100.

Example 1

Convert into percentages:

(a) $\frac{7}{10}$ (b) $\frac{3}{5}$ (c) $\frac{9}{20}$

(a) $\quad \frac{7}{10} = \frac{7 \times 10}{10 \times 10} = \frac{70}{100} = 70\%$

(b) $\quad \frac{3}{5} = \frac{3 \times 20}{5 \times 20} = \frac{60}{100} = 60\%$

(c) $\quad \frac{9}{20} = \frac{9 \times 5}{20 \times 5} = \frac{45}{100} = 45\%$

Example 2

(a) Change 0.3 to a percentage.

$$0.3 = \frac{3}{10} = \frac{3 \times 10}{10 \times 10} = \frac{30}{100} = 30\%$$

We see that the same result would be obtained if we multiplied the decimal number by 100. Thus

$$0.3 = 0.3 \times 100\% = 30\%$$

(b) Change 0.49 to a percentage.

We see that

$$0.49 = 0.49 \times 100\% = 49\%$$

From Example 2 we see that *to change a decimal number into a percentage multiply it by 100.*

You should learn by heart the following fractions as decimals and percentages.

Fraction	Decimal	Percentage
$\frac{1}{2}$	0.5	50
$\frac{1}{4}$	0.25	25
$\frac{3}{4}$	0.75	75
$\frac{1}{5}$	0.2	20
$\frac{2}{5}$	0.4	40
$\frac{3}{5}$	0.6	60
$\frac{4}{5}$	0.8	80
$\frac{1}{10}$	0.1	10
$\frac{3}{10}$	0.3	30
$\frac{7}{10}$	0.7	70
$\frac{9}{10}$	0.9	90

Not all percentages are whole numbers. For instance

$$\frac{3}{8} = 0.375 = 0.375 \times 100\% = 37.5\%$$

$$\frac{2}{3} = 0.667 = 0.667 \times 100\% = 66.7\%$$

Conversion of Percentages into Fractions and Decimals

To convert a percentage into a fraction or a decimal divide it by 100.

Example 3

(a) Change 5% into a fraction in its lowest terms.

$$5\% = \frac{5}{100} = \frac{5 \div 5}{100 \div 5} = \frac{1}{20}$$

(b) Change 18% into a fraction in its lowest terms.

$$18\% = \frac{18}{100} = \frac{18 \div 2}{100 \div 2} = \frac{9}{50}$$

Example 4

(a) Convert 86% into decimal numbers.

$$86\% = \frac{86}{100} = 0.86$$

The same result is obtained if we divide the percentage by 100.

$$86\% = 86 \div 100 = 0.86$$

(b) Convert 73.4% into decimal numbers.

$$73.4\% = 73.4 \div 100 = 0.734$$

Exercise 7.1

Convert the following fractions into percentages:

1. $\frac{9}{10}$ 2. $\frac{2}{5}$ 3. $\frac{7}{20}$

4. $\frac{19}{50}$ 5. $\frac{5}{8}$

Convert the following decimal numbers into percentages:

6. 0.6   7. 0.87   8. 0.03

9. 0.562   10. 0.917

Convert the following percentages into fractions in their lowest terms:

11. 8%   12. 15%   13. 40%

14. 95%   15. 32%

Convert the following percentages into decimal numbers:

16. 44%   17. 9%   18. 8.3%

19. 95.2%   20. 33.3%

Below is a table of corresponding fractions, decimal numbers and percentages. Copy the table and write in the figures which should be placed in each of the spaces marked with a question mark. Put fractions in their lowest terms.

	Fraction	Decimal	Percentage
	$\frac{1}{4}$	0.25	25
21.	$\frac{11}{20}$	?	?
22.	$\frac{17}{50}$	?	?
23.	$\frac{7}{8}$	?	?
24.	?	0.76	?
25.	?	0.08	?
26.	?	0.375	?
27.	?	?	15
28.	?	?	27
29.	?	?	45
30.	?	?	62

Percentage of a Sum of Money

When working out problems on percentages of sums of money percentages should be changed into decimals. A calculator can then be used to work out the answer.

Example 5

(a) Find 23% of £600.

$$23\% = 23 \div 100 = 0.23$$

$$23\% \text{ of } £600 = £600 \times 0.23 = £138$$

(Note that the word 'of' means multiply.)

(b) Find 3% of £15.

$$3\% = 3 \div 100 = 0.03$$

$$3\% \text{ of } £15 = £15 \times 0.03 = £0.45 = 45\text{p}$$

Find the values of

1. 5% of £80
2. 12% of £200
3. 40% of £90
4. 75% of £24
5. 60% of £25
6. 3% of £15
7. 8% of £30
8. 27% of £27
9. 49% of £38
10. 62% of £128

Increase and Decrease in a Sum of Money

Example 6

(a) A travel firm charges £180 for a holiday abroad but there is a fuel surcharge of 5% on top of the basic cost. What is the total cost of the holiday?

$$\text{Fuel surcharge} = 5\% \text{ of } £180$$

$$= 0.05 \times £180$$

$$= £9$$

$$\text{Total cost of holiday} = £180 + £9$$

$$= £189$$

(b) A coat has been reduced by 20% in a sale. The old price was £60. What is the sale price?

$$\text{Reduction in price} = 20\% \text{ of } £60$$

$$= 0.2 \times £60$$

$$= £12$$

$$\text{Sale price of coat} = £60 - £12$$

$$= £48$$

Exercise 7.3

1. In a sale an armchair has been reduced by 25%. The old price was £200. Calculate the sale price of the armchair.

2. The weekly wage of a craftsman is £180. He is given an increase of 5%. What is his new weekly wage?

3. The price of petrol was £2 per gallon. This price is to be increased by 10%. What will be the new price?

4. A price reduction of 17% is made on a three-piece suite costing £800. How much does the three-piece suite cost?

5. A holiday firm charges £260 for a fortnight's holiday in Portugal but a fuel surcharge of 8% is payable in addition. What is the total cost of the holiday?

Comparing Fractions, Decimals and Percentages

Since percentages and fractions are different ways of writing decimal numbers, it is possible to compare their sizes by converting them into decimal numbers.

Example 7

Write in order of size, beginning with the smallest, 0.65, 67% and $\frac{2}{3}$.

$$67\% = 67 \div 100 = 0.67$$

$$\frac{2}{3} = 2 \div 3 = 0.666$$

Putting these decimals in order of size, smallest first, we have 0.65, 0.666, 0.67. So the order is 0.65, $\frac{2}{3}$, 67%.

Exercise 7.4

Write in order of size, smallest first

1. $\frac{1}{2}$, 55%, 0.48
2. 23%, $\frac{1}{4}$, 0.24
3. $\frac{3}{5}$, 35%, 0.53
4. 0.33, $\frac{1}{3}$, 34%
5. 0.67, $\frac{2}{3}$, 66%
6. 0.43, $\frac{3}{4}$, 34%
7. $\frac{1}{8}$, 8%, 0.8
8. $\frac{5}{8}$, 63%, 0.62

Miscellaneous Exercise 7

Section A

1. Write in order of size, smallest first, 25%, $\frac{1}{5}$ and 0.22.

2. Change to percentages:
   (a) 0.65   (b) $\frac{3}{5}$

3. Change into fractions in their lowest terms:
   (a) 35%   (b) 48%

4. Find 34% of £220.

5. The cost of a table has been reduced by 20% in a sale. The old price was £120. What is the new price?

6. Is 0.4 greater or less than 45%?

7. Is $\frac{3}{20}$ greater or less than 0.12?

Section B

1. Change to fractions in their lowest terms:
   (a) 24%   (b) 6%

2. Change to decimal numbers:
   (a) 64%   (b) 7%

3. Opposite is a table of corresponding fractions, decimals and percentages. Write down the figures which should be placed in the spaces marked a, b, c, d, e and f, expressing fractions in their lowest terms.

Fraction	Decimal	Percentage
$\frac{1}{2}$	0.5	50
a	0.85	b
$\frac{7}{20}$	c	d
e	f	12

4. Find the value of
   (a) 9% of £20,   (b) 38% of £70.

5. A coat has been reduced by 15% in a sale. The old price was £80.

   (a) How much was the reduction in price?

   (b) How much will a customer pay for the coat?

6. In a class of 30 children, 40% are boys.

   (a) How many boys are there in the class?

   (b) What percentage of the class are girls?

7. Put in order of size, smallest first, 85% 0.875, $\frac{5}{6}$.

Multi-Choice Questions 7

1. 30% of £240 is
   A £8   B £72   C £720   D £800

2. Written as a fraction in its lowest terms, 45% is
   A $\frac{45}{100}$ B $\frac{9}{20}$ C 0.45   D 0.9

3. £1440 was divided between two people so that one person was given 45% of the total. How much money did the other person receive?
   A £628   B £696   C £744   D £792

4. 20.45% is the same as
   A 20.45   B 2.045
   C 0.2045   D 0.020 45

5. 0.305 is the same as
   A 0.305%   B 3.05%
   C 30.5%   D 35.0%

6. $\frac{12}{25}$ is the same as
   A 12%   B 13%   C 25%   D 48%

7. Written as a fraction in its lowest terms, 35% is
   A $\frac{35}{100}$ B $\frac{7}{20}$ C 0.35   D $\frac{35}{10}$

Mental Test 7

Try to answer the following questions without writing anything else down except the answer.

1. Express $\frac{4}{5}$ as a percentage.

2. What is 19% as a fraction?

3. Write 38% as a decimal number.

4. Change 0.89 to a percentage.

5. What is 30% of £80?

6. Find the value of 7% of £50.

7. Write 0.612 as a percentage.

8. Write 25% as a fraction in its lowest terms.

9. Is 32% less than 28%?

10. Is 0.5 greater than 60%?

Wages and Salaries

Introduction

Everyone who works for an employer receives a wage or salary in return for his or her labour. The payments, however, may be made in several different ways. **Wages** are usually paid weekly and **salaries** monthly.

Payment by the Hour

Many people are paid a certain amount of money for each hour that they work. They usually work a fixed number of hours called the **basic week**. It is this basic week which determines the hourly (or basic) rate of pay.

Example 1

(a) A man works a basic week of 35 hours and his weekly wage is £175. What is his hourly rate?

$$\text{Hourly rate of pay} = \frac{\text{Weekly wage}}{\text{Basic week}}$$

$$= \frac{£175}{35}$$

$$= £5.00$$

(b) A woman works a basic week of 38 hours. If her hourly rate of pay is £3.50, calculate her weekly wage.

$$\text{Weekly wage} = \text{Basic week} \times \text{Hourly rate}$$

$$= 38 \times £3.50$$

$$= £133$$

1. A woman works a basic week of 40 hours and her rate of pay is £4.00 per hour. Calculate her weekly wage.

2. £200 is paid to a mechanic per 40-hour week. What is the hourly rate of pay?

3. Calculate the weekly wage for a 35-hour week if the basic rate is

   (a) £2.00 **(b)** £2.50 **(c)** £3.00

4. A typist works a 38-hour week for which she is paid £152. What is her hourly rate of pay?

5. If the basic rate is £3.92 and the basic week is 40 hours, what is the weekly wage?

Overtime

Hourly paid workers are usually paid extra money for working more hours than the basic week requires. These extra hours of work are called **overtime**.

Overtime is usually paid at one of the following rates:

time-and-a-quarter ($1\frac{1}{4}$ times the basic rate)

time-and-a-half ($1\frac{1}{2}$ times the basic rate)

double-time (twice the basic rate)

Example 2

A girl is paid a basic rate of £2 per hour. Calculate her overtime rates when this is paid at **(a)** time-and-a-quarter, **(b)** time-and-a-half, **(c)** double time.

(a) Overtime rate at time-and-a-quarter

$$= 1\tfrac{1}{4} \times \text{Basic rate}$$

$$= £(\tfrac{5}{4} \times \tfrac{2}{1})$$

$$= £2\tfrac{1}{2}$$

$$= £2.50$$

(b) Overtime rate at time-and-a-half

$$= 1\tfrac{1}{2} \times \text{Basic rate}$$

$$= £(\tfrac{3}{2} \times \tfrac{2}{1})$$

$$= £3$$

(c) Overtime rate at double-time

$$= 2 \times \text{Basic rate}$$

$$= £(2 \times 2)$$

$$= £4$$

Example 3

Tom Kite is paid an hourly rate of £4 per hour. He works a basic week of 40 hours and, in addition, he works six hours' overtime for which he is paid time-and-a-half. Calculate his total wage for the week.

Overtime rate

$$= £(1\tfrac{1}{2} \times 4)$$

$$= £6 \text{ per hour}$$

Total wage

$$= 40 \times £4 + 6 \times £6$$

$$= £160 + £36$$

$$= £196$$

Exercise 8.2

1. A man is paid a basic rate of £5 per hour. Calculate his overtime rate at
   (a) time-and-a-quarter
   (b) time-and-a-half
   (c) double-time.

2. A woman's basic hourly wage is £3.20. What is her overtime rate when this is paid
   (a) at time-and-a-half
   (b) double-time?

3. Peter Cook works a basic week of 40 hours. His basic rate of pay is £5 per hour. During a certain week he worked 8 hours on Sunday for which he was paid double-time. Calculate his total wage for that week.

4. A man is paid £160 for a 40-hour week. He works 8 hours overtime during a certain week. If the overtime was paid at time-and-a-quarter, find
   (a) the basic hourly rate of pay
   (b) the overtime rate of pay
   (c) the amount earned in overtime
   (d) the man's total wage for the week.

5. A woman is paid £3 per hour for a 35-hour week. All overtime is paid at time-and-a-half. If she worked 5 hours of overtime, calculate
   (a) her basic weekly wage
   (b) her overtime rate
   (c) the amount earned in overtime
   (d) her total wage for the week.

Piecework

Some workers are paid a fixed amount for each article or piece of work that they make. Often, if they make more than a certain number of articles they are paid a bonus.

Example 4

A pieceworker received 5p for each button-hole sewn up to a limit of 200 per day. After that she is paid 7p per buttonhole. Calculate the amount she earned on a day when she sewed 250 buttonholes.

> Amount earned on first 200 = 200 × 5p = £10
>
> Amount earned on next 50 = 50 × 7p = £3.50
>
> Total amount earned = £10.00 + £3.50 = £13.50

Exercise 8.3

1. A woman is paid 14p for each pound of sweets that she packs up to a limit of 250 pounds. After that she earns a bonus of 2p per pound. Calculate how much she earns if she packs 400 lb of sweets.

2. A man is paid 22p for each mug he makes up to a limit of 100. After that he earns a bonus of 5p per mug. Calculate how much he earns if he makes 250 mugs.

3. A pieceworker is paid 8p for each handle he fixes to saucepans up to a limit of 300. After that he is paid 13p per handle. Find the amount he earns if he fixes 350 handles.

4. A woman is paid 50p for each plate she decorates up to a limit of 100 plates. After that she is paid 60p for each plate. How much does she earn if she decorates 150 plates?

5. A man is paid 8p for each spot-weld that he makes up to a limit of 300 per day. After that he is paid a bonus of 1p per spot-weld. If he makes 450 spot-welds in a day, calculate his earnings.

Commission

Sales people are usually paid a **commission** on top of their weekly wage. This is calculated as a small percentage of the value of the goods that they have sold.

Example 5

A salesman is paid a commission of 2% on the value of goods which he has sold. Calculate his commission if he sells goods to the value of £6000.

$$\text{Commission} = 2\% \text{ of } £6000$$
$$= £\left(\frac{2}{100} \times \frac{6000}{1}\right)$$
$$= £120$$

Example 6

A shop assistant is paid a weekly wage of £80. In addition she is paid 3% commission on the value of the goods which she sells. During a certain week she sells goods to the value of £3000. Calculate her total wages for that week.

$$\text{Commission} = 3\% \text{ of } £3000$$
$$= £\left(\frac{3}{100} \times \frac{3000}{1}\right)$$
$$= £90$$
$$\text{Total wages} = £80 + £90$$
$$= £170$$

Exercise 8.4

1. Find the commission at 2% on goods worth

   (a) £200   (b) £350   (c) £800

2. A car is sold for £9000 and the sales-man gets 5% commission. How much commission does the salesman get?

3. A furniture salesman is paid a basic weekly wage of £90. In addition he is paid 3% commission on goods which he sells. During a certain week he sold furniture to the value of £2500. How much were his total earnings for that week?

4. A woman earns £75 per week and gets 2% commission on goods sold to the value of £2000. How much did she earn that week?

5. Commission of 4% is paid on a sale worth £5000. How much is this commission?

Salaries

Salaried workers such as teachers and civil servants earn a fixed amount each year. The money is usually paid monthly. They are not paid overtime or commission.

Example 7

A civil servant has an annual salary of £9600. How much is he paid per month?

Monthly salary = £9600 ÷ 12 = £800

Exercise 8.5

Below are given the annual salaries of five people. How much is each paid monthly?

1. £5316   2. £6000   3. £8400

4. £6096   5. £7440

Deductions from Earnings

The wages and salaries discussed previously are not the take-home pay or salary of the workers. A number of deductions are made first. The most important of these are

(1) income tax

(2) national insurance

(3) pension fund payments

Before deductions the wage (or salary) is known as the **gross wage** (or **gross salary**).

After the deductions have been made the wage (or salary) is known as the **net wage** (or **net salary**) or as the **take-home pay**.

National Insurance

Both employees and employers pay national insurance. Employees pay a certain percentage of their gross wage (9% in 1986). Married women and widows can pay at a reduced rate (3.85% in 1986). Men over 65, women over 60 and employees earning less than £34 per week do not pay national insurance.

The rates for national insurance are fixed by the Chancellor of the Exchequer and they vary from year to year. National insurance pays for such things as sick-pay and unemployment benefit.

Example 8

A man earns £12 000 per annum. His deductions for national insurance are 9% of his gross salary. How much per month does he pay in national insurance?

$$\text{Monthly salary} = £12\,000 \div 12$$
$$= £1000$$

Amount of national insurance

$$= 9\% \text{ of } £1000$$
$$= £\left(\frac{9}{100} \times \frac{1000}{1}\right)$$
$$= £90$$

Income Tax

Taxes are levied by the Chancellor of the Exchequer to produce money to pay for such items as the armed services, the Civil Service and motorways. The largest producer of revenue is income tax.

Every person who earns more than a certain amount pays income tax. Tax is not paid on the entire income. Certain allowances are made. In 1988-9 these were as follows:

(1) Single person's allowance   £2605

(2) Married couple's allowance   £4095

(3) Wife's earned income allowance   £2605

In certain cases other allowances such as an additional personal allowance for one parent families and blind persons are available.

The residue of the income left over after all the allowances have been deducted is called the **taxable income**.

$$\text{Taxable income} = \text{Gross income} - \text{Allowances}$$

The amount of income tax payable per annum

$$= \text{Taxable income} \times \frac{\text{Rate of taxation}}{100}$$

In 1988-9 for taxable incomes up to £19 300 the rate of taxation was 25% but for any taxable income above £19 301 the rate of taxation was 40%.

Example 9

A woman's annual salary is £12 000. Her taxable income is calculated by deducting a personal allowance of £2605 and pension fund payments of £700 from her annual salary. She then pays income tax at 25% on her taxable income. How much does she pay in income tax per annum?

$$\text{Taxable income} = £12\,000 - £2605 - £700$$
$$= £8695$$

$$\text{Tax payable} = 25\% \text{ of } £8695$$
$$= £2173.75$$

Example 10

A man has a taxable income of £40 000. Work out the amount of income tax payable if the tax bands are 25% up to £19 300 and 40% above £19 301.

$$25\% \text{ of } £19\,300 = £4825$$
$$\text{Residue of taxable income} = £40\,000 - £19\,300$$
$$= £20\,700$$
$$40\% \text{ of } £20\,700 = £8280$$
$$\text{Total tax payable p.a.} = £4825 + £8280$$
$$= £13\,105$$

PAYE

Most people pay their income tax by a method called **Pay-As-You-Earn** or **PAYE** for short. The tax is deducted from their wages or salaries before they receive them by the employer. The taxpayer and his or her employer receive a notice of coding which gives the employee's allowances (based upon tax forms which he or she has previously completed) and sets a code number. The employer then knows from tax tables the amount of tax to deduct from the wage or salary of an employee.

Pension Funds

Many firms operate their own private pension scheme to provide a pension in addition to that provided by the state. Usually both employer and employee contribute, the employee's share being of the order of 5 or 6% of the gross annual wage or salary.

The amount of pension received depends upon the length of service and the amount of earnings at the time of retirement.

Example 11

An employee earns £9000 per annum. His pension fund payments are 5% of his gross annual salary. How much are his pension fund payments?

$$\text{Pension fund payments} = 5\% \text{ of } £9000$$
$$= £450 \text{ per annum}$$

Exercise 8.6

1. A man has a taxable income of £4000 and pays income tax at the rate of 30%. How much does he pay in income tax per annum.

2. From a gross income of £9000, allowances of £4325 are deducted.

   (a) Work out the taxable income.

   (b) If tax is to be paid at a rate of 30%, calculate the amount to be paid in income tax per annum.

3. A woman has a gross annual salary of £7000. Her tax-free allowances amount to £4000.

   (a) What is her taxable income?

   (b) Tax is levied at 29%. How much should she pay in income tax for the year?

4. How much national insurance is paid on a gross annual salary of £12 000 when it is levied at a rate of 9%?

5. Employees of a certain company contribute 6% of their gross annual wages to the pension fund. A woman earns £7500 per annum. Calculate her monthly payments to the pension fund.

6. A man earns £185 per week. His deductions are national insurance at 9% and pension fund payments of 5% of his gross wage. In addition he pays £36.45 in income tax. Work out his weekly take-home pay.

7. An employer pays 14% of a woman's salary in national insurance. How much does the employer pay if her salary is £8250 per annum.

Miscellaneous Exercise 8

Section A

1. Margaret worked for 40 hours. She is paid £2.90 per hour. What was her wage?

2. A man's taxable income is £3000. He pays income tax at 25%. How much tax did he pay?

3. A woman's total income is £8000 per annum. Her tax-free allowances amount to £3250. How much is her taxable income?

4. A man is paid £200 for a 40-hour week. Calculate the man's basic hourly rate of pay.

5. A girl is paid £2.50 per hour. Overtime is paid at time-and-a-half. What is her overtime rate?

6. A salesman is paid commission of 3% on his weekly sales. During one week he sold goods to the value of £8000. How much commission did he earn?

7. A woman is paid 12p for every article she makes. During one week she made 750 articles. How much money did she make?

8. A trainee earned £78 for a week's work. He paid £9.36 in income tax. What percentage of his weekly wage was paid in income tax?

Section B

1. A man works a basic week of 40 hours and he is paid £4.50 per hour. What is his weekly wage?

2. £137.67 is paid for a 39-hour week. Calculate the hourly rate of pay.

3. A garage mechanic is paid an hourly rate of £5.60. Any overtime that he works is paid at time-and-a-half. If he works 6 hours' overtime how much will he earn in overtime?

4. A man pays 8% of his salary to his firm's pension fund. If his gross annual salary is £7500, how much will he contribute per annum?

5. Miss Fairbairn earns an annual salary of £7200. She pays £632 in national insurance and £360 in pension fund payments. In addition she pays £1453 in income tax.

   (a) Work out the monthly deductions from her salary.

   (b) What is her monthly take home pay?

6. A man's gross annual salary is £12 000. His total tax-free allowances to be set against pay are £5650.

   (a) What is his taxable income?

   (b) Income tax is levied at a rate of 30%. How much does he pay in income tax for the year?

7. A pieceworker is paid 20p for each article that she makes up to a limit of 40. She is then paid a bonus of 5p per article. Work out her earnings if she makes 50 articles.

8. A salesman is paid a basic wage of £60 per week. In addition he is paid a commission of 3% on any sales that he makes. During a certain week his sales amounted to £6000.

   (a) Calculate the amount of his commission?

   (b) What is his gross wage for that week?

Multi-Choice Questions 8

1. A man's taxable income is £3600. If tax is levied at 30%, the amount paid in income tax is

   A £97.20   B £972

   C £1080   D £3240

2. An agent is paid 6% commission on each sale. If goods to the value of £1500 are sold, his commission is

   A £60   B £90   C £600   D £900

A woman works a 40-hour week and is paid £3 per hour. Overtime is paid at time-and-a-half. Use this information to answer questions 3, 4, 5 and 6.

3. How much money does she earn in a 40-hour week?

   A £30   B £45   C £120   D £180

4. What is her overtime rate of pay?

   A £1.50   B £3

   C £4.50   D £6

5. How much money will she earn if she works 52 hours in a certain week?

   A £116   B £120   C £156   D £174

6. The woman gains an 8% pay rise. How much would she then earn for a 40-hour week?

   A £120   B £129.60

   C £144   D £145

7. A man's basic wage for a 35-hour week is £136.50. His overtime rate is 60p per hour more than his basic rate. If he works 6 hours overtime in a certain week, his total wages for that week will be

   A £140.10   B £158.10

   C £163.50   D £168.00

8. A man's gross wage was £156 per week. He paid £7.80 in income tax. The percentage of his wage paid in income tax was

   A 2%   B 5%   C 20%   D 50%

Mental Test 8

1. A man is paid £200 for a 40-hour week. What is his hourly rate?

2. A woman's hourly rate is £3 per hour. If the basic week is 40 hours, calculate her weekly wage.

3. A firm pays its employees £4 per hour. All overtime is paid at time-and-a-half. Work out the overtime rate of pay.

4. A woman pieceworker is paid 10p for each article she makes. If she makes 80 articles, how much will she be paid?

5. Commission of 3% is paid to a salesman. How much commission will he earn on sales of £5000?

6. A teacher has a salary of £7200 per annum. Work out her monthly pay.

7. A man has a gross weekly wage of £200. His deductions total £70. How much is his take-home pay?

8. A woman has a gross annual salary of £10 000. Her allowances to be set against pay amount to £4000. What is her taxable income?

9. A man's taxable income is £5000. Income tax is levied at 30%. How much does the man pay in income tax per annum?

10. A woman contributes 5% of her annual salary to a pension fund. If she earns £8000 per annum, how much does she pay?

Interest

Introduction

If you invest money with a bank or a building society then **interest** is paid to you for lending the money. On the other hand if you borrow money from it you will be charged interest. Interest is given as a percentage per annum.

Example 1

I invest £2000 for 1 year at 8% interest per annum. How much interest will I be paid at the end of the year?

$$\text{Amount of interest} = 8\% \text{ of } £2000$$
$$= 0.08 \times £2000$$
$$= £160$$

Simple Interest

With **simple interest** the amount of interest earned is the same every year. If the money is invested for 2 years the amount of interest will be doubled; for 3 years the amount of interest will be 3 times as much. For 6 months (i.e. half a year) the amount of interest will be half as much.

Example 2

A man borrows £5000 for 6 months. If the rate of interest is 20% per annum, how much interest will he pay?

Amount of interest for 1 year
$$= 20\% \text{ of } £5000$$
$$= 0.2 \times £5000$$
$$= £1000$$

Amount of interest for 6 months
$$= \tfrac{1}{2} \times £1000$$
$$= £500$$

In practice, for periods over 1 year, simple interest is virtually unknown.

Exercise 9.1

Find the interest for 1 year on

1. £500 at 5% per annum

2. £1000 at 8% per annum

3. £4000 at 10% per annum

4. £8000 at 12% per annum

5. £15 000 at 20% per annum.

Find the amount of simple interest on

6. £2000 at 10% per annum for 6 months

7. £500 at 12% per annum for 4 months

8. £2400 at 20% per annum for 3 months.

Compound Interest

Compound interest is different from simple interest in that the interest is added to the amount invested (or borrowed) and hence also attracts interest. Thus after 2 years there will be more interest than after 1 year because there is more capital to attract interest.

Example 3

£500 is invested for 2 years at 10% compound interest. Calculate the value of the investment after this period.

	£	
Amount invested	500	
First year's interest	50	(i.e. 10% of £500)
Value of investment after 1 year	550	(i.e. £500 + £50)
Second year's interest	55	(i.e. 10% of £550)
Value of investment after 2 years	605	(i.e. £550 + £55)

When income is derived from money invested with a bank or a building society it attracts income tax which is deducted from the interest before you receive it. If you look in a newspaper you will often see advertisements like this:

9.75% (net compounded annual rate)

= 13.92% (gross compounded annual rate)

If there was no tax payable you would receive 13.92% compound interest on your investment. Since the tax is deducted at source, you receive only 9.75% on your investment. For anyone who is not a taxpayer (like children and some old-age pensioners) this is a bad investment because there is no way in which the tax can be reclaimed. A better investment is one in which the tax is not deducted at source (such as the National Savings Income Bonds).

Compound Interest Tables

Compound interest tables like the one below are available. They give the amount to which an investment of £1 grows.

Years	5%	6%	7%	8%	9%
1	1.050	1.060	1.070	1.080	1.090
2	1.103	1.124	1.145	1.166	1.188
3	1.158	1.191	1.225	1.260	1.295
4	1.216	1.262	1.311	1.360	1.412
5	1.276	1.338	1.403	1.469	1.539
6	1.340	1.419	1.501	1.587	1.677
7	1.407	1.504	1.606	1.714	1.828
8	1.477	1.594	1.718	1.851	1.993
9	1.551	1.689	1.838	1.999	2.172
10	1.629	1.791	1.967	2.159	2.367

Years	10%	11%	12%	13%	14%
1	1.100	1.110	1.120	1.130	1.140
2	1.210	1.232	1.254	1.277	1.300
3	1.331	1.368	1.405	1.443	1.482
4	1.464	1.518	1.574	1.603	1.689
5	1.611	1.685	1.762	1.842	1.925
6	1.772	1.870	1.974	2.082	2.195
7	1.949	2.076	2.211	2.353	2.502
8	2.144	2.304	2.476	2.658	2.853
9	2.358	2.558	2.773	3.004	3.252
10	2.594	2.839	3.106	3.395	3.707

Example 4

Using the compound interest table above, find the value of £6000 at the end of 5 years when the rate of compound interest is 12%.

From the table, £1 becomes £1.762 at the end of 5 years. To find to what £6000 grows we multiply £1.762 × 6000. Therefore

Value of the investment = £1.762 × 6000

= £10 572

Exercise 9.2

Use a calculator to work out the value of the following investments:

1. £300 invested for 2 years at 5% per annum compound interest.

2. £1000 invested for 3 years at 10% per annum compound interest.

3. £5000 invested for 2 years at 15% per annum compound interest.

Using the compound interest tables work out the value of the following:

4. £400 invested for 3 years at 5% per annum compound interest.

5. £1000 invested for 6 years at 8% per annum compound interest.

6. £500 invested for 10 years at 12% per annum compound interest.

7. £2000 invested for 8 years at 14% per annum compound interest.

Depreciation

When the value of an article decreases with age this is called **depreciation**. Examples of articles which depreciate with age are cars, machinery, typewriters, etc.

Problems with depreciation are very similar to those on compound interest. The difference is that with compound interest the interest is added to the value of the investment whereas with depreciation the interest is subtracted from the value of the article.

Example 5

A small business buys a computer for £2000. The rate of depreciation is 20% per annum. What is the value of the computer after 2 years?

	£	
Cost of computer	2000	
1st year's depreciation	400	(i.e. 20% of £2000)
Value at end of 1st year	1600	(i.e. £2000 − £400)
2nd year's depreciation	320	(i.e. 20% of £1600)
Value at end of 2nd year	1280	(i.e. £1600 − £320)

Hence after 2 years the value of the computer is only £1280.

Exercise 9.3

1. A firm buys a machine for £1000. If its value depreciates by 10% per annum, calculate its value at the end of 2 years.

2. The rate of depreciation of office machinery bought for £20 000 is 15% per annum. Calculate the value of the machinery at the end of 3 years.

3. A lorry costs £30 000 when new. It depreciates in value at a rate of 20% per annum. Calculate its value at the end of 3 years.

4. A typewriter cost £500 when bought. If it depreciates at a rate of 25% per annum, calculate its value at the end of 2 years.

Miscellaneous Exercise 9

Section A

1. £5000 is invested for 1 year at 10% per annum interest. How much interest will it earn at the end of the year?

2. £300 is invested for 6 months at 12% per annum simple interest. How much will the investment be worth at the end of this period?

3. £100 is invested at 10% compound interest for 2 years. How much is the investment worth at the end of this period?

4. A firm buys a machine for £5000. If it depreciates at a rate of 20% per annum, what will its value be at the end of 3 years?

5. A person wishes to invest £3000. He can buy Savings Bonds which pay simple interest at 12% per annum, or he can open a savings account which pays 11% compound interest per annum. Calculate the difference between the values of the two schemes at the end of 2 years.

Section B

1. What is the interest for 1 year on £800 at 10% per annum?

2. What is the interest for six months on £500 at 8% per annum?

3. £200 is invested for 2 years at 12% per annum simple interest:

   (a) What is the interest at the end of the first year?

   (b) How much is the interest at the end of the two years?

4. A woman invests £200 for 2 years at 10% compound interest:

   (a) How much interest does the money earn in the first year?

   (b) At the end of the second year, how much interest has been earned?

   (c) How much will her investment be worth at the end of the second year?

5. What sum of money must be invested to produce simple interest of £100 at the end of 1 year? The rate is 20% per annum.

Mental Test 9

1. £100 was invested for one year. The investment was then worth £110. How much interest has the investment attracted?

2. After one year an investment attracted £50 in simple interest. How much simple interest will there be after 2 years?

3. After a period of six months an investment attracted £30 in simple interest. What amount of simple interest will be payable after 1 year?

4. £1000 is invested at 10% per annum simple interest. Work out the amount of interest payable at the end of the year.

5. A machine was bought for £5000. After 3 years its value was £3000. By how much has it depreciated?

6. Find the interest on £200 invested at 5% for 1 year.

7. What is the simple interest on £100 invested at 10% for six months?

8. How much does £100 become when it is invested at 8% per annum for 1 year?

Money in Business and the Community

Profit and Loss

When a dealer buys or sells goods, the **cost price** is the price at which he buys the goods, and the **selling price** is the price at which he sells the goods. The **profit** is the difference between the selling price and the cost price. That is

$$\text{Profit} = \text{Selling price} - \text{Cost price}$$

Example 1

A dealer buys a table for £80 and sells it for £130. Calculate his profit.

$$\text{Profit} = \text{Selling price} - \text{Cost price}$$
$$= £130 - £80$$
$$= £50$$

The percentage profit is usually calculated on the cost price and

$$\text{Profit \%} = \frac{\text{Profit}}{\text{Cost price}} \times \frac{100}{1}$$

Example 2

A shopkeeper buys an article for £5 and sells it for £6. Calculate his percentage profit.

We are given:

$$\text{Cost price} = £5$$
$$\text{and} \quad \text{Selling price} = £6;$$
$$\text{Profit} = £6 - £5$$
$$= £1$$
$$\text{Profit \%} = \frac{1}{5} \times \frac{100}{1}$$
$$= 20\%$$

If a loss is made, the cost price is greater than the selling price and

$$\text{Loss} = \text{Cost price} - \text{Selling price}$$

Example 3

A motor bike is bought for £300 and sold for £250. Calculate the loss.

$$\text{Loss} = £300 - £250$$
$$= £50$$

As with percentage profit, the percentage loss is usually calculated on the cost price. That is

$$\text{Loss \%} = \frac{\text{Loss}}{\text{Cost price}} \times \frac{100}{1}$$

Example 4

A man buys a car for £3200 and sells it for £2400. Calculate his percentage loss.

We are given

$$\text{Cost price} = £3200$$
$$\text{and} \quad \text{Selling price} = £2400$$
$$\text{Loss} = £3200 - £2400$$
$$= £800$$
$$\text{Loss \%} = \frac{800}{3200} \times \frac{100}{1}$$
$$= 25\%$$

Exercise 10.1

1. A shopkeeper buys an article for £50 and sells it for £70. What is the amount of his profit?

2. An article is bought for 60p and sold for 90p. Calculate the profit.

3. A grocer buys a tin of beans for 20p and sells it for 30p. Work out his percentage profit.

4. An article is bought for £5 and sold for £4. What is the loss?

5. A second-hand car is bought for £3000 and sold for £2500. What is the loss?

6. An article is bought for £10 and sold for £8. Calculate the percentage loss.

7. A dealer buys a chair for £30 and sells it for £40:
   (a) What is the amount of his profit?
   (b) Calculate his percentage profit.

8. Goods are bought for £50 and sold for £55. Calculate the percentage profit.

Discount

Sometimes a dealer will deduct a percentage of the selling price, for example, if the customer is prepared to pay cash. This is called **discount**.

Example 5

A wardrobe is offered for sale at £270. A customer is offered a discount of 10% for a cash sale. How much will the customer actually pay for the wardrobe?

Method 1

Amount of discount $= 10\%$ of £270

$$= \frac{10}{100} \times \frac{270}{1}$$

$$= £27$$

Amount actually paid $= £270 - £27$

$$= £243$$

Method 2

Since a discount of 10% is offered, the customer pays only 90% (100% − 10%) of the selling price. Hence

Amount actually paid $= 90\%$ of £270

$$= \frac{90}{100} \times \frac{270}{1}$$

$$= £243$$

Exercise 10.2

1. An armchair is offered for sale at £200. The trader offers a discount of 20% for cash:
   (a) Work out the amount of the discount.
   (b) How much does a cash paying customer actually pay for the chair?

2. A shop offers a 15% discount on a table on sale for £120:
   (a) How much discount is allowed?
   (b) What is the table sold for?

3. During a sale a discount of 25% is allowed on all the items in a shop. A coat is marked at £48. How much will it be sold for?

4. A furniture shop offers a 12% discount for cash. How much discount will be allowed on furniture costing £2000?

5. A grocer offers a 3% discount to his customers provided they pay their bills within a week. If a bill of £60 is paid within one week, how much will the customer actually pay?

Value Added Tax

Value Added Tax, or VAT for short, is a tax on goods and services. Some goods and services bear no VAT, for instance food, books and insurance. The rate varies from time to time but at the moment it is levied at 15%.

Example 6

A man buys a lawn mower which is priced at £80 plus VAT at 15%. How much will the mower actually cost him?

$$VAT = 15\% \text{ of } £80$$

$$= \frac{15}{100} \times \frac{80}{1}$$

$$= £12$$

Amount actually paid = £80 + £12

$$= £92$$

Exercise 10.3

In the following questions take the rate of VAT as 15%.

1. A TV set is priced at £300 exclusive of VAT. How much VAT will be charged?

2. A woman buys a washing machine which is priced at £240 plus VAT. How much will the woman actually pay for the machine?

3. A housewife buys a refrigerator for £160 but on top of this VAT is charged. How much does the housewife pay altogether?

4. A telephone bill is £60 excluding VAT. How much is the bill when VAT is added on?

5. Calculate the amount of VAT which will be charged on a garage bill of £160.

Rates

Every property in a town or city is given a **rateable value** which is fixed by the local district valuer. This rateable value depends upon the size, condition and location of the property.

The rates are levied at so much in the pound of rateable value, for instance 90p in the pound. The money raised by the rates is used to pay for such things as libraries, police and education.

Annual rates = Rateable value × Rate in £1

Example 7

The rateable value of a house is £240. If the rate levied is £1.25 in the pound, how much must the householder pay in rates?

$$\text{Annual rates} = 240 \times £1.25$$

$$= £300$$

Most local authorities state, on their rate demand, the product of a penny rate. This is the amount that would be raised if the rate levied was 1p in the pound, that is, £0.01 in the pound.

Example 8

The rateable value of a town is £9 350 000. What is the product of a penny rate?

Product of penny rate = 9 350 000 × £0.01

$$= £93 500$$

Example 9

The cost of highways and bridges in a local authority area is £1 500 000. If the product of a penny rate is £80 000, calculate the rate in the pound needed to cover the cost of highways and bridges.

$$\text{Rate in the pound required} = \frac{1\,500\,000}{80\,000}$$

$$= 18.75p$$

Exercise 10.4

1. Copy and complete the following table:

	Rateable value	Rate in the £	Annual rates
(a)	£200	80 p	
(b)	£200	£1.20	
(c)	£350	£1.50	
(d)	£400	90 p	
(e)	£500	£1.30	

2. The rateable value of all the property in a city is £8 000 000. What is the product of a penny rate?

3. Calculate the total income of a local authority if the total rateable value is £4 500 000 and the rates are levied at £1.08 in the pound.

4. The total rateable value of a small town is £900 000. Calculate the total cost of the public library if a rate of 5.2 p in the pound must be levied to cover its cost.

5. In another town the cost of running the public library is £300 000 per annum. If the product of a penny rate is £15 000, find the rate in the pound needed for the upkeep of the library.

Insurance

Our future is something which is far from certain. We could become too ill to work or we could be badly injured or even killed in an accident. Our house could be burgled or burnt down. We might be involved in a car accident and be liable for injuries and damage. How do we take care of these eventualities? The answer is to take out insurance policies. The insurance company collects **premiums** from thousands of people who wish to insure themselves. It invests this money to earn interest and this money is then available to pay people who claim for their loss.

Example 10

A house and its contents are valued at £35 000. An insurance company charges a premium of £3 per £1000 of insurance. How much is the annual premium?

$$\text{Annual premium} = £\frac{3 \times 35\,000}{1000}$$

$$= £105$$

Car Insurance

By law a motor vehicle must be insured. That is, a policy must be taken out in case someone is injured, or damage is caused by an accident which is the fault of the policy holder. Car owners who have had no accidents are usually allowed a **no-claims** bonus which means that they pay less for their insurance policy.

Example 11

The insurance on a car is £150 but the owner is allowed a 40% no-claims bonus. What premium does the motorist pay for the year?

$$\text{No-claims bonus} = 40\% \text{ of } £150$$

$$= £60$$

$$\text{Annual premium} = £150 - £60$$

$$= £90$$

Cost of Running a Car

When calculating the cost of running a car, the costs of tax, insurance, petrol, maintenance and depreciation should all be taken into account. Every motor vehicle requires a Road Fund Licence which has to be paid when the vehicle is first registered and then renewed periodically. The vehicle will need maintaining and its value will depreciate with age.

Example 12

During one year a car depreciated by £1000 and during this year it used petrol costing £750. Insurance cost £87, tax £100 and maintenance £350.

(a) Work out the cost of running the car for 1 year.

(b) During the year the car did 20 000 km. What was the total cost per kilometre?

   (a) Cost for 1 year is

$$\begin{aligned}
\text{Depreciation} &= \text{£1000} \\
\text{Petrol} &= \text{£750} \\
\text{Insurance} &= \text{£87} \\
\text{Tax} &= \text{£100} \\
\text{Maintenance} &= \underline{\text{£350}} + \\
\text{Total} &= \underline{\text{£2287}}
\end{aligned}$$

   (b) Cost per kilometre travelled

$$\begin{aligned}
&= \text{£2287} \div 20\,000 \\
&= \text{£0.114} \\
&= 11.4\text{p}
\end{aligned}$$

Example 13

A car does 30 miles to 1 gallon of petrol which costs £2 per gallon. During one year the motorist travels 33 000 miles. How much does petrol cost the motorist for a year's motoring?

$$\begin{aligned}
\text{Number of gallons used} &= 33\,000 \div 30 \\
&= 1100
\end{aligned}$$

$$\begin{aligned}
\text{Cost of petrol} &= \text{Number of gallons used} \\
&\qquad \times \text{Price per gallon} \\
&= 1100 \times \text{£2} \\
&= \text{£2200}
\end{aligned}$$

Life Assurance

With this type of **assurance** a sum of money, depending upon the size of the premium, etc. is paid to the dependants (wife or husband and children) of the policy holder upon his or her death. The premiums depend upon:

(1) The age of the policy holder (the younger the policy holder is, the less he or she pays in premiums because there is less chance of him or her dying suddenly).

(2) The amount of money the policy holder wants the dependants to receive (the larger the amount the larger the premium).

Example 14

A man aged 35 years wishes to assure his life for £7000. The insurance company quotes an annual premium of £13.90 per £1000 assured. Calculate the amount of the monthly premium.

$$\begin{aligned}
\text{Annual premium} &= \text{£13.90} \times \frac{7000}{1000} \\
&= \text{£13.90} \times 7 \\
&= \text{£97.30} \\
\text{Monthly premium} &= \text{£97.30} \div 12 \\
&= \text{£8.11}
\end{aligned}$$

Endowment Assurance

This is very similar to whole life assurance but the policy holder can decide for how long he or she is going to pay the premiums. At the end of the chosen period a lump sum is paid to the policy holder. If, however, the policy holder dies before the end of the chosen period, the assured sum of money will be paid to his or her dependants.

Some endowment and whole life policies are 'with profits'. This means that a bonus based on the profits that the insurance company makes will be paid in addition to the sum assured.

Example 15

Guaranteed death benefits during 15-year term					
Your age now		Monthly premium			
Male	Female	£4.95	£7.95	£10.95	£14.95
20	20–24	14050	27631	41212	
21	25	13843	27222	40600	
22	26	13520	26583	39645	Not
23	27	13091	25735	38379	Available
24	28	12572	24710	36848	
25	29	11982	23544	35106	50522
26	30	11329	22256	33182	47750
27	31	10625	20864	31104	44757
28	32	9880	19395	28910	41597
29	33	9108	17872	26636	38321
30	34	8329	16336	24343	35018
31	35	7554	14808	22062	31733
32	36	6800	13322	19843	28538
33	37	6081	11904	17728	25492
34	38	5408	10578	15748	22642
35	39	4812	9405	13997	20120
36	40	4295	8385	12475	17929
37	41	3840	7490	11139	16005
38	42	3440	6702	9963	14312
39	43	3086	6005	8923	12814
40	44	2771	5385	7998	11482
41	45	2492	4834	7177	10301
42	46	2244	4347	6451	9255
43	47	2025	3916	5808	8330
44	48	1831	3536	5241	7513
45	49	1660	3199	4738	6790
46	50	1508	2902	4295	6152
47	—	1375	2639	3903	5589
48	—	1256	2406	3557	5090
49	—	1152	2201	3250	4650
50	—	1058	2018	2977	4256

Look at the table above which is for a 15-year Endowment Plan and find the guaranteed death benefit:

(a) if a man aged 31 years pays monthly premiums of £7.95

(b) if a woman aged 39 pays monthly premiums of £10.95.

    (a) The death benefit for a man aged 31 years paying £7.95 per month is £14 808.

    (b) The death benefit for a woman aged 39 paying £10.95 per month is £13 997.

Exercise 10.5

1. A householder wishes to insure his house for £30 000. His insurance company charge an annual premium of £1.50 per £1000 insured. Find the householder's annual premium.

2. An insurance company charges £2.50 per £1000 insured. Calculate the premium to insure for £8000.

3. The premium on a car is £120 but a 30% no-claims bonus is allowed. How much is the annual premium?

4. The insurance on a car is £180 but the owner is allowed a 20% no-claims bonus. How much premium does the motorist pay for the year?

5. A car is bought for £4200 and used for one year. It is then sold for £3500. During the year it used petrol costing £900. Insurance cost £120, tax £100 and maintenance £280. Work out the cost of running the car for one year.

6. During one year the depreciation on a car was £750 and maintenance and repairs cost a further £250. The car travelled 30 000 km and used petrol costing £880. Third-party insurance cost £80 and tax cost £120. How much did it cost the motorist for his year's driving?

7. During one year a motorist travelled 30 000 km. The car did 15 km per litre of petrol which cost 50p per litre. What was the cost of petrol for a year's motoring?

8. After one year's use the depreciation of a motor bike was £200. The owner travelled 10 000 miles on it during that year and fuel consumption was, on the average, 80 miles per gallon. If petrol cost £2 per gallon, tax cost £60 and insurance £90 calculate:

    (a) the cost of a year's motorcycling

    (b) the cost per mile.

9. Look at the table on page 95 and answer the following questions:

   (a) What is the death benefit for a man aged 30 years if he pays £4.95 per month in premiums?

   (b) How much will a woman aged 25 years get at the end of 15 years if she pays premiums of £7.95 per month?

10. A man aged 36 years takes out an endowment policy for which he is quoted an annual premium of £2.02 per £1000 assured. If he assures himself for £8000, how much will he pay yearly?

11. For a monthly premium of £50 a woman aged 40 is guaranteed £6624 at the end of 10 years. However, the insurance company states that the maturity value of the policy is likely to be £11 241.

    (a) Work out how much the woman will pay in premiums over the 10-year period.

    (b) Calculate the amount of profit she is likely to make on the policy.

Miscellaneous Exercise 10

Section A

1. The premium for insuring a car is £200 but the owner is allowed a no-claims bonus of 30%. How much does he actually pay for the insurance?

2. Mrs Jones insures her jewellery for £1800. The premium is 25p per £100 insured. What is her total premium?

3. A home computer costs £200 plus 15% VAT. Work out the total price of the computer including VAT.

4. A dealer buys a table for £200 and sells it for £250. Work out:

   (a) his profit

   (b) his profit per cent.

5. A tea set is offered for sale at £120. A customer is given a discount of 5% for a cash sale. How much does the customer actually pay?

6. The rateable value of a house is £200. How much must be paid in rates when these are levied at £1.30 in the pound.

7. A man aged 30 years wishes to assure his life for £5000. The annual premium is £10 per £1000 assured. Work out the amount he will pay monthly.

8. The costs involved in running a car are: depreciation £230, fuel £780, tax £120, maintenance £240, insurance £90.

   (a) Work out the cost of running the car for a year.

   (b) If the owner travelled 9000 miles during that year, how much did the car cost per mile?

Section B

1. The cost price of a table is £120 and it is sold for £200:

   (a) How much is the profit?

   (b) What is the percentage profit?

2. A motor cycle is bought for £300 and sold for £150:

   (a) How much loss is made?

   (b) What is the percentage loss?

3. A dealer offers a discount of 25% on all the clothes in his shop. A coat is priced at £80. How much will a customer pay when he buys the coat?

4. Value added tax is charged at 15%. An electric fire is priced at £50 exclusive of tax:

   (a) How much tax will be charged?

   (b) How much will the customer pay for the fire?

5. A house has a rateable value of £400. Rates are levied at £1.25 in the pound. How much will the householder pay in rates per year?

6. A man wishes to insure his house for £30 000. The insurance company quotes him £3 per £1000 insured. How much will the man pay in premiums per annum?

7. The insurance on a car is £120 but there is a 30% no-claims bonus. How much is the premium for the year?

Multi-Choice Questions 10

1. When a dealer sells an article for £18 he makes a profit of £3. His percentage profit is therefore

   A 14.3%   B 16.7%

   C 20%   D 25%

2. The rateable value of a house is £150. The rates are £1.60 in the pound. The amount of rates paid anually is

   A £24   B £240   C £280   D £300

3. A car owner is quoted a premium of £100 to insure his car. If his no-claims bonus is 40%, the amount he pays for insurance is

   A £40   B £60   C £100   D £140

4. When an article is sold for £90, a loss of 10% on its cost price is recorded. The cost price of the article is

   A £81   B £95   C £99   D £100

5. A furniture shop decides to give 10% discount on all purchases made in the shop. How much would a customer actually pay for a bedroom suite that was originally priced at £480?

   A £452   B £436   C £432   D £216

6. A car is priced at £3600 plus VAT at 15%. The price including VAT is

   A £4140   B £3615

   C £3515   D £3060

7. The annual premium on a whole life policy is £8 per £1000 of insurance. If a man insures his life for £8000, his annual premium is

   A £64   B £80   C £640   D £800

Mental Test 10

1. The total cost of running a car for a year is £600. If it travels 6000 km in the year, what is the cost per kilometre?

2. VAT is charged at 15%. How much VAT is payable on an article priced at £200 plus VAT?

3. The rateable value of a house is £500. Rates are levied at £2 in the pound. How much must the householder pay in rates?

4. A shopkeeper made a profit of £10 on an article which he bought for £50. What is his percentage profit?

5. An article is priced at £200 but a discount of 10% is offered for a cash sale. How much discount will a cash customer receive?

6. A car owner suffered a loss of £50 on selling his car which he bought for £500. What is his percentage loss?

7. A car is insured for £100 but a no-claims bonus of 35% is allowed. How much does the car owner pay for his insurance?

8. A house is insured for £30 000. The premium is £5 per £1000 of insurance. How much must the householder pay per annum?

Household Finance

Rent

Rent is a charge for accommodation. It is usually paid weekly or monthly to the landlord who owns the property. The biggest owners of rented property are the local authorities.

The landlord may include the rates in his charge for the accommodation but if he does not it is the tenant's responsibility to see that these are paid.

Example 1

A landlord charges £18 per week for the rent of a flat. The rates are £104 per annum. What charge should the landlord make for rent and rates per week?

$$\text{Rates per week} = £104 \div 52$$
$$= £2$$

$$\text{Charge for rent and rates per week} = £18 + £2$$
$$= £20$$

Mortgages

If a person buying a house cannot pay out-right, he or she arranges a loan called a **mortgage** from a building society. The society generally requires a deposit of 5 or 10% but sometimes 100% mortgages are available. The balance of the loan plus interest is paid back over a number of years, perhaps 20 or 25 years. The interest rates charged by the building societies vary from time to time.

Example 2

A man wishes to buy a house costing £30 000. He pays a deposit of 10% to the building society. How much mortgage will he require?

$$\text{Deposit} = 10\% \text{ of } £30\,000$$
$$= £3000$$

$$\text{Mortgage required} = £30\,000 - £3000$$
$$= £27\,000$$

Example 3

A building society states that the monthly repayments on a mortgage are £11.80 per £1000 borrowed. Calculate the monthly repayments on a mortgage of £30 000.

$$\text{Monthly repayments} = \frac{30\,000}{1000} \times £11.80$$
$$= 30 \times £11.80$$
$$= £354$$

Exercise 11.1

Copy and complete the following table:

	House price	Deposit at 10%	Mortgage required
1.	£20 000		
2.	£25 000		
3.	£30 000		
4.	£40 000		
5.	£50 000		

Copy and complete the following table:

	Amount of mortgage	Monthly repayments per £1000 borrowed	Monthly repayments to the building society
6.	£10 000	£12.00	
7.	£20 000	£10.00	
8.	£25 000	£11.50	
9.	£40 000	£10.80	
10.	£50 000	£11.70	

Copy and complete the following table:

	Weekly rent	Annual rates	Rates per week	Rent + rates per week
11.	£20.00	£156.00		
12.	£30.00	£208.00		
13.	£17.50	£109.20		
14.	£45.00	£364.00		
15.	£63.00	£483.60		

Hire-Purchase

When goods are bought and paid for by instalments they are said to have been bought on **hire-purchase**. Usually the purchaser pays a deposit. The remainder of the purchase price (called the **outstanding balance**) plus interest is repaid by a number of instalments.

Example 4

The cash price of a refrigerator is £200. A woman buys it on hire-purchase. She pays a deposit of £50 and 12 equal monthly instalments of £20.

(a) How much is the outstanding balance?

(b) Work out the amount paid for the refrigerator when it is bought on hire-purchase.

(c) What is the difference between the hire-purchase price and the cash price?

$$\text{(a)} \quad \text{Outstanding balance} = £200 - £50$$
$$= £150$$

(b) Amount paid in instalments
$$= 12 \times £20$$
$$= £240$$

Total hire-purchase price
$$= £240 + £50$$
$$= £290$$

(c) Difference between hire-purchase price and cash price
$$= \text{Hire-purchase price} - \text{Cash price}$$
$$= £290 - £200$$
$$= £90$$

Example 5

A man buys a lawn mower for £200. He pays a deposit of 15%. What is his outstanding balance?

$$\text{Deposit} = 15\% \text{ of } £200$$
$$= £30$$
$$\text{Outstanding balance} = \text{Price} - \text{Deposit}$$
$$= £200 - £30$$
$$= £170$$

Example 6

The cash price of a motor bike is £800. When bought on hire-purchase interest at 20% is charged.

(a) If no deposit is paid calculate the hire-purchase price.

(b) The cash price plus interest is to be paid in 12 equal monthly payments. How much is each instalment?

(a) Interest = 20% of £800

$$= £160$$

Hire-purchase price = Cash price + Interest

$$= £800 + £160$$

$$= £960$$

(b) Amount of each instalment

$$= £960 \div 12$$

$$= £80$$

In Example 6, 20% would only be the true rate of interest if all of the outstanding balance was paid at the end of the year. However, as each instalment is paid the actual outstanding balance is reduced and hence a larger amount of each successive instalment is interest. The true rate of interest is about 38%.

Exercise 11.2

1. The hire-purchase payments on a washing machine are £25 per month. How much will be paid at the end of two years?

2. The cash price of a suite of furniture is £600. If bought on hire-purchase the terms are 12 monthly instalments of £60. Work out the difference between the cash price and the hire-purchase price.

3. A young man buys a second-hand car on hire-purchase. The hire-purchase payments are 24 monthly payments of £25:

    (a) How much will the young man pay for the car?

    (b) The cash price was £400. Work out the difference between the cash price and the hire-purchase price.

4. The cash price of some kitchen equipment is £400. To buy on hire-purchase a deposit of £80 followed by 10 equal instalments of £50 is required:

    (a) What is the total hire-purchase price of the equipment?

    (b) Calculate the difference between the cash price and the hire-purchase price.

5. The cash price of a motor bike is £800. To buy it on hire-purchase a deposit of 20% followed by 20 equal payments of £45 is required:

    (a) How much is the deposit?

    (b) What is the balance owing after the deposit has been paid?

    (c) What is the total hire-purchase price, i.e. the amount of 20 instalments plus the deposit?

    (d) Work out the difference between the cash and hire-purchase prices.

Bank Loans

Many people take out personal loans from a bank. The bank will calculate the interest for the whole period of the loan; the loan plus interest is usually repaid in equal monthly instalments.

Example 7

A man borrows £500 from a bank. The bank charges 18% interest for the whole period of the loan. If the repayments are in 12 equal monthly instalments, calculate the amount of each instalment.

$$\text{Interest} = 18\% \text{ of } £500$$

$$= £90$$

Total amount to be repaid = £500 + £90

$$= £590$$

Amount of each instalment = £590 $\div$ 12

$$= £49.17$$

Exercise 11.3

1. A man borrows £500 from a bank and interest amounts to £100. He repays the loan plus interest in 10 equal instalments. What is the amount of each instalment?

2. A bank lends a woman £1000. They charge interest at 20% per annum. She plans to pay back the loan plus interest in 12 equal monthly instalments. Work out:

   (a) the amount of interest payable

   (b) the total amount ot be paid to the bank

   (c) the amount of each monthly instalment.

3. An advertisement in a newspaper states that if you borrow £1000 for a period of 10 years the repayments will amount to £20.70 per month:

   (a) A man borrows £5000 over 10 years. Calculate the amount of his monthly repayments.

   (b) How much will the man pay in total over the 10-year period?

4. A man borrows £600 from his bank. To repay the loan he makes 12 monthly payments of £65:

   (a) How much does the man pay the bank in total?

   (b) What is the difference between the amount of the loan and the amount he pays back to the bank?

5. A bank lends a woman £2000. She agrees to repay the loan by making 30 payments of £90:

   (a) How much money does she repay in total?

   (b) Find the difference between the amount of the loan and the amount paid back.

Gas Bills

Gas is charged according to the number of **therms** used (1 therm = 100 cubic feet approximately). To find the cost of the gas used, multiply the number of therms used by the charge per therm. A standing charge may also be made.

Example 8

A customer uses 90 therms of gas during one quarter. She is charged 30p per therm plus a standing charge of £10. How much is her gas bill?

$$\text{Cost of gas used} = 90 \times 30\,\text{p} = £27$$

$$\text{Gas bill} = \text{Cost of gas used} + \text{Standing charge}$$

$$= £27 + £10 = £37$$

Reading Gas Meter Dials

The amount of gas used is always measured in cubic feet (often in hundreds of cubic feet) and this is changed into therms. In the ordinary way your gas meter will be read at regular intervals. However, there may be occasions when you wish to check the gas consumption yourself.

Fig. 11.1 shows the dials on a gas meter. To read the dials copy down the readings in the order in which they appear. Adjacent dials revolve in opposite directions. Where the hand stands between two figures, write down the lower figure. However, if the hand stands between 9 and 0 you should write down 9.

Fig. 11.1

Dial 1 pointer has passed 7, reading is 7   Dial 2 pointer has passed 5, reading is 5   Dial 3 pointer has passed 1, reading is 1   Dial 4 pointer has passed 9, reading is 9

The dials in the diagram show 7519 hundred cubic feet. If the previous reading was 7491 then

$$\text{Amount of gas used} = 7519 - 7491$$

$$= 28 \text{ hundred cubic feet}$$

$$= 28 \text{ therms approx.}$$

This is the amount of gas used since the meter was last read.

Exercise 11.4

1. Copy and complete the following table:

Charge per therm	Number of therms used	Cost of gas used	
		in pence	in pounds
30p	100		
40p	80		
45p	120		
35.3p	200		
37.5p	300		

2. In a certain quarter a customer uses 100 therms of gas which is charged at 35p per therm. If the standing charge is £10, how much is the gas bill?

3. The present reading on a householder's gas meter is 8765 and the previous reading was 8413 hundreds of cubic feet:

   (a) How many hundreds of cubic feet have been used?

   (b) If 1 therm = 100 cubic feet, how many therms have been used?

   (c) If gas is charged at 40p per therm, what is the cost of the gas consumed?

4. Copy the set of dials shown in Fig. 11.2 three times and draw in hands to show

   (a) 2400   (b) 8345   (c) 4568

Fig. 11.2

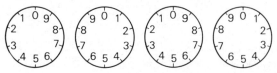

5. Look at the gas bill shown opposite and answer the following questions:

   (a) What is the cost of gas used?

   (b) What is the total amount of the bill?

6. Gas meters are read every quarter. If the reading was 350 at the beginning of the quarter and 593 hundreds of cubic feet at the end of the quarter

   (a) How many hundreds of cubic feet of gas have been used?

   (b) How many therms is this (1 therm = 100 cubic feet)?

   (c) If the charge is 37p per therm, calculate the cost of the gas used.

   (d) If the standing charge is £8 per quarter, work out the total amount of the bill.

7. Write down how many cubic feet the meters in Fig. 11.3 show:

Fig. 11.3

(a)

(b)

(c)

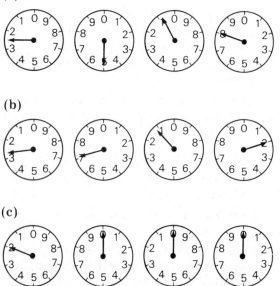

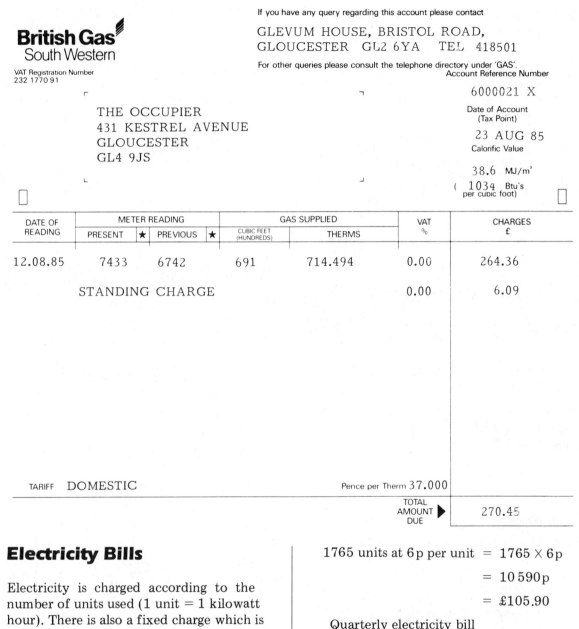

British Gas South Western

VAT Registration Number
232 1770 91

If you have any query regarding this account please contact

GLEVUM HOUSE, BRISTOL ROAD,
GLOUCESTER   GL2 6YA   TEL 418501

For other queries please consult the telephone directory under 'GAS'.

Account Reference Number

6000021 X

THE OCCUPIER
431 KESTREL AVENUE
GLOUCESTER
GL4 9JS

Date of Account
(Tax Point)

23 AUG 85

Calorific Value

38.6 MJ/m³

(1034 Btu's
per cubic foot)

| DATE OF READING | METER READING | | | | GAS SUPPLIED | | VAT % | CHARGES £ |
	PRESENT	★	PREVIOUS	★	CUBIC FEET (HUNDREDS)	THERMS		
12.08.85	7433		6742		691	714.494	0.00	264.36
STANDING CHARGE							0.00	6.09

TARIFF   DOMESTIC

Pence per Therm 37.000

TOTAL AMOUNT DUE ▶   270.45

Electricity Bills

Electricity is charged according to the number of units used (1 unit = 1 kilowatt hour). There is also a fixed charge which is added to the bill.

Example 9

A household uses 1765 units of electricity in a quarter. If the standing charge is £7 and the price per unit is 6p, find the amount of the quarterly electricity bill.

$$1765 \text{ units at } 6\text{p per unit} = 1765 \times 6\text{p}$$
$$= 10\,590\text{p}$$
$$= £105.90$$

Quarterly electricity bill
$$= \text{Cost of units used} + \text{Standing charge}$$
$$= £105.90 + £7.00$$
$$= £112.90$$

Reading Electricity Meters

When reading an electricity meter remember that adjacent dials revolve in opposite directions. The five dials are read from left to right. The method of reading an electricity meter is shown in Fig. 11.4.

Fig. 11.4

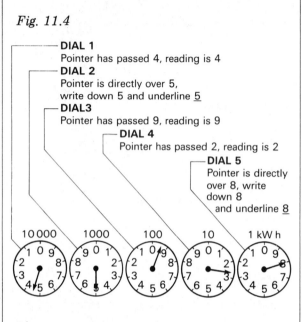

DIAL 1
Pointer has passed 4, reading is 4
DIAL 2
Pointer is directly over 5,
write down 5 and underline 5
DIAL 3
Pointer has passed 9, reading is 9
DIAL 4
Pointer has passed 2, reading is 2
DIAL 5
Pointer is directly
over 8, write
down 8
and underline 8

10 000   1000   100   10   1 kW h

The reading in the diagram is 45928. Now look at the figures underlined.

If one of these is followed by a 9, reduce the previous figure by 1. Thus the corrected reading is 44928.

Example 10

Fig. 11.5 shows the dials of an electricity meter.

(a) Read the meter.

(b) If the previous meter reading was 53 432, find the number of units used.

(c) If the charge per unit is 5.93p and the standing charge is £9.80, find the amount of the quarterly electricity bill.

Fig. 11.5

10 000   1000   100   10   1 kW h

(a) The meter reads 56 378.

(b) Number of units used

$$= 56\,378 - 53\,432$$

$$= 2946$$

(c) 2946 units at 5.93p $= 2946 \times 5.93$p

$$= £174.70$$

Cost of electricity

$= $ Cost of units used

     $+$ Standing charge

$$= £174.70 + £9.80$$

$$= £184.50$$

Exercise 11.5

Find the total amounts payable for these electricity bills:

1. Standing charge £7.40; 1000 units at 5p per unit.

2. Standing charge £5.90; 850 units at 6p per unit.

3. Standing charge 8.00; 759 units at 5.73p per unit.

4. Standing charge £6.82; 958 units at 5.17p per unit.

Read the electricity meters shown in Fig. 11.6:

Fig. 11.6

5.

6.

7.

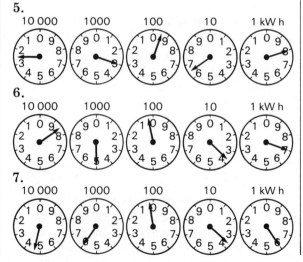

8. Copy the set of dials shown in Fig. 11.7 three times and put in arrows to show readings of

   (a) 81 372 units

   (b) 56 325 units

   (c) 73 452 units.

Fig. 11.7

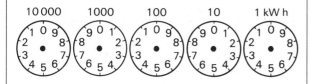

10 000   1000   100   10   1 kW h

9. A meter reading on 15th February was 009 870 and on 17th May it was 009 986:

   (a) How many units of electricity have been used?

   (b) If each unit costs 5.93 p and the standing charge is £8.70, find out how much the electricity bill will be.

10. An economy tariff charges for electricity as follows: day rate 5.70 p per unit; night rate 2.04 p per unit:

    (a) 2430 units are used during the day. What is the cost of these units?

    (b) 932 units were used at night. Work out the cost of these units.

    (c) If the standing charge is £8.90 find the total amount of the electricity bill.

11. A typical electricity bill is shown overleaf. Answer the following questions about it:

    (a) How much does the customer have to pay?

    (b) How many units of electricity were used?

    (c) What is the cost per unit.

    (d) What is the standing charge?

    (e) Is the bill correct?

    (f) On what date was the meter read?

Telephone Bills

The amount of a telephone bill depends on the number of units used and the charge for the rental of the telephone itself. The number of units for a dialled call depends upon the destination of the call, the time of day and the length of the call. The bills are paid quarterly and VAT at 15% is charged.

Example 11

A householder used 500 units during a quarter and the quarterly rental charge was £20. If each unit was charged at 5p and VAT is levied at 15%, find the total cost of the householder's telephone bill.

$$500 \text{ units at } 5p \text{ each} = 500 \times £0.05$$
$$= £25$$
$$\text{Cost less VAT} = £25 + \text{Rental charge}$$
$$= £25 + £20$$
$$= £45$$
$$\text{VAT at } 15\% = 15\% \text{ of } £45$$
$$= £6.75$$
$$\text{Total cost} = \text{Cost less VAT} + \text{VAT at } 15\%$$
$$= £45 + £6.75$$
$$= £51.75$$

Exercise 11.6

In the following questions VAT is charged at 15%.

Work out the amount payable for the following telephone bills:

1. Rental charge £15; 200 units at 5p each; plus VAT.

2. Rental charge £20; 180 units at 4p each; plus VAT.

3. Standing charge £17.60; 180 units at 4.8p each; plus VAT.

MIDLANDS   ELECTRICITY

ACCOUNT DATE	CONSUMER REFERENCE No.
21.8.85	4834/02689/21

MR T SMITH
431 KESTREL AVENUE
GLOUCESTER
GL4 9JS

CD36   VAT REGISTRATION No. | 238 | 5679 | 21 |

METER READINGS			VAT REGISTRATION No. 238 5679 21	AMOUNT £
PREVIOUS	PRESENT			
20013	20520	TARIFF D1 - DOMESTIC SUPPLY		
		STANDING CHARGE		6.09
		UNIT CHARGE 507 AT 5.520P		27.98

DATE OF READING	E INDICATES ESTIMATED READING, SEE REVERSE OF ACCOUNT	TOTAL AMOUNT NOW DUE FOR PAYMENT	£	34.07
21.08.85	C INDICATES CONSUMERS OWN READING. THE DATE SHOWN ON THE LEFT IS THE DATE WE RECEIVED YOUR READING R INDICATES METER REMOVED			

PAYMENTS
BY POST   to   MIDLANDS ELECTRICITY BOARD, P.O. BOX No. 18, KINGSWINFORD,
WEST MIDLANDS, DY6 8BL. (further advice on methods of payment is given overleaf)

ENQUIRIES   to   M.E.B. HAMMOND WAY, BARNWOOD, GLOUCESTER, GL4 7HQ
TEL.   GLOS. 69221

A RECEIPT when given
by machine to be printed here

4. Standing charge £16.15; 250 units at 5.3p each; plus VAT.

5. The telephone meter was read on 15th May as 009 768 and on 17th August as 009 975:

(a) How many units have been used during this period?

(b) If each unit is charged at 4.8p per unit, what is the cost of the units used?

(c) If the rental charge is £17.30, what is the total bill (exclusive of VAT)?

(d) How much is the VAT which is charged on the total bill?

(e) How much will the subscriber actually pay?

Holidays

Holidays whether at home or abroad can prove quite costly and some provision out of weekly or monthly income must be made to cover the cost. The main costs are accommodation, miscellaneous expenses like ice-cream for the children and drinks for both children and adults, and the cost of travelling, whether by car, rail or coach. Some people buy **package holidays** and the package will then consist of accommodation and cost of travelling.

Example 12

A family consisting of two adults and two children, under 11, decide to book a week's seaside holiday. The cost of the boarding house is £30 per week for adults and £18 per week for children under 11. The rail fare per adult is £23 and half-price for children. Miscellaneous expenses are expected to total £110.

(a) Work out the total cost of the holiday.

(b) The parents reckon they need to save for 40 weeks. How much a week do they need to save?

    (a) Cost of accommodation = £96.00
      (2 × £30 + 2 × £18)

      Rail fare   = £69.00
      (2 × £23 + 2 × £11.50)

      Miscellaneous expenses = £110.00

      Total cost of holiday   = £275.00

    (b) Weekly savings needed = £275 ÷ 40

                     = £6.88

Example 13

The table opposite shows the charges made by the Wintersun Holiday Company for skiing holidays to Switzerland. The prices shown are in pounds per person.

Date of departure	12th Dec	19th Dec	8th Jan & 15th Jan	22nd Jan	29th Jan & 5th Feb
7 nights	195	220	305	287	198
14 nights	370	400	585	550	460
Reductions	Children aged 2 to 11 years, half-price.				

(a) Miss Thomson plans to spend 1 week in Switzerland. How much will it cost her if she departs on 15th January?

(b) How much will she save if she departs on 19th December instead?

(c) A family consisting of two adults and one child, aged 9 years, depart on 19th December for 14 nights. How much will the holiday cost them?

    (a) Cost when departing on 15th January = £305.

    (b) Cost when departing on 19th December = £220.

    Saving = £305 − £220 = £85

    (c) Cost of adults = £800 (2 × £400)

      Cost of child = £200 ($\frac{1}{2}$ × £400)

  Total cost = £1000 (£800 + £200)

Exercise 11.7

1. A family estimates that their holiday will cost them £600. If they save for 40 weeks how much per week do they need to save?

2. A camping holiday in France is advertised for £245 all in. Other expenses are estimated to be £200. If a family save for 11 months how much do they need to put away each month?

3. A woman decides to take a coach touring holiday. The cost of her travel, food and hotels is £230. In addition she needs pocket money amounting to £70. To pay for this holiday she saves for 30 weeks. How much should she save per week?

4. The table below has been taken from the brochure of the Wonderful Holiday Company! Answer the following questions about it:

    (a) Which is the cheapest 7-night holiday with half board?

    (b) How much is the air fare from Gatwick to Costa del Sol on 6th February?

    (c) From the figures given, estimate the cost of food (half board) and accommodation only for 7 nights in Majorca flying from Bristol.

    (d) A family decide to have a holiday on the Algarve flying from Gatwick on 7th February. The duration of the holiday is 7 nights. What is the price of the holiday per person? How much would they pay if they decided on flight only?

5. A hotel offers bed, breakfast and evening meal for £25 per person per day. Children under 14 years old are allowed a discount of 30%. VAT is charged at 15% on the total bill:

    (a) A family consisting of 3 children (under 14 years old) and 2 adults stay for 7 days. What is the cost excluding VAT?

    (b) How much VAT is payable?

    (c) What is the amount of the bill including VAT?

Miscellaneous Exercise 11

Section A

1. The rent of a flat is £80 per month and the annual rates are £180. How much should the landlord charge for rent and rates per month?

2. A building society charges £16.00 per month per £1000 borrowed. How much would a borrower pay per year if he borrowed £5000?

3. The price of a house is £20 000. The person who is buying it puts down a deposit of 15%:

    (a) How much was the deposit?

    (b) How much mortgage does the person require?

Destination	Nights	Dates	Airport	Hotel grade & board	Price	Air fare only
Majorca	4 nts	3, 10 Feb	Gatwick	HB	£79	£59
Tunisia	7 nts	28 Jan, 4 Feb	Gatwick	HB	£125	—
Costa del Sol	7 nts	6 Feb	Gatwick	HB	£139	£59
Tunisia	14 nts	28 Jan, 1, 4, 8, 11 Feb	Gatwick	HB	£169	—
Algarve	7 nts	7 Feb	Gatwick	HB	£189	£69
Gambia	7 nts	12 Feb	Gatwick	HB	£329	—
Malta	7 nts	30 Jan, 6, 13 Feb	Luton	SC	£99	£49
Majorca	7 nts	31 Jan	Luton	HB	£115	£49
Malta	14 nts	30 Jan, 6, 13 Feb	Luton	SC	£125	£49
Costa del Sol	7 nts	30 Jan, 6 Feb	Luton	HB	£129	£59
Tunisia	7 nts	28 Jan, 1, 4, 11 Feb	Luton	HB	£139	—
Estoril Coast	7 nts	3 Feb	Luton	BB	£149	£49
Majorca	7 nts	31 Jan	Bristol	HB	£119	£49
Majorca	7 nts	31 Jan	Cardiff	HB	£105	£49
Costa Blanca	7 nts	1 Feb	Birmingham	FB	£109	£59
Costa del Sol	7 nts	30 Jan, 6 Feb	Birmingham	HB	£129	£49
Majorca	7 nts	31 Jan	East Midlands	HB	£129	£49

Abbreviations: HB = half board; SC = self catering; FB = full board

4. The cash price for a suite of furniture is £1000. A woman buys it on hire-purchase. She pays a deposit of £100 and ten monthly payments of £110:

   (a) How much is there left to pay after the deposit?

   (b) Work out the hire-purchase price of the suite.

   (c) Find the difference between the hire-purchase price and the cash price.

5. A man borrows £2000 from a bank which charges interest at 20% for the whole period of the loan:

   (a) How much in interest does the bank charge?

   (b) If the loan plus interest is to be repaid in 12 equal monthly instalments, work out the amount of each payment.

6. Fig. 11.8 shows the dials of a gas meter. Put hands on the dials to show a reading of 4637 cubic feet (hundreds).

Fig. 11.8

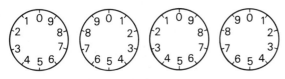

7. What is the reading on the gas meter shown in Fig. 11.9?

Fig. 11.9

8. A seven-day holiday in Norway is advertised at £170 per person with a reduction of 30% for children under 11 years old:

   (a) How much is the price for a child 10 years old?

   (b) A family consists of man and wife and a 10-year-old child. Work out the total cost of this holiday.

9. An electricity meter was read on 15th May as 007 852 units. On 17th February the meter read 007 328:

   (a) How many units of electricity have been used?

   (b) If each unit is charged at 6p and the standing charge is £8, calculate the amount of the electricity bill.

10. A householder used 400 units during a certain quarter. If each unit is charged at 5p and the quarterly rental charge is £15, calculate the amount of the telephone bill when VAT is charged at 15%.

Section B

1. Mrs Evans insured her jewellery for £2400. The premium is 25p per £100 of cover. What is her total premium?

2. A small computer is priced at £300 plus VAT at 15%:

   (a) How much is the VAT?

   (b) What is the total cost of the computer including VAT?

   (c) Tindalls offer the computer with £30 off its total price. Lyons offer the computer with 20% off the total price. Find the difference between the two offers.

   (d) Thompsons offer the computer on hire-purchase as follows: Deposit £90 plus 12 equal monthly payments of £24.75. What is the hire-purchase price of the computer?

3. (a) Fig. 11.10 shows the positions of an electricity meter at the beginning of a quarter. Write down the meter reading.

Fig. 11.10

(b) If during this quarter 760 units of electricity are used write down the meter reading.

(c) Copy the dials shown in Fig. 11.11 and show the new meter reading on them.

Fig. 11.11

(d) If electricity is charged at 5.7p per unit plus a standing charge of £8.50, calculate the total cost of the 760 units in the quarter.

4. Copy and complete the following telephone bill:

Quarterly rental charge = £9.75

Dialled units 600 at 4.7p per unit =

Total (exclusive of VAT) =

VAT at 15% =

Total payable =

5. Fig. 11.12 shows the positions of the dials of a gas meter at the end of a quarter. The reading gives the number of hundreds of cubic feet.

(a) What is the reading on the meter?

(b) If 1 therm = 100 cubic feet, how many therms have been used?

(c) If the previous reading was 1013 therms, how many therms have been used this quarter?

(d) Gas is charged at 34.9p per therm plus a standing charge of £9.25 per quarter. Calculate the total amount of the gas bill.

Fig. 11.12

Multi-Choice Questions 11

1. For her holidays a woman saved 10% of her weekly wage of £150 for 40 weeks in the year. How much did she save for her holiday?

   A £150   B £300   C £45   D £600

2. A debt of £2000 is to be repaid in equal instalments of £20. How many instalments are required?

   A 5   B 8   C 10   D 100

3. A telephone subscriber is charged 5p for each dialled unit she uses. If she used 400 units and the standing charge is £15, how much is her telephone bill?

   A £20   B £35   C £200   D £235

4. A householder is charged for electricity at 8p per unit. If he uses 600 units in a quarter, the amount of his bill is

   A £4.80   B £6   C £48   D £60

5. A building society charges £78 per year per £1000 borrowed. A man borrows £20 000. His annual repayments are

   A £78   B £1560

   C £1666.67   D £2340

6. A woman wishes to buy a house for £30 000. She pays a 10% deposit. How much mortgage does she require?

   A £2700   B £3000

   C £27 000   D £30 000

7. A householder is charged 4.8p for each unit of electricity used plus a standing charge of £14.40. If he used 330 units during one quarter, the amount of his electricity bill was

   A £15   B £17.04

   C £30.24   D £32.20

Mental Test 11

1. The annual rates for a house are £240. How much should the landlord charge the tenant per month for rates?

2. The rent of a flat is £20 per week and the rates are £2 per week. How much should the landlord charge the tenant for rent and rates per week?

3. The rent of a small house is £480 per annum. What is the monthly rent?

4. A building society charges £10 per month per £1000 borrowed. A man takes out a mortgage for £10 000. How much does he pay per month?

5. A man wishes to buy a house for £20 000. He pays a deposit of 10%. How much deposit does he pay?

6. A man buys a table on hire-purchase. He makes 10 payments of £8. How much does he pay for the table?

7. A woman borrows £1000 from her bank which charges 20% interest. How much interest does she pay?

8. A customer uses 200 therms of gas during a quarter. The gas is charged at 40p per therm. How much does the gas cost?

9. SW Gas make a standing charge of £10. If the gas used costs a further £25, how much is the gas bill?

10. An electricity board makes a standing charge of £6. If a customer uses 500 units charged at 6p per unit, how much is her electricity bill?

The table below has been taken from the brochure of the Remarkable Holiday Company! The abbreviations are B/B (bed and breakfast), S/C (self catering), H/B (half board, i.e. breakfast and evening meal).

Answer the following questions about it.

11. Which is the cheapest holiday available?

12. On what date does the cheapest holiday to Tenerife occur?

13. What is the most expensive holiday?

14. What is the destination for the cheapest self-catering holiday and on what date does it depart?

15. Why is a 1-week holiday more than half a 2-week holiday?

16. What is the destination for the special offers?

Date	Resort	Accommodation	Rating	Board	1 Wk	2 Wks
22/1	Israel – Eilat	Sunsaver	3	B/B	£169	£265
23/1	Lanzarote	Arena Dorada Apt.	3+	S/C	£155	—
23/1	Lanzarote	Sunsaver	—	S/C	£149	£195
24/1	Tenerife	Palmeras Pl. Apt.	3	*H/B	£169	£238
24/1	Tenerife	Sunsaver	—	H/B	£142	£189
25/1	Costa del Sol	Stella Polaris TM	3	H/B	£99	—
27/1	Gran Canary	Walhalla Apt.	3	S/C	£159	—
29/1	Israel – Eilat	Sunsaver	3	B/B	£189	£269
29/1	Israel – Eilat	Caesar	3+	B/B	£287	£371
30/1	Lanzarote	La Penita Apt.	2+	S/C	£165	—
31/1	Tenerife	Palmeras Pl. Apt.	3	*H/B	£175	£238
31/1	Tenerife	El Cortijo Apt.	2+	S/C	£175	—
2/2	Algarve	Clube Praia Apt.	3	S/C	£145	—
3/2	Gran Canary	Sol Y Mar	3	S/C	£175	—

Apt. prices based on two sharing — discounts for 3/4 occupancy available.
Prices are for Gatwick and include all taxes and surcharges.
Manchester departures and flight only also available.

★ Special Offer ★   ★ Special Offer ★

Time, Distance and Speed

Measurement of Time

In the winter, time in the United Kingdom is measured using Greenwich Mean Time (GMT). In summer the clocks are brought forward by 1 hour and this is called British Summer Time.

The units of time are:

$$60 \text{ seconds (s)} = 1 \text{ minute (min)}$$
$$60 \text{ minutes (min)} = 1 \text{ hour (h)}$$
$$24 \text{ hours (h)} = 1 \text{ day (d)}$$
$$7 \text{ days (d)} = 1 \text{ week (wk)}$$
$$28, 30, 31 \text{ days} = 1 \text{ calendar month}$$
$$365 \text{ days} = 1 \text{ year}$$
$$366 \text{ days} = 1 \text{ leap year}$$
$$52 \text{ weeks} = 1 \text{ year}$$
$$12 \text{ months} = 1 \text{ year}$$
$$13 \text{ weeks} = 1 \text{ quarter}$$

When the number denoting a year is exactly divisible by 4, the year is called a **leap year** (unless it is a century year, e.g. 1900: then it is a leap year only if the first two figures are divisible by 4). So 1908, 1936 and 1984, for instance, were all leap years, as will be 2000, 2004, etc. A leap year has 366 days, the extra day being 29th February.

Example 1

(a) Change 720 s into minutes.

$$720 \text{ s} = 720 \div 60 \text{ min}$$
$$= 12 \text{ min}$$

(b) Change 9 h into minutes.

$$9 \text{ h} = 9 \times 60 \text{ min}$$
$$= 540 \text{ min}$$

(c) How many seconds are there in 1 h?

$$1 \text{ h} = 60 \text{ min}$$
$$= 60 \times 60 \text{ s}$$
$$= 3600 \text{ s}$$

(d) How many weeks are there in 3 quarters?

$$3 \text{ quarters} = 3 \times 13 \text{ weeks}$$
$$= 39 \text{ weeks}$$

Exercise 12.1

1. How many minutes are there in $3\frac{1}{4}$ hours?

2. How many seconds are there in $5\frac{1}{2}$ minutes.

3. Change 420 min into hours.

4. How many seconds are there in 4 hours?

5. How many weeks are there in 3 quarters?

6. A woman earns £9000 per annum. How much does she earn per month?

7. Taking 4 weeks = 1 month, find the number of weeks corresponding to:
   (a) 5 months   (b) 8 months
   (c) 11 months.

8. An electricity bill was £78 per quarter. How much is this per week?

9. Milk costs 20p per pint and a householder uses 2 pints per day. How much is his weekly bill for milk?

10. A television programme consists of plays, music and cartoons and it lasts for $2\frac{1}{2}$ hours. 20% of the time is devoted to cartoons and 50% to plays. Work out the time devoted to each part.

11. Which of the following were leap years?

    (a) 1784   (b) 1907   (c) 1931

    (d) 1942   (e) 1940

The Clock

There are two ways of showing the time:

(1) With a 12-hour clock (Fig. 12.1), in which there are two periods each of 12 hours' duration during each day.

The period between midnight and noon is called **a.m.** whilst the periods between noon and midnight is called **p.m.** Thus 8.45 a.m. is a time in the morning whilst 8.45 p.m. is a time in the evening.

Example 2

Find the length of time between 2.15 a.m. and 8.30 p.m.

    2.15 a.m. to 12 noon is   9 h 45 min

    12 noon to 8.30 p.m. is   8 h 30 min +

       Length of time is   18 h 15 min

Fig. 12.1

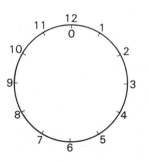

(2) With a 24-hour clock (Fig. 12.2) in which there is one period of 24 hours. This clock is used for railway and airline timetables. Times between midnight and noon are given times of 0000 hours to 1200 hours whilst times between noon and midnight are given times of 1200 hours to 2400 hours.

Fig. 12.2

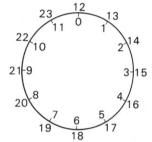

3.30 a.m. is written 0330 hours whilst 3.30 p.m. is written 1530 hours and this avoids confusion between the two times.

Note that 24-hour-clock times are written with four figures. 1528 hours should be read as 'fifteen twenty-eight hours' and 1300 hours as 'thirteen hundred hours'.

Example 3

Find the length of time between 0425 hours and 1812 hours.

    1812 hours is   18 h 12 min

    0425 hours is   4 h 25 min −

    Length of time is   13 h 47 min

Exercise 12.2

Find the length of time in hours and minutes from the first time given to the second.

1. 3.49 a.m. and 8.53 a.m.

2. 2.35 p.m. and 7.42 p.m.

3. 1.46 a.m. and 5.34 p.m.

4. 0050 hours and 1120 hours

5. 0342 hours and 1848 hours.

6. An aeroplane starts off from an airport at 0830 hours and flies for 50 minutes to a second airport. At what time does it reach its destination?

7. A train leaves London for Newcastle at 9.35 a.m. and travels for 3 hours and 50 minutes to York, where it stops for 10 minutes. It then travels on to Newcastle, which it reaches in a further 1 hour and 35 minutes. At what time does it arrive at Newcastle?

8. A car leaves Bristol at 1035 hours and arrives in Hereford $1\frac{1}{4}$ hours later. What time did it arrive in Hereford?

Timetables

Timetables are usually written using the 24-hour clock. Here is a simplified portion of a railway timetable.

Hester	0730	0830	0930	1030	1100	1230	1430	1600
Manton	0735	....	0935	1047	....	1236	1447	1606
Turley	0755	....	0955	1058	....	1255	1503	1627
Denton	0810	0900	1010	1115	1128	1310	1518	1640

Exercise 12.3

Using the timetable given above, answer the following questions:

1. I want to arrive in Turley at 1058. Which train must I catch from Hester? How long does the journey take?

2. Which train from Hester to Denton is the fastest? Which train is the slowest?

3. How long must I wait for the first train from Turley after 10.00 a.m.?

4. I arrive at Manton station at 2.15 p.m. How long do I have to wait for a train to Denton?

5. How long does it take the fastest train to travel between Hester and Turley?

Shown below is part of a railway timetable. Use it to answer the following questions:

6. Which is the first train to Paddington from Worcester after 10.00 a.m.?

7. How long does it take the 1358 train from Gloucester to reach Swindon?

8. What time must I arrive at Stroud station in order to catch the 0943 train from Cheltenham?

9. What time train must I catch from Kemble to make sure of arriving at Reading by 2.00 p.m.?

10. What time does the 1256 train from Stonehouse arrive at Paddington? How long does the journey take?

WORCESTER · CHELTENHAM · GLOUCESTER · SWINDON · LONDON

Mondays to Fridays

WORCESTER, Shrubb Hill	....	0653	0805	0915	....	1107	....	1317	....	1508	1554	....	1757	2022	....	....	....
CHELTENHAM. . .	0620	0722	0828	0943	1041	1135	1230	1344	1437	1536	1624	1718	1826	2051	2138	....	....
Gloucester	0633	0741	0847	0956	1055	1148	1244	1358	1451	1550	1638	1732	1843	2105	2152	....	....
Stonehouse	0645	0755	0902	1008	1107	1200	1256	1410	1503	1602	1650	1744	1855	2117	2204	....	....
Stroud	0650	0802	0909	1013	1112	1206	1302	1415	1508	1607	1656	1749	1900	2123	2209	....	....
Kemble.	0706	0819	0926	1029	1128	1222	1318	1431	1524	1623	1713	1805	1916	2140	2225	....	....
SWINDON	0723	0836	0943	1046	1145	1239	1335	1448	1541	1640	1730	1824	1933	2157	2242	....	....
Reading	0803	0915	1017	1143	1245	1353	1443	1545	1643	1718	1853	1908	2010	2245	....	....	....
LONDON, Paddington	0834	0938	1053	1216	1308	1426	1517	1618	1717	1747	1920	1937	2039	2314	....	....	....

The Calendar

1986 CALENDAR

JANUARY
Mon		6	13	20	27
Tue		7	14	21	28
Wed	1	8	15	22	29
Thu	2	9	16	23	30
Fri	3	10	17	24	31
Sat	4	11	18	25
Sun	5	12	19	26

FEBRUARY
Mon		3	10	17	24
Tue		4	11	18	25
Wed		5	12	19	26
Thu		6	13	20	27
Fri		7	14	21	28
Sat	1	8	15	22
Sun	2	9	16	23

MARCH
Mon		3	10	17	24	31
Tue		4	11	18	25
Wed		5	12	19	26
Thu		6	13	20	27
Fri		7	14	21	28
Sat	1	8	15	22	29
Sun	2	9	16	23	30

APRIL
Mon		7	14	21	28
Tue	1	8	15	22	29
Wed	2	9	16	23	30
Thu	3	10	17	24
Fri	4	11	18	25
Sat	5	12	19	26
Sun	6	13	20	27

MAY
Mon		5	12	19	26
Tue		6	13	20	27
Wed		7	14	21	28
Thu	1	8	15	22	29
Fri	2	9	16	23	30
Sat	3	10	17	24	31
Sun	4	11	18	25

JUNE
Mon		2	9	16	23	30
Tue		3	10	17	24
Wed		4	11	18	25
Thu		5	12	19	26
Fri		6	13	20	27
Sat		7	14	21	28
Sun	1	8	15	22	29

JULY
Mon		7	14	21	28
Tue	1	8	15	22	29
Wed	2	9	16	23	30
Thu	3	10	17	24	31
Fri	4	11	18	25
Sat	5	12	19	26
Sun	6	13	20	27

AUGUST
Mon		4	11	18	25
Tue		5	12	19	26
Wed		6	13	20	27
Thu		7	14	21	28
Fri	1	8	15	22	29
Sat	2	9	16	23	30
Sun	3	10	17	24	31

SEPTEMBER
Mon	1	8	15	22	29
Tue	2	9	16	23	30
Wed	3	10	17	24
Thu	4	11	18	25
Fri	5	12	19	26
Sat	6	13	20	27
Sun	7	14	21	28

OCTOBER
Mon		6	13	20	27
Tue		7	14	21	28
Wed	1	8	15	22	29
Thu	2	9	16	23	30
Fri	3	10	17	24	31
Sat	4	11	18	25
Sun	5	12	19	26

NOVEMBER
Mon		3	10	17	24
Tue		4	11	18	25
Wed		5	12	19	26
Thu		6	13	20	27
Fri		7	14	21	28
Sat	1	8	15	22	29
Sun	2	9	16	23	30

DECEMBER
Mon	1	8	15	22	29
Tue	2	9	16	23	30
Wed	3	10	17	24	31
Thu	4	11	18	25
Fri	5	12	19	26
Sat	6	13	20	27
Sun	7	14	21	28

Exercise 12.4

1. How many days were there altogether in February, March and April, 1986?

2. On which day did 7th October fall in 1986?

3. How many Saturdays were there in November, 1986?

4. Shrove Tuesday is 40 days before Palm Sunday. If Palm Sunday was 23rd March, what date was Shrove Tuesday?

5. The summer holiday in a certain area lasts for seven weeks. A school broke up on 18th July, 1986. On what date did it reopen?

6. How many weekdays (i.e. Monday to Friday) fell between 5th May and 13th June?

7. In 1986, 31st December fell on a Wednesday. On what day did 9th January, 1987 fall?

Average Speed

The road from Campton to Burley is 50 miles long. It takes me two hours to drive from Campton to Burley. Obviously I do not drive at the same speed all the time. Sometimes I drive slowly and sometimes I have to stop at crossroads and traffic lights.

However, the time it takes me is exactly the same as if I were driving along a straight road at 25 miles per hour, i.e. 50 miles ÷ 2 hours. So we say that my **average speed** is 25 miles per hour.

$$\text{Average speed} = \frac{\text{Total distance travelled}}{\text{Total time taken}}$$

The unit of speed depends upon the unit of distance and the unit of time. If the distance is measured in kilometres and the time in hours then the average speed will be measured in kilometres per hour (km/h). If the distance is measured in miles and the time in hours then the average speed will be measured in miles per hour (mile/h). If the distance is measured in metres and the time in seconds then the average speed will be measured in metres per second (m/s).

Note carefully that

$$\text{Distance travelled} = \text{Average speed} \times \text{Time taken}$$

$$\text{Time taken} = \frac{\text{Distance travelled}}{\text{Average speed}}$$

Example 4

(a) A car travels a total distance of 400 km in 5 hours. What is the average speed?

$$\text{Average speed} = \frac{\text{Distance travelled}}{\text{Time taken}}$$

$$= \frac{400}{5} \text{km/h}$$

$$= 80 \text{ km/h}$$

(b) A lorry travels at 35 miles per hour for 3 hours. How far does it travel?

$$\text{Distance travelled} = \text{Average speed} \times \text{Time taken}$$

$$= 35 \times 3 \text{ miles}$$

$$= 105 \text{ miles}$$

(c) A train travels 120 km at 40 km/h. How long does the journey take?

$$\text{Time taken} = \frac{\text{Distance travelled}}{\text{Average speed}}$$

$$= \frac{120}{40}$$

$$= 3 \text{ hours}$$

Remember that average speed is defined as *total distance travelled divided by total time taken.*

Example 5

A car travels 120 miles at a speed of 40 miles per hour. It then travels 60 miles at 30 miles per hour. Calculate its overall average speed. For questions like this, which consist of two parts, we must find the **total** distance travelled and the total time taken.

$$\text{Total distance} = 120 + 60$$

$$= 180 \text{ miles}$$

Since

$$\text{Time taken} = \frac{\text{Distance}}{\text{Speed}}$$

$$\text{Time for first part} = \frac{120}{40}$$

$$= 3 \text{ hours}$$

$$\text{Time for second part} = \frac{60}{30}$$

$$= 2 \text{ hours}$$

$$\text{Total time taken} = 3 \text{ hours} + 2 \text{ hours}$$

$$= 5 \text{ hours}$$

$$\text{Overall average speed} = \frac{\text{Total distance}}{\text{Total time}}$$

$$= \frac{180}{5}$$

$$= 36 \text{ miles per hour}$$

Exercise 12.5

Copy the table below and work out the values of the **average speeds**:

	Distance	Time	Average speed
1.	8 miles	2 h	
2.	20 km	5 h	
3.	80 m	4 s	
4.	40 ft	5 s	
5.	120 km	2 h	
6.	100 miles	4 h	
7.	27 km	3 h	
8.	150 miles	$2\frac{1}{2}$ h	

Copy the table below and work out the values of the distances:

	Average speed	Time	Distance
9.	5 miles/h	2 h	
10.	8 miles/h	3 h	
11.	20 km/h	5 h	
12.	30 m/s	4 s	
13.	12 ft/s	5 s	
14.	35 km/h	8 h	
15.	50 miles/h	$1\frac{1}{2}$ h	

Copy the table below and work out the values of the times:

	Distance	Average speed	Time
16.	20 km	5 km/h	
17.	80 miles	20 miles/h	
18.	6 m	3 m/s	
19.	50 ft	25 ft/s	
20.	300 miles	30 miles/h	
21.	400 km	50 km/h	
22.	10 miles	4 miles/h	

23. A motorist leaves Taunton at 1155 hours and arrives in Bath $1\frac{1}{2}$ hours later. The distance between Bath and Taunton is 45 miles:

   (a) At what time did the motorist arrive in Bath?

   (b) What was his average speed?

24. The distance between Gloucester and Manchester is 120 miles:

   (a) How long will it take a motorist to make the journey if he travels at an average speed of 40 miles per hour?

   (b) At what average speed must the motorist travel if he is to complete the journey in $2\frac{1}{2}$ hours?

25. (a) A motorist travelled for 2 hours at an average speed of 30 miles per hour. How far did he travel?

   (b) He then increased his average speed to 40 miles per hour and continued at this speed for a further 3 hours. How far did he travel at 40 miles per hour?

   (c) Using the answers to parts (a) and (b), calculate the average speed for the complete journey.

26. (a) A motorist drives for 3 hours at an average speed of 50 miles per hour. How far has he travelled?

   (b) He then travels at 60 miles per hour for 2 hours. What is the total distance he has now travelled?

   (c) Work out the total time taken for the complete journey.

   (d) What is his overall average speed?

27. A motorist travels 160 miles at an average speed of 40 miles per hour and spends 2 hours travelling the remaining 50 miles.

   (a) What is the total distance travelled?

   (b) What is the total time taken?

   (c) What is the overall average speed?

28. An aeroplane flies for 3 hours at a speed of 500 km/h and for 2 hours at a speed of 650 km/h.

   (a) What is the total distance travelled?

   (b) What is the overall average speed of the aeroplane?

Miscellaneous Exercise 12

Section A

1. How many seconds are there in 2 hours?

2. How many minutes are there in one day?

3. A car leaves Sheffield at 9.55 a.m. If it arrives in Birmingham at 1208 hours, how long did the journey take?

4. Part of a bus timetable is shown below. Use it to answer the following questions:

(a) At what time does the 1150 bus from Cirencester arrive at White Horse Inn?

(b) How long does it take the 1450 bus from Cirencester to reach Stroud?

(c) What time must I arrive at Bourne Bridge in order to catch the bus that arrives at 1239 in Stroud?

CIRENCESTER, Bus Station . . .	0950	1150	1450
Daglingworth	1000	1200	1500
Sapperton	1010	1210	1510
White Horse Inn	1017	1217	1517
Chalford, Marle Hill	1022	1222	1522
Bourne Bridge	1027	1227	1527
Thrupp, Brewery Lane	1031	1231	1531
STROUD, Bus Station	1039	1239	1539

5. A college closed on Friday, 11th July, 1986 for the long vacation which lasted for 65 days. On what date did the college reopen?

6. A motorist travelled a total distance of 300 miles in $5\frac{1}{2}$ hours. Calculate his average speed to the nearest mile per hour.

7. (a) A car travels for 3 hours at an average speed of 40 miles per hour. Calculate the distance travelled during this time.

(b) The car then increases its speed to 45 miles per hour. It travelled a distance of 90 miles at this speed. How long did it take to cover the 90 miles?

(c) How long did the complete journey take?

(d) Calculate the overall average speed of the car.

Section B

1. (a) A clock at a bus station shows the departure of an evening bus to be 8.15. How would this time be written in a timetable which uses the 24-hour-clock system?

(b) The bus arrived at its destination at 2045 hours. For how long was the bus travelling?

2. An aeroplane left town A at 1535 hours taking 2 h 40 min for its flight to town B. At what time did the aeroplane reach town B?

3. A car leaves Baxton at 1325 and reaches Flighton at 1555 hours. The distance between the two towns is 125 km:

(a) How long did the journey take?

(b) What was the average speed of the car?

4. A boat leaves a harbour at 0235 hours and arrives at its destination at 0405 hours:

(a) How long did the journey take?

(b) If the distance travelled was 12 miles, calculate the average speed of the boat?

5. The distance between Cardiff and London is 150 miles. A car leaves Cardiff at 9.35 a.m. and arrives in London 3 hours later.

(a) What time was it when the car arrived in London?

(b) What was the average speed of the car?

Multi-Choice Questions 12

1. An aeroplane flies non-stop for $2\frac{1}{4}$ hours and travels 1620 km. Its average speed in kilometres per hour is

A 364.5   B 720
C 800   D 3645

2. A cyclist covers a distance of 66 km at an average speed of 12 km/h. How long does the journey take?

A 5 h 30 min   B 5 h 20 min
C 5 h 5 min   D 5 h 3 min

3. A motorist travels a distance of 80 km at an average speed of 50 km/h. The time, in minutes, taken for the journey is

   A 40   B 96   C 150   D 360

4. A boy takes 15 minutes to walk to school at a speed of 5 km/h. How far did he walk?

   A 0.8 km   B 1.25 km
   C 20 km   D 75 km

5. How many seconds are there in 3 hours?

   A 180   B 1080
   C 3600   D 10 800

6. The length of time between 3.32 a.m. and 9.19 p.m. is

   A 5 h 47 min   B 8 h 28 min
   C 12 h 51 min   D 17 h 47 min

7. A car travels 50 km at 50 km/h and 70 km at 70 km/h. Its average speed, in kilometres per hour, is

   A 58   B 60   C 62   D 65

Mental Test 12

1. Change 120 seconds into minutes.

2. Change 5 minutes into seconds.

3. Change 240 minutes into hours.

4. Change 3 hours into minutes.

5. How many hours are there in 5 days?

6. What is the length of time between 8.00 a.m. and 3.00 p.m.?

7. Find the length of time between 0330 hours and 1600 hours.

8. A motorist travels 200 miles in 4 hours. What is his average speed?

9. A car travels for 5 hours at an average speed of 30 miles/h. How far does it travel?

10. A train travels 400 km at an average speed of 50 km/h. How long does the journey take?

Using the bus timetable shown below answer the following questions:

11. How long does it take for the bus to go from

    (a) Marcle to Parton

    (b) Wroughton to Parton

    (c) Marcle to Trumpton.

12. By the fastest bus, how long does it take from Wroughton to Trumpton?

Wroughton	0900	0930	1000	1030	1100
Marcle	0915	0945	....	1045	1115
Parton	0925	0955	....	1055	1125
Trumpton	0945	1015	1035	1115	1145

Basic Algebra

Introduction

In **algebra** letters are used instead of numbers.

$m + n$ means a number m added to a second number n.

$m - n$ means a number n subtracted from another number m.

$m \times n$ means a number m multiplied by a second number n.

$m \div n$ means a number m divided by a second number n.

Example 1

(a) Put into symbols six times a number p.

Six times a number p is $6 \times p$

In algebra, multiplication signs are usually left out, so that $6 \times p$ would be written as $6p$.

(b) Put into symbols eight multiplied by r multiplied by s.

Eight multiplied by r multiplied by s

$= 8 \times r \times s$

$= 8rs$

(c) Put into symbols three times a number y minus four.

Three times y is $3 \times y = 3y$

Three times y minus four is $3y - 4$

Put each of the following into symbols:

1. seven times a number y

2. five times a number b minus seven

3. five times a number p plus three times a number q

4. the numbers x, y and z, multiplied together

5. eight times a number r times a number s

6. twice the number m divided by the number n

7. nine times the number a minus three times the number b

8. the numbers p and q multiplied together and divided by a third number r.

Algebraic Terms

$a + a + a + a + a$ is $5 \times a$. In algebra it is written as $5a$ (i.e. with the multiplication sign left out). $5a$ is called an **algebraic term**.

In the same way

$3y = 3 \times y = y + y + y$

Exercise 13.2

Write down each of the following as an algebraic term:

1. $m + m + m$ 2. $x + x$

3. $t + t + t + t$ 4. $r + r + r$

5. $p + p + p + p + p$

6. $q + q + q + q + q + q$

7. $w + w + w$

8. $y + y + y + y + y$

Substitution

Substitution means putting numbers in place of the letters.

Example 2

Find the value of $3x + 5y - 2z$ when $x = 6$, $y = 7$ and $z = 4$.

Remembering that $3x$ means $3 \times x$, $5y$ means $5 \times y$ and $2z$ means $2 \times z$ we have

$$3x + 5y - 2z = 3 \times 6 + 5 \times 7 - 2 \times 4$$
$$= 18 + 35 - 8$$
$$= 45$$

(Remember that we always multiply before adding and subtracting.)

Exercise 13.3

If $x = 3$ and $y = 4$, find values for the following expressions. Remember that multiplication and division must always be done before addition and/or subtraction.

1. $x + 5$ 2. $3y - 7$

3. $6x + y$ 4. $x + 3y$

5. $2x + 7$ 6. $3x - 5$

7. $3x - y$ 8. $3y - 3x$

9. $7x + 2y - 9$ 10. $x + 3y + 8$

Like and Unlike Terms

Like terms are terms which contain the same letter. Thus $3x$, $2x$ and $5x$ are three like terms.

Terms which do not contain the same letter are called **unlike terms**. Thus $7a$, $3b$ and $8c$ are three unlike terms.

Only like terms can be added or subtracted.

Example 3

(a) $2x + 3x = (x + x) + (x + x + x)$
$$= (2 + 3)x$$
$$= 5x$$

(b) $7b - 5b = (7 - 5)b$
$$= 2b$$

(c) $3y - 2y = (3 - 2)y$
$$= y$$

(In algebra, $1 \times y$ or $1y$ is written as y.)

(d) $9b + b + 3b = (9 + 1 + 3)b$
$$= 13b$$

Although expressions like $x + y$ and $3b - 5c$ cannot be made any simpler it is possible for an expression to contain several sets of like terms.

The expression can then be made simpler by adding or subtracting each set of like terms.

Example 4

Simplify $8p + 2q - 5p + 3q + 4p$.

$$8p + 2q - 5p + 3q + 4p$$
$$= (8 - 5 + 4)p + (2 + 3)q$$
$$= 7p + 5q$$

Exercise 13.4

Simplify if possible:

1. $3x + 5x$ 2. $7y + 3y$

3. $6c + 3c$ 4. $5a + 11a$

5. $d + 7d$ 6. $9y - 3y$

7. $6x - 5x$ 8. $4q - q$

9. $3t - 5t + 7t$ 10. $5m + 8m - 3m$

11. $6x - 9x + 4x$ 12. $8p + 3q$

13. $3a + 4b - 5c$ 14. $3x + 2x - 5y$

15. $5x - 4y + 3x + 7y - 6x$

16. $15y + 11 - 3y - 8$

17. $9p - 2q + 3q - 8p$

18. $6b + 3a - 8 + 2a - 3b + 9$

19. $a - 3b - 2c + 5a - 6c + 8b$

20. $5x - 4y + 2z - 4x + 5y - z$

Powers

In algebra, $a \times a \times a$ is written a^3.

The number 3 which shows the number of a's to be multiplied together is called the **index** (plural indices). We say that a has been raised to the third **power**.

$$b^5 = b \times b \times b \times b \times b$$

Notice carefully that a^2 is not the same as $2a$.

$$a^2 = a \times a \quad \text{but} \quad 2a = a + a$$

Example 5

Find the value of y^6 when $y = 2$.

$$y^6 = y \times y \times y \times y \times y \times y$$
$$= 2 \times 2 \times 2 \times 2 \times 2 \times 2$$
$$= 64$$

Example 6

Find the value of $a^2 - b^2$ when $a = 3$ and $b = 2$.

$$a^2 - b^2 = (a \times a) - (b \times b)$$
$$= (3 \times 3) - (2 \times 2)$$
$$= 9 - 4$$
$$= 5$$

Exercise 13.5

Write the following in index form:

1. $b \times b$ 2. $c \times c \times c$

3. $n \times n \times n \times n$ 4. $p \times p \times p$

5. $q \times q \times q \times q \times q \times q$

6. $r \times r \times r \times r \times r$

If $p = 2$, $q = 3$ and $r = 5$ find values for each of the following:

7. p^3 8. q^2

9. p^4 10. $p^2 + q^2$

11. $r^2 - q^2$ 12. $p^3 + q^3$

13. $q^2 - r^2$ 14. $r^2 - p^2$

15. $p^3 + r^2$

When dealing with terms like $5ab^4$ note that it is only the b which is raised to the fourth power.

$$5ab^4 = 5 \times a \times b \times b \times b \times b$$

Example 7

Find the value of $3mn^3$ when $m = 3$ and $n = 2$.

$$5mn^3 = 5 \times m \times n \times n \times n$$
$$= 5 \times 3 \times 2 \times 2 \times 2$$
$$= 120$$

Exercise 13.6

Find the values of each of the following when $p = 2$, $q = 3$ and $r = 4$:

1. pq^2
2. p^2q
3. p^2q^2
4. p^3r
5. $3qr^2$
6. $5p^2r^2$
7. $3pq^2r$
8. p^3qr^2

Multiplication

When terms are multiplied together we can write them in a shorter way by missing out the multiplication sign.

For example

$$a \times b \text{ can be written as } ab$$

Because multiplication can be done in any order

$$5a \times 3b = 5 \times a \times 3 \times b$$
$$= 5 \times 3 \times a \times b$$
$$= 15ab$$

When multiplying terms containing the same letter, indices are used.

$$p \times p \text{ can be written as } p^2$$
$$2m \times 5m = 2 \times 5 \times m \times m$$
$$= 10m^2$$

In writing algebraic terms it is usual to write the numbers first followed by the letters in alphabetical order. It is not wrong to write terms in any other order but it looks better to write, for example, $5xyz$ rather than $x5zy$.

Exercise 13.7

Make each of the following into single terms:

1. $y \times x$
2. $2p \times q$
3. $2y \times 3p$
4. $8p \times 3q$
5. $3a \times 2b \times c$
6. $2z \times 3y \times x$
7. $3p \times 5n$
8. $3p \times 5n \times 2m$

Brackets

As in arithmetic, brackets are used to group terms together.

$$3(a + b) = 3 \times (a + b)$$
$$= 3 \times a + 3 \times b$$
$$= 3a + 3b$$

So when 'removing' a bracket each term within the bracket is multiplied by the term outside the bracket.

Example 8

$$5(2m + 3p) = 5 \times 2m + 5 \times 3p$$
$$= 10m + 15p$$

When simplifying expressions containing brackets first remove the brackets. Then add and/or subtract the terms containing the same letters.

Example 9

(a) Simplify $5(3a + 4b) + 3(2a - 5b)$

$$5(3a + 4b) + 3(2a - 5b)$$
$$= 5 \times 3a + 5 \times 4b + 3 \times 2a - 3 \times 5b$$
$$= 15a + 20b + 6a - 15b$$
$$= (15 + 6)a + (20 - 15)b$$
$$= 21a + 5b$$

(b) Find the value of $4(x + 3y) - 2(3x - 4y)$ when $x = 5$ and $y = 2$.

$$4(x + 3y) - 2(3x - 4y)$$
$$= 4(5 + 3 \times 2) - 2(3 \times 5 - 4 \times 2)$$
$$= 4(5 + 6) - 2(15 - 8)$$
$$= 4 \times 11 - 2 \times 7$$
$$= 44 - 14$$
$$= 30$$

Exercise 13.8

Remove the brackets and simplify if possible:

1. $3(a + b)$
2. $2(x + 3y)$
3. $5(2x - y)$
4. $2(3x - 4y)$
5. $3(m + 2n)$
6. $5(2x - 3y)$
7. $4(p + 2q) + 5(3p - 4q)$
8. $3(a - 2b) + 5(2a + 3b)$
9. $3(x + 3y) + 8x$
10. $15p + 3(p - 8q)$

Given that $a = 2$, $b = 3$ and $c = 5$, work out the values of the following expressions:

11. $2(a + 3b)$
12. $4(3a - b)$
13. $2(5a - 2b)$
14. $2(a + 3b)$
15. $15 - 2(a + b)$
16. $5(2a + 3b) - c$
17. $4c - 2(a + b)$
18. $2(3a + 2b) + 3(5a - 2b)$
19. $3(a + 2c) - 2(5a - c)$
20. $2(3a + 5b) + 3(5a - 2b)$

Division

When dividing cancelling between the top and bottom parts is often possible, but always remember that cancelling means dividing the top and bottom parts by the same quantity.

Example 10

(a)
$$\frac{12y}{4} = \frac{\overset{3}{\cancel{12}} \times y}{\underset{1}{\cancel{4}}}$$
$$= 3 \times y$$
$$= 3y$$

(b)
$$\frac{pq}{p} = \frac{\cancel{p} \times q}{\cancel{p}}$$
$$= q$$

Exercise 13.9

Simplify each of the following:

1. $\dfrac{12x}{6}$
2. $\dfrac{3a}{a}$
3. $\dfrac{10x}{5x}$
4. $\dfrac{4ab}{2a}$
5. $\dfrac{7ab}{b}$
6. $\dfrac{6xy}{3x}$
7. $\dfrac{12abc}{3ab}$
8. $\dfrac{6xyz}{3xz}$

Making Algebraic Expressions

It is important to be able to translate written or verbal information into symbols.

Example 11

(a) Write down an expression which will give the total weight of a box containing y articles, if the empty box has a weight of 7 kilograms and each article weighs 2 kilograms.

$$\text{Weight of } y \text{ articles} = 2 \times y$$
$$= 2y \text{ kg}$$

Total weight of box and contents
$$= (2y + 7) \text{ kg}$$

(b) How many minutes are there in
   (i)   3 hours   (ii) t hours
   (iii) 120 seconds   (iv) m seconds?

(i)   3 hours $= 3 \times 60$ min
$$= 180 \text{ min}$$

(ii) t hours $= t \times 60$ min
$$= 60t \text{ min}$$

(iii)   120 seconds $= 120 \div 60$
$$= 2 \text{ min}$$

(iv) m seconds $= m \div 60$
$$= \frac{m}{60} \text{ min}$$

Exercise 13.10

1. How many articles are there in
   (a) 3 dozen   (b) n dozen?

2. A rocket has a speed of 1200 miles per hour. How far will it travel in
   (a) 4 hours   (b) t hours
   (c) 2 minutes   (d) m minutes?

3. An empty bag weighs 1 kg. Find the weight of the bag and contents when w kg of groceries and m kg of meat are placed in it.

4. (a) 12 apples are shared equally between 2 boys and 4 girls. How many apples does each child receive?
   (b) If k apples are shared equally between x girls and y boys, how many apples does each child receive?

5. (a) A pail holds 4 pints. How many times can it be filled from a tank holding 20 gallons?
   (b) A pail holds x pints. How many times can it be filled from a tank holding k gallons?

6. A paper costs y pence. How much do 5 papers cost?

7. A man works x hours each weekday (i.e. Monday to Friday). If he works y hours on Saturday and z hours on Sunday, how many hours did he work in the week?

Miscellaneous Exercise 13

Section A

1. A boy is n years old now. How old will he be in 10 years' time?

2. Take the number x. Square it and add 5. What is the result?

3. Take the two numbers x and y. Indicate three times their sum.

4. If $p = 7$, $q = 8$ and $r = 0$, work out the value of each of the following:
   (a) $q - p + r$ (b) $pq - qr$
   (c) $p(q + r)$

5. If $x = 3$, $y = 2$ and $z = 4$, find the values of
   (a) $x + 3(y + z)$
   (b) $xy(x + z)$

6. Simplify each of the following:
   (a) $5x + 9 - 2x - 7$
   (b) $3x + 5y + 2x - 3y$
   (c) $3x \times 4y$
   (d) $2x \times 5x$
   (e) $8y \div 2$

7. Remove the brackets and simplify:
   (a) $2(3x + 4y) + 3(x - 3y)$
   (b) $5(p - q) + 4(2p - q)$

Section B

1. What is the value of $x^2 + x + 8$ when $x = 4$?

2. Express $\dfrac{a - b}{a + b}$ as a fraction in its lowest terms given that $a = 5$ and $b = 3$.

3. If $a = 0.5$ and $b = 37$, what is the value of $\dfrac{b}{a}$?

4. If $p = 3xy$, what is the value of p when $x = 2$ and $y = 5$?

5. An approximate way of converting degrees Celsius into degrees Fahrenheit is $F = 2C + 30$, where F is the temperature in degrees Fahrenheit and C is the temperature in degrees Celsius. Convert a temperature of 12°C into degrees Fahrenheit.

6. If $a = 3$ and $b = 6$, find the value of $5a^2b$.

7. Work out the value of $a + 2b$ when $a = 1$ and $b = \frac{1}{2}$.

Multi-Choice Questions 13

1. In its simplest form $(2x + y) + (x - 2y)$ is

   A $3x + y$ B $3x - y$

   C $x - y$ D $x - 3y$

2. If $x = 2$, the value of $2x^2 + 3$ is

   A 11   B 19   C 14   D 24

3. Which of the following is **not** equal to $\frac{1}{2}pq$?

   A $\dfrac{pq}{2}$ B $\dfrac{1}{2p} \times q$

   C $\frac{1}{2}qp$ D $q \times \dfrac{p}{2}$

4. The value of $2^2 + 3^3$ is

   A 13   B 25   C 31   D 36

5. 5^4 has the value of

   A 9   B 20   C 125   D 625

6. Evaluate $13^2 - 12^2$. It is

   A 25   B 5   C 2   D 1^2

7. What is the value of $x^2 + x - 6$ when $x = 3$?

   A 0   B 3   C 6   D 18

8. If $x = 0.2$ and $y = 4.8$ calculate the value of $\dfrac{y}{x}$. It is

   A 240   B 24   C 2.4   D 0.24

Mental Test 13

1. Translate into algebraic symbols: 8 times a number b.

2. Translate into algebraic symbols a number a divided by a number b.

3. Find the value of $2y$ when $y = 3$.

4. Find the value of $3x - 5$ when $x = 10$.

5. Find the value of 2^3.

6. Simplify $3a \times 2a$.

7. Simplify $3x + 5x$.

8. Remove the brackets from $3(x + 2y)$.

9. Simplify $3x \div x$.

10. Simplify $10x \div 2$.

Simple Equations and Inequalities

Introduction

As an example of a simple equation consider

$$x + 2 = 7$$

This means that some number x plus 2 equals 7.

Because 5 is the only number which when added to 2 gives 7 then x must be equal to 5.

The solution of the equation $x + 2 = 7$ is $x = 5$.

Example 1

(a) Solve the equation $x + 3 = 5$.

Since 2 is the only number which when added to 3 gives 5, the solution of the equation $x + 3 = 5$ is $x = 2$.

(b) Solve the equation $x - 4 = 3$.

The only number that gives 3 when 4 is subtracted from it is 7. So the solution of the equation $x - 4 = 3$ is $x = 7$.

Exercise 14.1

Using the methods shown in Example 1 solve the following equations:

1. $x + 3 = 8$ 2. $x + 2 = 9$

3. $x + 7 = 10$ 4. $x + 5 = 7$

5. $x + 4 = 12$ 6. $x + 2 = 5$

7. $x - 2 = 6$ 8. $x - 3 = 8$

9. $x - 7 = 2$ 10. $x - 5 = 9$

Harder Equations

In Example 1 we solved equations by the method of trial and error. Sometimes though we get more difficult equations to solve and trial and error methods can then become tedious. Hence we now look for a second method of solving simple equations.

Fig. 14.1 shows a pair of scales which are in balance. That is each scale pan contains exactly the same number of kilograms. Therefore the balance represents the equation

$$x + 2 = 7$$

Fig. 14.1

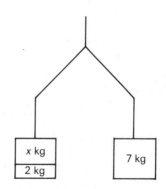

To solve this equation we have to find a value for x so that the scales remain in balance. This can be done by adding or subtracting the same amount from each pan.

If we take 2 kilograms from the left-hand pan then we are left with x kilograms in this pan. To keep the balance we must also take 2 kilograms from the right-hand pan leaving 5 kilograms in this pan. We see that the solution of the equation $x + 2 = 7$ is $x = 5$.

Let us now take a second example as shown in Fig. 14.2. In the left-hand scale pan we have 4 packets each having a weight of x kilograms. In the right-hand scale pan we have a weight of 8 kilograms. The balance represents the equation

$$4x = 8$$

Fig. 14.2

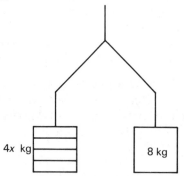

4x kg   8 kg

If we divide each of the quantities in the pans by 4 we will still keep the scales in balance.

Dividing $4x$ by 4 gives x and dividing 8 by 4 gives 2. Hence the solution of the equation $4x = 8$ is $x = 2$ (that is, each packet has a weight of 2 kilograms).

From these two equations we can say that:

(1) An equation shows balance between two sets of quantities. That is, the equals sign means that one side of the equation is equal to the other side.

(2) The balance will still be kept if we

  (a) add the same quantity to each side of the equation

  (b) take away the same quantity from each side of the equation

  (c) multiply each side of the equation by the same quantity

  (d) divide each side of the equation by the same quantity.

Example 2

(a) Solve the equation $x + 3 = 7$.

Taking 3 away from both sides we get

$$x + 3 - 3 = 7 - 3$$

The $+3$ and the -3 'cancel each other out' and we are left with

$$x = 4$$

(b) Solve the equation $x - 6 = 2$.

Adding 6 to each side we get

$$x - 6 + 6 = 2 + 6$$

The -6 and the $+6$ 'cancel each other out' and we are left with

$$x = 8$$

(c) Solve the equation $\dfrac{x}{2} = 3$.

Multiplying both sides by 2 we get

$$\frac{x}{2} \times \frac{2}{1} = 3 \times 2$$

$$x = 6$$

(d) Solve the equation $5x = 10$.

Dividing each side by 5 we get

$$\frac{5x}{5} = \frac{10}{5}$$

$$x = 2$$

Exercise 14.2

Solve the following equations for x:

1. $x + 3 = 9$ 2. $x + 3 = 7$

3. $x + 5 = 8$ 4. $x + 7 = 11$

5. $x - 5 = 7$ 6. $x - 3 = 6$

7. $x - 2 = 8$ 8. $x - 4 = 5$

9. $3x = 6$ 10. $7x = 14$

11. $8x = 16$ 12. $2x = 12$

13. $\dfrac{x}{2} = 4$ 14. $\dfrac{x}{3} = 5$

15. $\dfrac{x}{7} = 3$ 16. $\dfrac{x}{5} = 2$

Example 3

Solve the equation $4x - 1 = 7$.

Adding 1 to each side we get

$$4x - 1 + 1 = 7 + 1$$
$$4x = 8$$

Dividing both sides by 4 we get

$$\frac{4x}{4} = \frac{8}{4}$$
$$x = 2$$

Although the method shown in Example 3 can always be used it takes up quite a lot of time. Provided you understand the method, you can use the quicker method which follows:

The operation of adding 1 to each side can be done by taking 1 to the RHS (right-hand side) and changing its sign:

$$4x - 1 = 7$$
$$4x = 7 + 1$$
$$4x = 8$$

Dividing both sides by 4 we get

$$x = 2$$

We now need a way of checking the solution of an equation to make sure that the solution is correct. We already know that the solution of an equation is that value of the unknown (x in Example 3) which makes the LHS equal to the RHS. Therefore to check the solution we substitute the value of the unknown into the original equation and look to see if the LHS is equal to the RHS. For Example 3 the check is:

$$\begin{aligned} \text{LHS} &= 4 \times 2 - 1 \\ &= 8 - 1 \\ &= 7 \\ &= \text{RHS} \end{aligned}$$

Therefore the solution is correct.

Example 4

Solve the equation $5x - 3 = 7$.

$$5x - 3 = 7$$
$$5x = 7 + 3 \quad \text{(adding 3 to both sides)}$$
$$5x = 10$$
$$x = 2 \quad \text{(dividing both sides by 5)}$$

Check:

$$\begin{aligned} \text{LHS} &= 5 \times 2 - 3 \\ &= 10 - 3 \\ &= 7 \\ &= \text{RHS} \end{aligned}$$

Therefore the correct solution is $x = 2$.

Exercise 14.3

Solve the following equations for x and check the solutions:

1. $2x - 3 = 7$ 2. $5x - 8 = 2$

3. $3x + 2 = 5$ 4. $2x + 3 = 7$

5. $3x - 4 = 8$ 6. $3x - 8 = 7$

7. $8x - 11 = 13$ 8. $5x - 9 = 6$

We can change the sides of an equation completely and we can do this at any stage of the working (as shown in Example 5 overleaf).

Example 5

(a) Solve the equation $7 = x + 4$.

Subtracting 4 from each side gives

$$7 - 4 = x$$

$$3 = x$$

By changing sides

$$x = 3$$

When writing the solution of an equation it is usual to write the letter on the LHS.

(b) Solve the equation $15 = 3x$.

$$3x = 15 \quad \text{(changing sides)}$$

$$x = \frac{15}{3} \quad \text{(dividing both sides by 3)}$$

$$x = 5$$

Example 6

Solve the equation $5x + 2 = 3x + 8$.

To solve this equation we need to get the terms containing x on the LHS and the numbers on the RHS.

Subtracting $3x$ from both sides gives

$$5x - 3x + 2 = 8$$

Subtracting 2 from each side

$$2x = 8 - 2$$

$$2x = 6$$

Dividing both sides by 2 gives

$$x = 3$$

Check:

$$\text{LHS} = 5 \times 3 + 2$$

$$= 15 + 2$$

$$= 17$$

$$\text{RHS} = 3 \times 3 + 8$$

$$= 9 + 8$$

$$= 17$$

Since the LHS equals the RHS the correct solution is $x = 3$.

Exercise 14.4

Solve the following equations and check the solutions:

1. $7 = 3 + x$
2. $12 = 5 + x$
3. $16 = x + 9$
4. $7 = x - 8$
5. $3 = 9 - x$
6. $4 = 7 - x$
7. $12 = 2x$
8. $16 = 4x$
9. $25 = 5x$
10. $36 = 9x$
11. $5x + 3 = 3x + 9$
12. $7x + 4 = 5x + 10$
13. $8x + 6 = 5x + 12$
14. $3x - 9 = x + 7$
15. $7x - 8 = 2x + 12$

Formulae

A **formula** is an equation which shows the relationship between two or more quantities. It may be stated in words or with letters. For example

Wages $=$ Rate per hour $\times$ Hours worked

$$A = l \times b$$

Example 7

(a) A woman works on piecework making dolls. She is paid according to the formula

Pay $=$ Number of dolls made

$\times$ Amount paid per doll

If she makes 50 dolls in a week and she is paid £2 per completed doll, work out her pay for the week.

To work out her pay we have to substitute figures for the words in the formula.

$$\text{Pay} = 50 \times 2$$

$$= 100$$

Hence her pay is £100.

(b) If $V = A \times h$, find the value of V when $A = 8$ and $h = 4$.

To work out the value of V we substitute the given figures for the letters. Thus

$$V = 8 \times 4$$
$$= 32$$

(c) If $E = IR$, find the value of E when $I = 5$ and $R = 20$.

$$E = I \times R$$
$$= 5 \times 20$$
$$= 100$$

Exercise 14.5

1. The cooking instructions for a turkey are given by: Cooking time = 25 min per kg + 20 min. Work out the time needed to cook a turkey weighing 7 kg.

2. The following formula gives the cooking time for roast beef: Time needed (min) = 50 × weight in kg + 20. Find the time required to cook a roast weighing three-and-a-half kilograms.

3. The distance round a rectangle is called its perimeter. It is found by: Perimeter = twice the length + twice the width. Calculate the perimeter of a rectangle whose length is 8 cm and whose width is 5 cm.

4. The speed of a car is found by using the formula: Speed = distance travelled ÷ time taken. Work out the speed of a car if it travelled 40 km in 2 hours.

5. An approximate conversion from degrees Celsius to degrees Fahrenheit is

$$F = 2C + 30$$

where F is the Fahrenheit temperature and C is the Celsius temperature. Use this formula to convert $20\,°C$ into degrees Fahrenheit.

6. A man's weekly wage is calculated from: Wage = number of hours × rate per hour. Calculate the wage when 40 hours are worked and the rate is £4 per hour.

7. **(a)** In income tax calculations the formula

$$T = G - A$$

is used where T is the taxable income, G is the gross income and A is the tax-free allowances. Calculate the taxable income when the gross income is £10 000 and the tax-free allowances are £3500.

(b) The amount of income tax payable is obtained from: Tax payable = 25% of taxable income. Work out the amount of tax payable.

8. When goods are bought on hire-purchase the hire-purchase price is given by

$$H = In + D$$

where I is the amount of each instalment, n is the number of instalments and D is the amount of the deposit. Work out the hire-purchase price when 24 instalments of £25 are paid plus a deposit of £50.

Inequalities

An **inequality** is a statement that one quantity is greater (or less) than a second quantity. For example

3 is less than 7 and £5 is greater than £3

Instead of writing 'is less than' the symbol $<$ is used and instead of writing 'is greater than' the symbol $>$ can be used. Thus

$$3 < 7 \quad \text{and} \quad £5 > £3$$

Note carefully that if the symbols $<$ and $>$ are used the arrow always points to the smaller quantity.

Exercise 14.6

Using the words 'is less than', 'is greater than' and 'equals', or the symbols $<$, $>$ and $=$, fill in the gap between the pairs of quantities below in order to make the statements true:

1. 2   5
2. 10 mm   1 cm
3. £1   80 p
4. 9   1
5. -3 -7
6. 2   4
7. $\frac{1}{2}$ $\frac{1}{3}$
8. $\frac{2}{3}$ $\frac{4}{9}$
9. 0.25 $\frac{1}{3}$
10. 0.3   30%

Flow Diagrams

A **flow diagram** can be used to show the order in which calculations are to be done. Flow diagrams may also be used to evaluate formulae.

Example 8

Draw a flow chart to represent the formula $y = 3p + 5$ and use it to find the value of y when $p = 4$.

Remembering that multiplication and division must be done before addition and subtraction the flow charts will look like Fig. 14.3.

Fig. 14.3

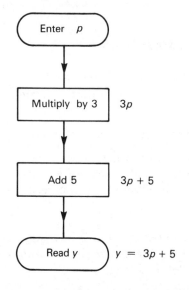

Start and read boxes are usually shaped like this

and instruction boxes like this

Fig. 14.4 shows how the flow diagram is used to calculate the value of y when $p = 4$.

Fig. 14.4

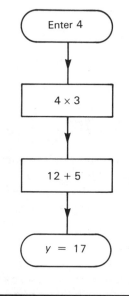

Exercise 14.7

1. Draw flow charts for the following calculations and, using a calculator, work out the answers:
    (a) $27 \times 14 + 18$
    (b) $36 + 28 \times 7$
    (c) $128 \div 8 - 7$
    (d) $56 \times 9 - 18 + 72$

2. Draw a flow chart to evaluate $5 \times (3 + 4) - 7$.

3. Draw a flow chart to represent the formula $p = 5q - 8$ and use it to find the value of p when $q = 3$.

4. Draw a flow chart for the formula $C = nM$ and use it to find the value of C when $n = 5$ and $M = 12$.

5. The flow chart shown in Fig. 14.5 is used to convert degrees Celsuis (C) into degrees Fahrenheit (F). Use the chart to convert the following Celsuis temperatures into degrees Fahrenheit:

   (a) 5   **(b)** 20

   (c) 0   **(d)** −15

Fig. 14.5

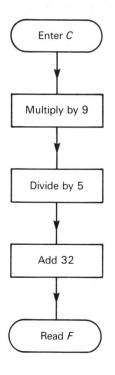

6. To find the amount of a gas bill the following formula is used:

   Bill = Number of therms

         × Price per therm ÷ 100

         + Standing charge

   Draw a flow chart to represent this formula and use it to work out the amount of the gas bill when the standing charge is £10 and 200 therms are used costing 40p each.

7. An approximate formula for converting degrees Fahrenheit into degrees Celsius is

$$C = (F - 30) \div 2$$

Draw a flow chart for this formula and use it to convert the following Fahrenheit temperatures into Celsius temperatures:

(a) 70   **(b)** 120   **(c)** 0

Miscellaneous Exercise 14

Section A

1. Solve the equation $7x - 9 = 5$.

2. Solve the equation $x - 8 = 6$.

3. Solve the equation $m - 2 = 3$.

4. Solve the equation $3x + 7 = 22$.

5. The following formula is used to work out the amount of a telephone bill without the VAT:

   Bill = Number of units

         × Price in pence per unit ÷ 100

         + Rental charge

   Work out the telephone bill when 200 units costing 4p per unit are used and the rental charge is £8.

6. Use the words 'less than', 'greater than' or 'equals' in each of the following to make them true statements:

   (a) $\frac{1}{2}$ $\frac{1}{4}$ **(b)** 100 cm   1 m

   (c) 50p   £1

7. Solve the equation $12 = 3x$.

8. When roasting lamb the following formula is used:

   Time (in minutes) = 25 min per lb

                     + 25 min

   Work out the time needed to cook a leg of lamb weighing 5 lb.

Section B

1. If $A = l \times b$, calculate the value of A when $l = 8$ and $b = 5$.

2. For cooking a stuffed chicken, the formula below is used:

$$T = 40W + 20$$

where T is the cooking time in minutes, and W is the weight of the chicken in kilograms:

(a) Find the time required, in minutes, to cook a chicken weighing $1\frac{1}{2}$ kilograms.

(b) If the chicken is frozen the defrosting time, using a microwave oven, is 17 minutes. How many minutes are needed to cook a frozen chicken weighing 2 kilograms?

3. Use one of the symbols $>$, $<$, $=$ in each of the following to make it a true statement:

(a) $5^2 - 3^2$ 4^2

(b) $9^2 - 3^2$ 3^2

(c) 0.5^3 0.5^2

4. An approximate formula for converting degrees Fahrenheit into degrees Celsius is

$$C = \tfrac{1}{2}(F - 30)$$

where C is the Celsius temperature and F the Fahrenheit temperature. The recommended temperature for cooking doughnuts is $360°F$. What is this temperature in degrees Celsius?

5. I think of a number. I double it and add five to the result. The answer is 21. What is the number I first thought of?

6. Solve each of the following equations for x:

(a) $5x - 8 = 12$ (b) $7 = \dfrac{x}{3}$

(c) $16 = 8x$

7. The formula $v = u + at$ is used in physics to calculate the speed of a body after a time t when the initial speed is u and the acceleration is a. Find the value of v when $u = 10$, $a = 2$ and $t = 3$.

Multi-Choice Questions 14

1. Jean had x pence. She spent 37 pence on sweets and 24 pence on foreign stamps. She then had 15 pence left. What is the value of x?

   A 76   B 66   C 61   D 39

2. When $\dfrac{x}{4} = 5$, the value of x is

   A 4   B 8   C 10   D 20

3. To convert degrees Celsius into degrees Fahrenheit the formula $F = 2C + 30$ is used. The temperature for cooking chips is $180°C$. This temperature in degrees Fahrenheit is

   A 210   B 300   C 360   D 390

4. To calculate the amount of a telephone bill the formula $C = \dfrac{Np}{100} + R$ is used where N is the number of units used, p is the cost per unit in pence and R is the rental charge in pounds. If 200 units costing 4p each are used and the rental charge is £5, the amount of the bill is

   A £8   B £9   C £13   D £2

5. Given that $x + 1 = 18$, the value of x is

   A 19   B 17   C 5   D 3

6. A tumbler used in the kitchen holds 250 grams of whole rice. A recipe asks for 875 grams of rice. The number of tumblers of rice the cook should use is

   A 3   B $3\frac{1}{2}$ C 4   D $4\frac{1}{2}$

7. The cooking time for a duckling is 20 minutes per pound plus 20 minutes. The time it will take to cook a duckling weighing $2\frac{1}{2}$ lb is

A 40 min   B 50 min

C 70 min   D 100 min

Mental Test 14

1. Construct a formula if a number a is equal to the numbers b and c added together.

2. When multiplied together, the numbers x and y equal the number p. Write this information as a formula.

3. Solve the equation $3x = 6$.

4. A young duck takes 50 min per kilogram to cook. How long will it take to cook one weighing 2 kilograms?

5. John had x pence and he spent 25 pence on chocolate. He had 20 pence left. What was the value of x?

6. If $P = UV$ find P when $U = 3$ and $V = 5$.

7. 3   6. Which symbol $>$ or $<$ should be used to make the statement true?

8. Solve the equation $x - 3 = 5$.

Angles and Straight Lines

Some Definitions

(1) A **point** has position but no size, although to show it on paper it obviously must be given some size.

(2) If two points A and B are chosen (Fig. 15.1) then only one **straight line** can contain them.

Fig. 15.1

(3) If the two points chosen are the end points of the line then AB is called a **line segment** (Fig. 15.2).

Fig. 15.2

A ●————————————————● B

Angles

When two lines meet at a point they form an **angle.** The size of the angle depends upon the amount of opening between the two lines. It does not depend upon the lengths of the lines forming the angle. In Fig. 15.3 the angle A is larger than the angle B despite the fact that the arms are shorter.

Fig. 15.3

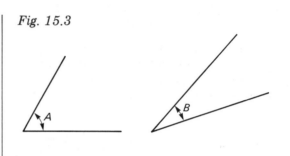

Measurement of Angles

An angle may be looked upon as the amount of rotation or turning. In Fig. 15.4 the line OA has been turned about O until it takes up the position OB. The angle through which the line OA has turned is the amount of opening between the lines OA and OB.

Fig. 15.4

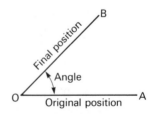

If the line OA is rotated past B until it returns to its starting place it will have completed one revolution. Therefore we can measure an angle as a fraction of a revolution.

Fig. 15.5 shows a circle divided into 36 equal parts. The first division is split into 10 equal divisions so that each small division is $\frac{1}{360}$ of a complete revolution. This small division is called a **degree**. Therefore

$$360 \text{ degrees} = 1 \text{ revolution}$$

which is written as

$$360° = 1 \text{ rev}$$

The small symbol ° stands for degrees.

Fig. 15.5

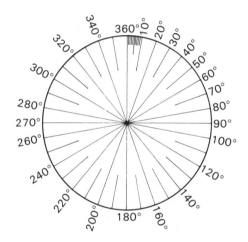

Most angles are stated in degrees and decimals of a degree. A typical angle might be 36.7° or 206.4°.

Example 1

(a) Find the angle, in degrees, corresponding to one-third of a revolution.

$$1 \text{ rev} = 360°$$
$$\tfrac{1}{3} \text{ rev} = \tfrac{1}{3} \times 360°$$
$$= 120°$$

(b) Find the angle, in degrees, corresponding to 0.7 of a revolution.

$$0.7 \text{ rev} = 0.7 \times 360°$$
$$= 252°$$

Example 2

(a) Add together 46.3° and 36.9°.

$$\begin{array}{r} 46.3° \\ 36.9° \; + \\ \hline 83.2° \end{array}$$

(b) Subtract 38.75° from 65.62°.

$$\begin{array}{r} 65.62° \\ 38.75° \; - \\ \hline 26.87° \end{array}$$

Exercise 15.1

Find the angle in degrees corresponding to each of the following:

1. $\frac{2}{3}$ of a revolution

2. $\frac{3}{8}$ of a revolution

3. 0.2 of a revolution

4. 0.9 of a revolution

5. 0.15 of a revolution.

Add the following angles:

6. 18.9° and 27.6°

7. 39.2° and 17.8°

8. 43.2°, 54.5° and 72.6°.

Subtract the following angles:

9. 18.3° from 54.6°

10. 37.9° from 46.1°.

Types of Angle

An **acute angle** is an angle of less than 90° (Fig. 15.6).

Fig. 15.6

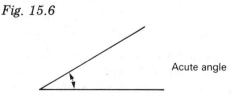

Acute angle

A **right angle** is an angle equal to $90°$ or $\frac{1}{4}$ of a revolution (Fig. 15.7). Note carefully how a right angle is marked.

Fig. 15.7

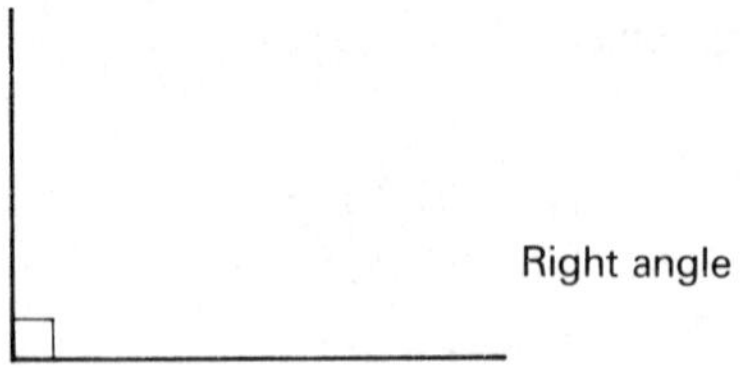

Right angle

An **obtuse angle** is an angle between $90°$ and $180°$ (Fig. 15.8).

Fig. 15.8

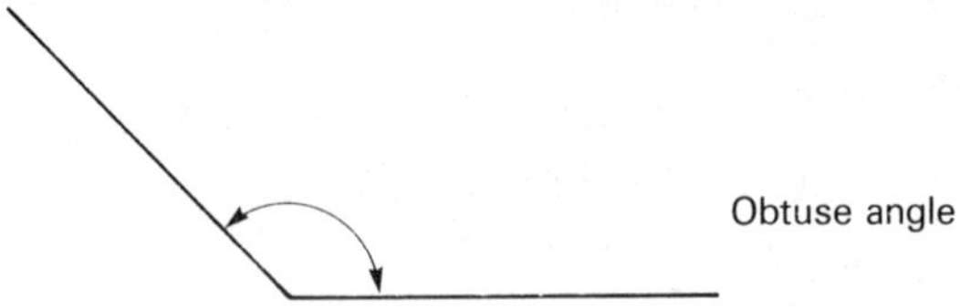

Obtuse angle

A **reflex angle** is an angle greater than $180°$ (Fig. 15.9).

Fig. 15.9

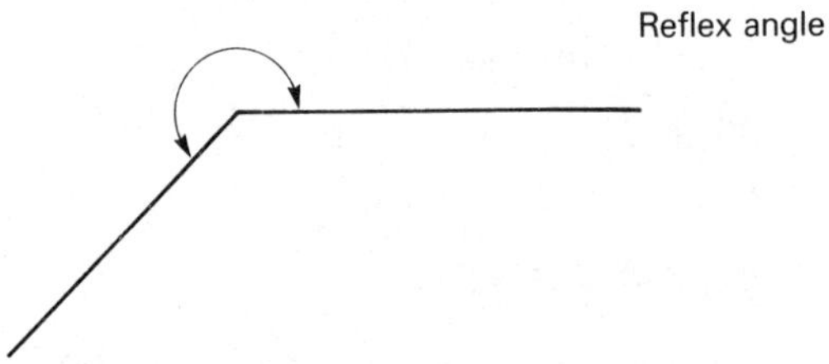

Reflex angle

Complementary angles are angles whose sum is $90°$. Thus $18°$ and $72°$ are complementary angles because $18° + 72° = 90°$.

Supplementary angles are angles whose sum is $180°$. Thus $103°$ and $77°$ are supplementary angles because $103° + 77° = 180°$.

1. Look at each of the angles in Fig. 15.10. State which are acute, which are obtuse and which are reflex.

Fig. 15.10

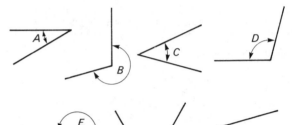

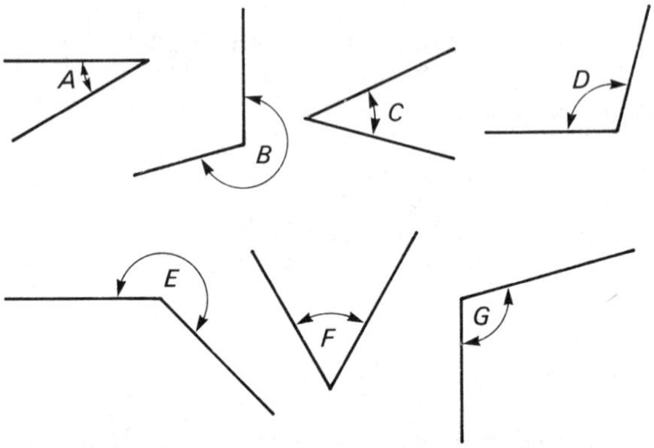

2. State the angle, in degrees, corresponding to each of the following:

   (a) $\frac{2}{3}$ of a right angle

   (b) $\frac{1}{4}$ of a right angle

   (c) 2 right angles

   (d) $1\frac{1}{2}$ right angles

   (e) 0.6 of a right angle

   (f) 0.35 of a right angle.

3. In each of the diagrams of Fig. 15.11, the angles at the centre are of equal size. Work out the number of degrees in each of the angles.

Fig. 15.11

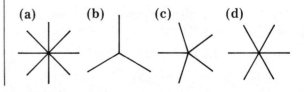

(a)   (b)   (c)   (d)

4. Find the number of degrees in each of the angles marked *x* in Fig. 15.12.

Fig. 15.12

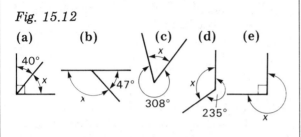

(a)   (b)   (c)   (d)   (e)

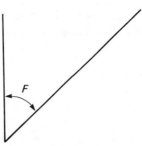

5. (a) Using a protractor, measure each of the angles shown in Fig. 15.13.

    (b) State which are acute, which are obtuse and which are reflex.

Fig. 15.13

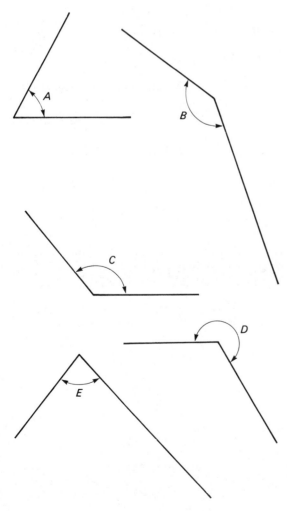

6. Make a table with three headings: acute, obtuse and reflex then write these angles in the appropriate columns:

27°, 195°, 220°, 165°, 75°, 245°, 173°, 64°, 126°, 280°, 153°, 82°, 178°, 325°, 30°, 340°, 15°, 110°, 98°, 220°.

7. (a) Two angles are complementary. One of them is 57°. What is the other?

    (b) Two angles are complementary. If one is 34°, find the other.

    (c) Two angles are supplementary. One is 27°, what is the other?

    (d) Angles *A* and *B* are supplementary. If *A* = 98°, what is the size of *B*?

Properties of Angles and Straight Lines

(1) The total angle on a straight line is 180°, i.e. two right angles. The angles *A* and *B* in Fig. 15.14 are called **adjacent angles** on a straight line and their sum is 180°.

Fig. 15.14

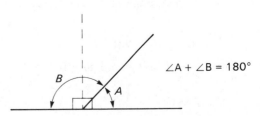

∠A + ∠B = 180°

(2) When two straight lines intersect, the **vertically opposite angles** are equal. (Fig. 15.15).

Fig. 15.15

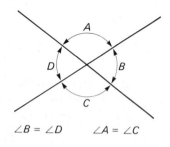

$$\angle B = \angle D \qquad \angle A = \angle C$$

Parallel Lines

Two lines in a plane that have no points in common, no matter how far they are produced, are called **parallel lines**:

(1) When two parallel lines are cut by a transversal (Fig. 15.16), the **corresponding angles** are equal. That is

$$a = l \quad b = m \quad c = p \quad d = q$$

Fig. 15.16

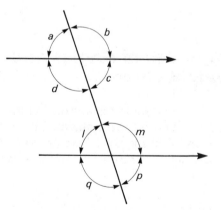

(2) The **alternate angles** are also equal. That is

$$d = m \quad \text{and} \quad c = l$$

(3) The **interior angles** are supplementary. That is

$$d + l = 180° \quad \text{and} \quad c + m = 180°$$

Example 3

In Fig. 15.17, AB and CD are two parallel straight lines. Find the size of the angles marked *a*, *b*, *c*, *d* and *e*.

Fig. 15.17

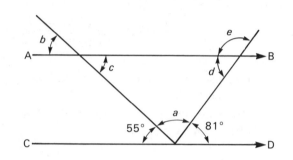

$a + 55° + 81° = 180°$ (sum of the angles on a straight line is 180°)

$a = 44°$

$b = 55°$ (AB ∥ CD, corresponding angles)

$c = b = 55°$ (vertically opposite angles)

$d = 81°$ (AB ∥ CD, alternate angles)

$e = 180° - 81°$ (sum of the angles on a straight line is 180°)

Exercise 15.3

1. Find the size of the angle *a* in Fig. 15.18.

Fig. 15.18

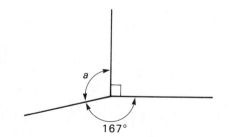

2. In Fig. 15.19, find the size of the angles marked x and y. AB and CD are parallel.

Fig. 15.19

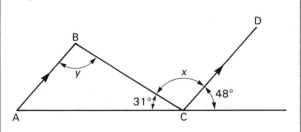

3. In Fig. 15.20, AB and CD are parallel. Calculate the size of the angles marked m, n, p and q.

Fig. 15.20

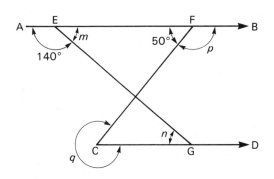

4. In Fig. 15.21, AB and CD are parallel lines crossed by the line ST. Find the size of the angle marked y.

Fig. 15.21

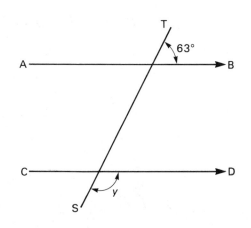

5. AB and CD (Fig. 15.22) are two parallel straight lines crossed by the straight line EF. Find the size of the angles marked x, y and z.

Fig. 15.22

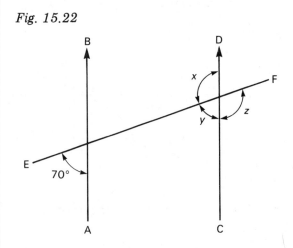

6. In Fig. 15.23, AB is parallel to DC. Find the size of the angles marked x and y.

Fig. 15.23

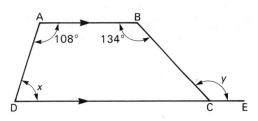

7. In Fig. 15.24, calculate the angle marked a.

Fig. 15.24

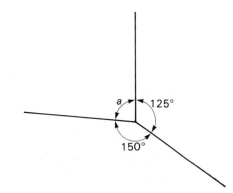

8. In Fig. 15.25, ABCDE and FGH are two parallel straight lines crossed by the straight lines BKG, FKCJ and GDJ. Find the size of the angles marked u, v, w, x, y and z.

Fig. 15.25

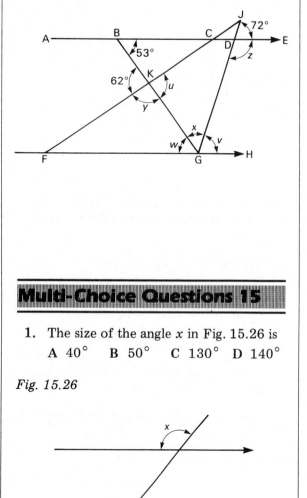

Multi-Choice Questions 15

1. The size of the angle x in Fig. 15.26 is

   A 40°   B 50°   C 130°   D 140°

Fig. 15.26

2. The size of the angle a in Fig. 15.27 is

   A 60°   B 90°   C 100°   D 110°

Fig. 15.27

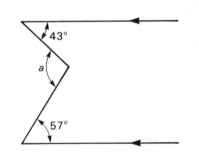

3. In Fig. 15.28, PQ and RS are parallel and the point T lies on RS. The size of the angle marked X is

   A 20°   B 40°   C 60°   D 70°

Fig. 15.28

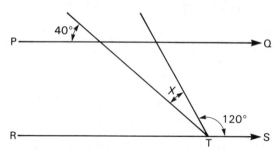

4. The size of the angle marked a in Fig. 15.29 is

   A 73°   B 107°   C 117°   D 146°

Fig. 15.29

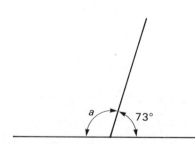

5. The size of the angle marked *b* in Fig. 15.30 is

A 60°   B 110°   C 120°   D 130°

Fig. 15.30

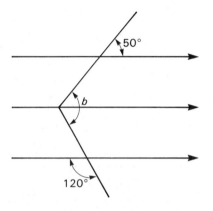

In Fig. 15.31, the lines AB and CD are parallel and the point E lies on CD. Use this diagram to answer questions 6, 7 and 8.

6. The size of angle *a* is

A 25°   B 50°   C 75°   D 80°

7. The size of angle *b* is

A 80°   B 100°   C 105°   D 150°

8. The size of angle *c* is

A 25°   B 50°   C 80°   D 135°

Fig. 15.31

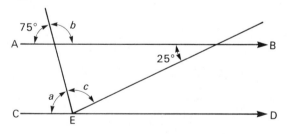

9. In Fig. 15.32, the value of *x* is

A 45°   B 67½°   C 90°   D 135°

Fig. 15.32

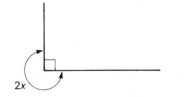

10. The angle shown in Fig. 15.33 is

A acute   B right

C reflex   D obtuse

Fig. 15.33

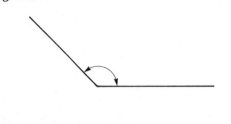

Mental Test 15

1. Express one degree as a fraction of a complete revolution.

2. Find the angle in degrees corresponding to one-sixth of a revolution.

3. Find the angle in degrees corresponding to 0.3 of a revolution.

4. What is the angle in Fig. 15.34 called?

Fig. 15.34

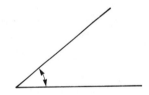

5. Two angles are complementary. One is 60°. What is the size of the other?

6. Two angles are supplementary. One is 130°. What is the size of the other?

7. Find the number of degrees in each of the angles marked *x* in Fig. 15.35.

Fig. 15.35

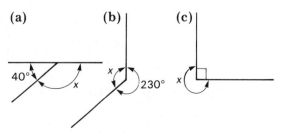

8. In Fig. 15.36, write down the size of the angles A and B.

Fig. 15.36

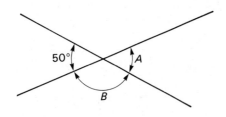

9. In Fig. 15.37, write down the size of the angles marked b, c, d, e, f, g and h.

Fig. 15.37

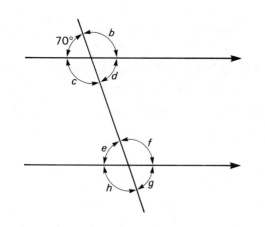

10. Add 35.3° and 44.7°.

11. Using a protractor measure each of the angles shown in Fig. 15.38.

Fig. 15.38

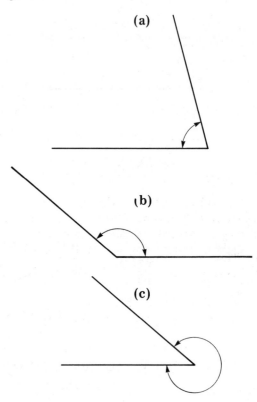

(a)

(b)

(c)

Symmetry

Lines of Symmetry

If you fold a piece of paper so that one half of the paper covers the other half exactly, then the fold is called a **line of symmetry**.

The shape shown in Fig. 16.1 is symmetrical only about the line AA'. The shape is said to have one line of symmetry.

Fig. 16.1

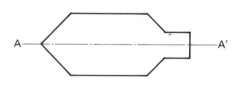

The shape shown in Fig. 16.2 is symmetrical about the lines XX' and YY'. The shape therefore has two lines of symmetry.

Fig. 16.2

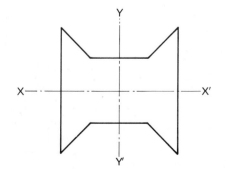

It is possible for a shape to have several lines of symmetry. For instance, the square shown in Fig. 16.3 has four lines of symmetry AA', BB', CC' and DD'.

Fig. 16.3

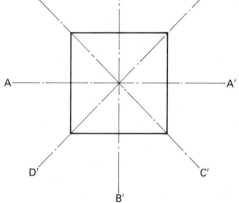

Some shapes have no lines of symmetry. The parallelogram shown in Fig. 16.4 is an example.

Fig. 16.4

From the examples given above it can be seen that a line of symmetry may be horizontal, vertical or oblique. Symmetry is also discussed on page 246.

Exercise 16.1

1. Each of the shapes shown in Fig. 16.5 has one line of symmetry. Copy the shapes on squared paper and then draw the line of symmetry.

Fig. 16.5

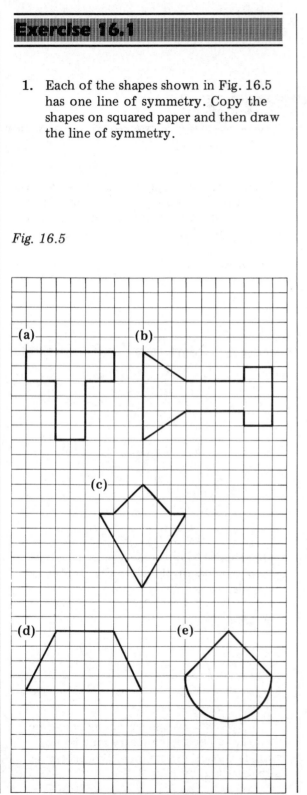

2. Fig. 16.6 shows a number of half-shapes with the line of symmetry indicated in chain dot. Draw the complete symmetrical shape using squared paper.

Fig. 16.6

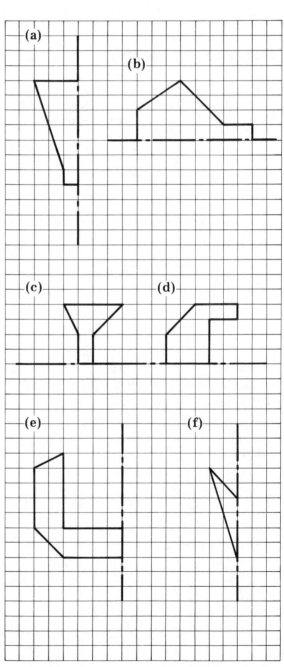

3. Each of the shapes shown in Fig. 16.7 has two lines of symmetry. Using squared paper copy each of these shapes and then draw the two lines of symmetry.

Fig. 16.7

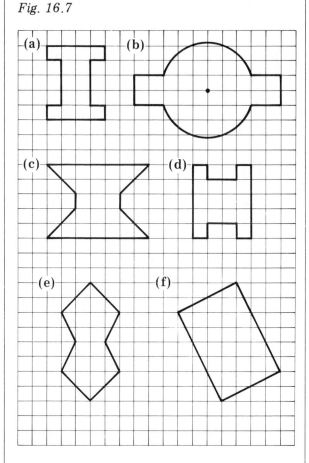

4. Each of the capital letters M, K, T, C, X and H has one or more lines of symmetry. Sketch the letters and draw all the lines of symmetry for each of the letters.

Rotational Symmetry

Fig. 16.8 shows a square ABCD whose diagonals intersect at O. If we rotate the square through 90°, 180° and 270° the square does not appear to have moved (unless we label the corners A, B, C and D, when the change is apparent). Because there are four positions where it appears not to have moved, we say that the square has **rotational symmetry of order four**.

Fig. 16.8

Original position and rotated through 360°

Rotated through 90°

Rotated through 180°

Rotated through 270°

Point Symmetry

Rectangles and parallelograms appear to be in the same position when rotated through 180°. They are said to have **point symmetry**.

The parallelogram (Fig. 16.9) does not appear to have moved after being rotated through 180° and hence it has point symmetry, but the trapezium (Fig. 16.10) appears upside down after a rotation of 180°. A trapezium therefore has no point symmetry.

Fig. 16.9

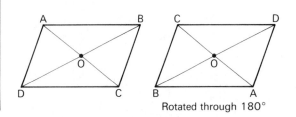

Rotated through 180°

Fig. 16.10

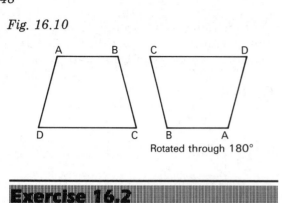

Rotated through 180°

Exercise 16.2

For each of the shapes shown in Fig. 16.11
write down

(a) the number of axes of symmetry

(b) the order of rotational symmetry

(c) whether the shape has point symmetry
    or not.

Fig. 16.11

1. Rhombus   **2**. Regular pentagon

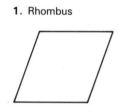 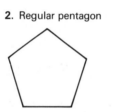

3. Ellipse   **4**. Regular octagon

5. Star   **6**. Letter X

7. Letter E   **8**. Equilateral triangle

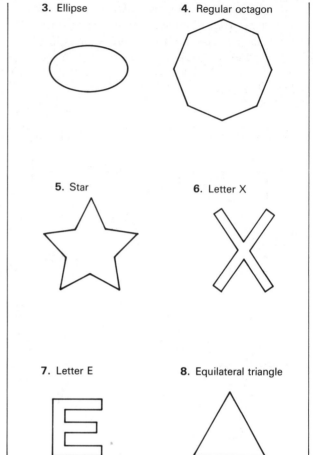

Plane Figures

Introduction

A **plane figure** is flat like a sheet of paper or metal. It is made up of lines called the sides of the figure. A triangle has three sides and a quadrilateral has four sides.

Types of Triangle

(1) An **acute-angled triangle** has every angle less than $90°$ (Fig. 17.1).

Fig. 17.1

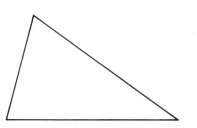

(2) A **right-angled triangle** has one of its angles equal to $90°$ (Fig. 17.2).

The side opposite to the right angle is the longest side and it is called the **hypotenuse**.

Fig. 17.2

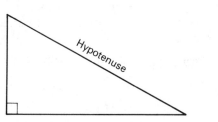

(3) An **obtuse-angled triangle** has one angle greater than $90°$ (Fig. 17.3).

Fig. 17.3

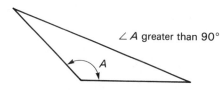

$\angle A$ greater than $90°$

(4) A **scalene triangle** has all three sides of different length and all three angles of different size.

(5) An **isosceles triangle** has two equal sides and two equal angles. The equal angles lie opposite to the equal sides (Fig. 17.4).

Fig. 17.4

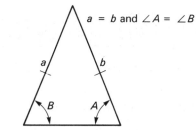

$a = b$ and $\angle A = \angle B$

(6) An **equilateral triangle** has all its sides equal in length and all its angles equal to $60°$ (Fig. 17.5).

Fig. 17.5

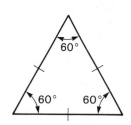

A small line drawn across the sides of a figure shows that the lengths of those sides are equal. In triangle ABC (Fig. 17.6) the sides AC and AB are equal. Sometimes a pair of lines are used to show equal sides (Fig. 17.7).

Fig. 17.6

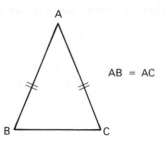

AB = AC

Fig. 17.7

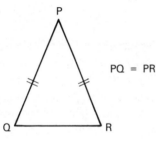

PQ = PR

Exercise 17.1

Look at the triangles shown in Fig. 17.8 and decide

1. which are equilateral triangles

2. which are obtuse-angled triangles

3. which are acute-angled triangles

4. which are isosceles triangles

5. which are right-angled triangles.

Fig. 17.8

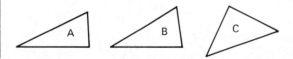

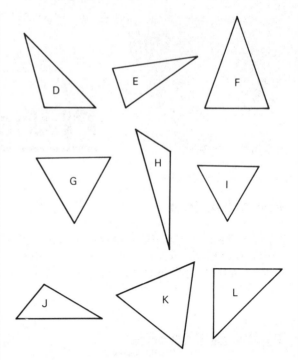

Angle Properties of Triangles

The angles of a triangle add up to $180°$ (Fig. 17.9).

Fig. 17.9

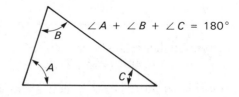

$\angle A + \angle B + \angle C = 180°$

Example 1

Two of the angles of a triangle are $35°$ and $72°$. What is the size of the third angle of the triangle?

$$\text{Third angle} = 180° - (35° + 72°)$$
$$= 180° - 107°$$
$$= 73°$$

Example 2

In Fig. 17.10, find the size of the angle *y*.

Fig. 17.10

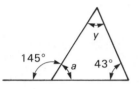

The angle on a straight line is 180°.

$$145° + a = 180°$$

$$a = 180° - 145°$$

$$a = 35°$$

The angles of a triangle add up to 180°.

Hence

$$43° + 35° + y = 180°$$

$$78° + y = 180°$$

$$y = 180° - 78°$$

$$y = 102°$$

1. In each of the triangles listed in the table below, the size of two angles is given. Find, by calculation, the size of the third angle.

Triangle	Given angles		Third angle
A	28°	67°	
B	37°	82°	
C	80°	80°	
D	90°	28°	
E	104°	63°	
F	60°	30°	

2. Find the size of each of the angles *x* and *y* shown in Fig. 17.11.

Fig. 17.11

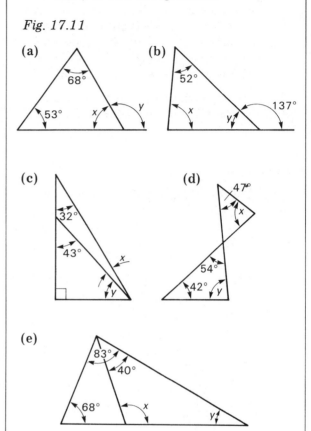

Constructing Triangles

To construct a triangle, given the lengths of the three sides, only a rule and compasses are needed.

Example 3

Construct accurately the triangle ABC which is shown in Fig. 17.12.

Fig. 17.12

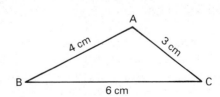

Draw BC = 6 cm (Fig. 17.12). Using your compasses set at 4 cm and centred at B draw an arc. Next set the compasses to 3 cm and with centre at C draw a second arc to cut the first arc at A. Finally join A and B and also A and C. ABC is then the required triangle.

To construct a triangle like the one shown in Fig. 17.13 a rule and protractor are needed.

Example 4

Construct the triangle ABC which is shown in Fig. 17.13.

Fig. 17.13

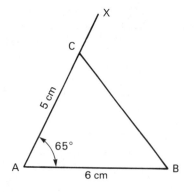

First draw AB 6 cm long. Then using a protractor draw AX so that ∠BAX = 65°. Along AX mark off AC = 5 cm. Finally join B and C, then ABC is the required triangle.

Exercise 17.3

Construct each of the triangles shown in Fig. 17.14. Then measure the lengths of the sides not given and, using a protractor, measure the sizes of the angles not given.

Fig. 17.14

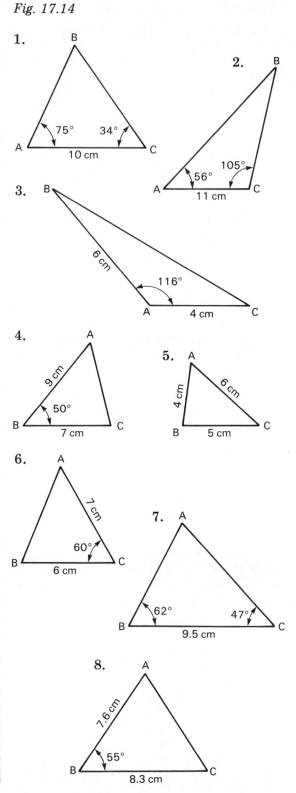

The Isosceles Triangle

It will be remembered that an isosceles triangle (Fig. 17.15) has two sides and two angles equal. The equal angles lie opposite to the equal sides.

Fig. 17.15

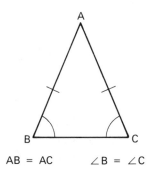

AB = AC   ∠B = ∠C

Example 5

In Fig. 17.16, find the size of the angles x and y.

Fig. 17.16

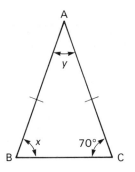

Since AB = AC, triangle ABC is isosceles.

Hence $x = 70°$

The angles of a triangle add up to $180°$.

Hence

$$70° + 70° + y = 180°$$
$$140° + y = 180°$$
$$y = 180° - 140°$$
$$y = 40°$$

If we cut out the isosceles triangle ABC shown in Fig. 17.17 and fold it along the line AD we will see that the one part fits exactly over the other part. So AD is a line of symmetry.

This means that:

Angle BAD = Angle DAC

Angle BDA is a right angle and so is angle ADC.

Length BD = Length DC.

Fig. 17.17

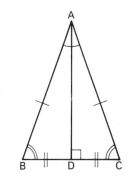

Example 6

In triangle ABC shown in Fig. 17.18, work out the size of the angles marked x and y and the length of the side BC.

Fig. 17.18

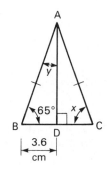

Since triangle ABC is isosceles

$$x = 65°$$
$$\angle BAC = 180° - 65° - 65°$$
$$= 180° - 130°$$
$$= 50°$$

Since AD is the line of symmetry

$$y = \tfrac{1}{2} \text{ of } \angle BAC$$
$$= \tfrac{1}{2} \text{ of } 50°$$
$$= 25°$$
$$BC = 2 \times BD$$
$$= 2 \times 3.6 \text{ cm}$$
$$= 7.2 \text{ cm}$$

The Equilateral Triangle

An equilateral triangle has all its sides the same length and all of its angles equal to 60°.

In Fig. 17.19 the equilateral triangle ABC has three axes of symmetry. They are AD, BE and CF.

Fig. 17.19

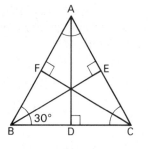

Example 7

In Fig. 17.20 the equilateral triangle ABC has each of its sides 8 cm long. Find the length of AD and the size of the angle ABD.

BD is a line of symmetry and so

$$AD = \tfrac{1}{2} \text{ of } AC$$
$$= \tfrac{1}{2} \text{ of } 8 \text{ cm}$$
$$= 4 \text{ cm}$$

Each angle of triangle ABC equals 60°. Hence

$$\text{Angle ABD} = \tfrac{1}{2} \text{ of } 60°$$
$$= 30°$$

Fig. 17.20

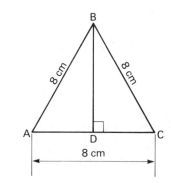

Exercise 17.4

Find the size of the angles marked x and y in Fig. 17.21.

Fig. 17.21

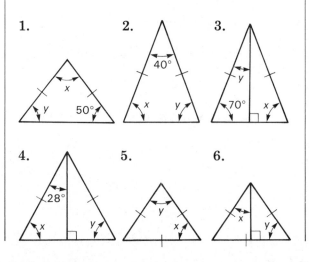

The Angles of a Quadrilateral

In Fig. 17.22 the figure ABCD has four sides and so it is a **quadrilateral.**

Fig. 17.22

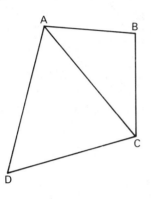

If we draw a straight line AC from one corner to the opposite corner the quadrilateral is divided into two triangles.

The angles of a triangle add up to 180° and so the four angles of ABCD add up to 360°.

Example 8

Fig. 17.23 shows a quadrilateral. Find the size of the angle marked y.

Fig. 17.23

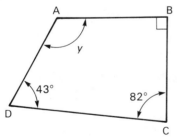

The angles of ABCD add up to 360°. So

$$y + 43° + 82° + 90° = 360°$$
$$y + 215° = 360°$$
$$y = 360° - 215°$$
$$y = 145°$$

The Rectangle

A **rectangle** is a quadrilateral with each of its angles equal to 90°.

A rectangle is symmetrical about the two axes shown in Fig. 17.24. It also has point symmetry about the point O.

Therefore its diagonals AC and BD are equal in length and they bisect each other, i.e. AO = CO and BO = DO.

Fig. 17.24

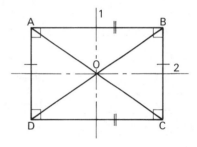

The Square

A **square** is a rectangle with all its sides equal in length.

It is symmetrical about the four axes shown in Fig. 17.25 and also about the point O. The diagonals of a square bisect at right angles, that is AO = CO, BO = DO and angle AOB is a right angle.

Fig. 17.25

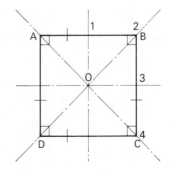

The Parallelogram

A **parallelogram** is, in effect, a rectangle pushed out of square (Fig. 17.26). It has both pairs of opposite sides parallel, i.e. AB is parallel to DC and BC is parallel to AD. Also the sides AB and CD are equal in length as are the sides BC and AD.

A parallelogram has no axes of symmetry but it has point symmetry about the point O. This means that the diagonals bisect each other, i.e. AO = CO and BO = DO.

Fig. 17.26

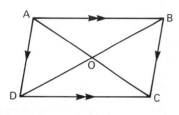

Example 9

In Fig. 17.27, ABCD is a parallelogram. Write down the size of the angles marked a, b and c.

Fig. 17.27

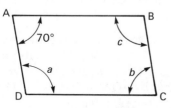

Because AB is parallel to CD

$$a = 180° - 70°$$
$$a = 110°$$

Because AD is parallel to BC

$$c = 180° - 70°$$
$$c = 110°$$
$$b = 180° - a$$
$$b = 180° - 110°$$
$$b = 70°$$

The Rhombus

A **rhombus** (Fig. 17.28) is a parallelogram with all its sides equal in length. Whereas a parallelogram has no axes of symmetry, a rhombus is symmetrical about each of its diagonals, i.e. about the lines AC and BD. This means that the diagonals bisect the angles through which they pass, i.e. BD bisects the angles ABC and ADC whilst AC bisects the angles DAB and BCD. Like a square the diagonals of a rhombus intersect at right angles, i.e. angle AOB is a right angle.

Fig. 17.28

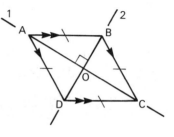

Example 10

In Fig. 17.29, ABCD is a rhombus. Work out the sizes of the angles marked w, x and y.

Fig. 17.29

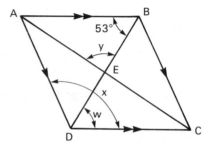

Because AB is parallel to DC

$$w = 53° \quad \text{(alternate angles)}$$

Because the diagonal BD bisects ADC

$$x = 2 \times w$$
$$= 2 \times 53°$$
$$= 106°$$

Because the diagonals AC and BD bisect at right angles

$$y = 90°$$

The Kite

Fig. 17.30 shows a **kite**. The sides AB and BC are equal in length and so are the sides AD and CD. The kite is symmetrical about the line BD and this means that the angles DAB and DCB are equal. Also, the diagonals AC and BD intersect at right angles.

Fig. 17.30

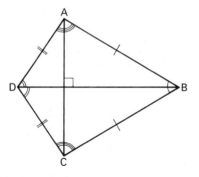

Example 11

Find the angles marked *a* and *b* for the kite shown in Fig. 17.31

Fig. 17.31

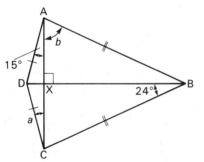

Because AD = CD, triangle ADC is isosceles, so

$$a = 15°$$

In triangle BXC,

$$XBC = 24° \quad \text{(given)}$$
$$AXB = 90°$$

Let $\angle XCB = x$ then

$$x + 24° + 90° = 180°$$
$$x + 114° = 180°$$
$$x = 180° - 114°$$
$$x = 66°$$

Because $\angle BAX = \angle XCB$,

$$b = x$$
$$= 66°$$

The Trapezium

A **trapezium** (Fig. 17.32) is a quadrilateral with one pair of sides parallel.

Fig. 17.32

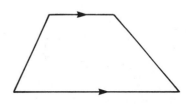

Example 12

Fig. 17.33 shows a trapezium. Calculate the size of the angles *x* and *y*.

Fig. 17.33

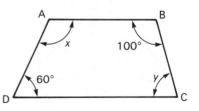

Because AB is parallel to CD

$$x = 180° - 60°$$
$$= 120°$$
$$y = 180° - 100°$$
$$= 80°$$

Exercise 17.5

1. Calculate the angle x in Fig. 17.34.

Fig. 17.34

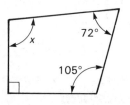

2. Fig. 17.35 shows a trapezium with AB parallel to CD. Find the size of the angle marked x.

Fig. 17.35

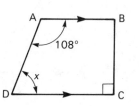

3. In Fig. 17.36, calculate the size of the angle x.

Fig. 17.36

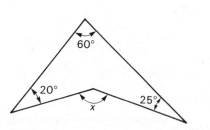

4. In Fig. 17.37, ABCD is a parallelogram. Work out the size of the angles w, x and y.

Fig. 17.37

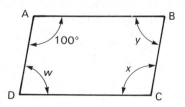

5. Name each of the plane figures shown in Fig. 17.38.

Fig. 17.38

(a)

(b)   (c)

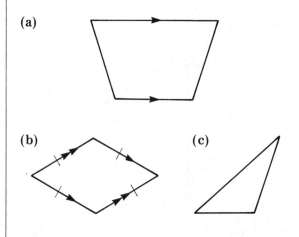

6. For the rhombus ABCD (Fig. 17.39), find the size of the angles w, x and y.

Fig. 17.39

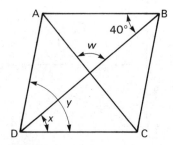

7. In Fig. 17.40, ABCD is a square. Find the size of the angles *a* and *b*.

Fig. 17.40

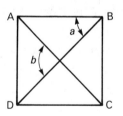

Polygons

A **polygon** is any plane shape whose sides are straight lines. A triangle is a polygon with three sides and a quadrilateral is a polygon with four sides.

A shape is **regular** if all its sides are the same length and all its angles are equal in size. An equilateral triangle and a square are examples of regular figures.

A regular pentagon (five-sided), a regular hexagon (six-sided) and a regular octagon (eight-sided) are shown in Fig. 17.41.

Fig. 17.41

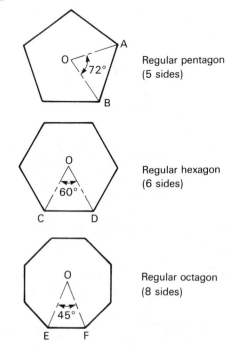

Regular pentagon (5 sides)

Regular hexagon (6 sides)

Regular octagon (8 sides)

All regular polygons have the number of axes of symmetry equal to the number of its sides. So a pentagon has five axes of symmetry (Fig. 17.42), a hexagon has six axes of symmetry and an octagon has eight axes of symmetry. In addition, all regular polygons have rotational symmetry.

Fig. 17.42

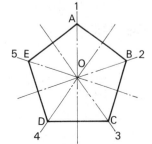

Therefore every regular polygon can be split up into a number of equal isosceles triangles.

For a pentagon there are five equal triangles all the same as AOB. Therefore

$$\text{Angle AOB} = 360° \div 5$$
$$= 72°$$

For a hexagon there are six equal triangles the same as OCD. Therefore

$$\text{Angle COD} = 360° \div 6$$
$$= 60°$$

Similarly an octagon can be split up into eight triangles like EOF. Therefore

$$\text{Angle EOF} = 360° \div 8$$
$$= 45°$$

We can use this information to find the angles of any regular polygon.

Example 13

What size is each angle of a regular pentagon?

Looking at Fig. 17.43 (overleaf):

$$\text{Angle AOB} = 360° \div 5$$
$$= 72°$$

Because triangle AOB is isosceles

Angle OBA $= (180° - 72°) \div 2$

$= 108° \div 2$

$= 54°$

Because triangles AOB and BOC are equal

Angle OBA $=$ Angle OBC

$= 54°$

Angle ABC $= 54° + 54°$

$= 108°$

The size of the exterior angle of a regular pentagon is $108°$

Fig. 17.43

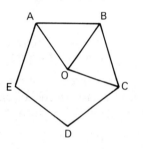

The Circle

The main parts of a circle are shown in Fig. 17.44.

Fig. 17.44

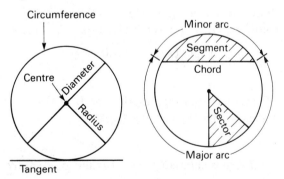

Every circle is symmetrical about any of its diameters and so it has an infinite number of axes of symmetry. It also has rotational symmetry and this means that:

The diameter $= 2 \times$ The radius

Exercise 17.6

1. Write down the number of sides which each of the following polygons possess:
    (a) rhombus
    (b) equilateral triangle
    (c) square
    (d) hexagon
    (e) isosceles triangle
    (f) trapezium
    (g) pentagon
    (h) kite
    (i) quadrilateral
    (j) octagon
    (k) scalene triangle.

2. What is the size of the angles of:
    (a) a regular hexagon
    (b) a regular octagon
    (c) an equilateral triangle.

3. Fig. 17.45 consists of equilateral triangles. Using the letters in the diagram, find
    (a) a rhombus   (b) a regular hexagon.

Fig. 17.45

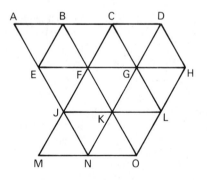

4. What are the shapes shown in Fig. 17.46 called?

Fig. 17.46

(a)   (b)

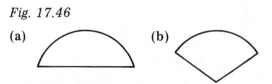

5. How many axes of symmetry has
   (a) a pentagon
   (b) a square
   (c) an equilateral triangle
   (d) an octagon?

6. For a regular pentagon, calculate the size of each internal angle.

7. ABCDEF is a regular hexagon drawn on top of a square ABXY. Work out the size of the angle marked y in the diagram (Fig. 17.47).

Fig. 17.47

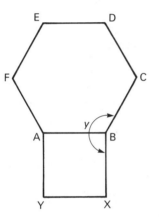

8. The diagrams in Fig. 17.48 show an equilateral triangle, a regular pentagon and a rhombus. Sketch the figures and show, by broken lines, all the lines of symmetry for each shape.

Fig. 17.48

(a)   (b)

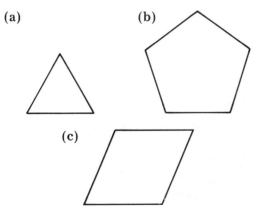

(c)

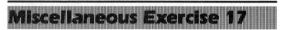

Miscellaneous Exercise 17

Section A

1. For the kite shown in Fig. 17.49 work out the angles x, y and z.

Fig. 17.49

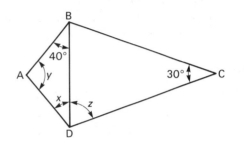

2. In Fig. 17.50 find the angles marked a, b, c, d and e. The figure is a rhombus.

Fig. 17.50

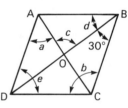

3. Find the angle marked x in Fig. 17.51.

Fig. 17.51

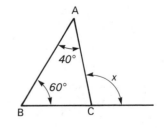

4. Using a rule and compasses only, draw accurately a triangle with sides 5 cm, 6 cm and 7 cm long. Measure and write down the vertical height of the triangle.

5. Fig. 17.52 shows dots arranged in squares. Copy the diagram on to squared paper and perform the following:

(a) Join four of the lettered dots to form a square. Write down the letters you used.

(b) Join four different dots to make a parallelogram and write down the letters you used.

(c) Join four of the dots to form a kite. Which letters did you use?

Fig. 17.52

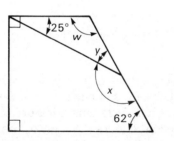

Section B

1. The radius of a big wheel at a fairground is 20 ft. What is its diameter?

2. Calculate the angles *w*, *x* and *y* in Fig. 17.53.

Fig. 17.53

3. Fig. 17.54 shows a parallelogram. Calculate the size of the angles *a* and *b*.

Fig. 17.54

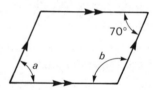

4. In Fig. 17.55 work out the size of the angles *x* and *y*.

Fig. 17.55

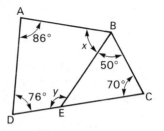

5. In Fig. 17.56 calculate the size of the angles

(a) *x*   (b) *y*   (c) BAC

Fig. 17.56

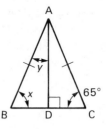

6. In Fig. 17.57 what is the size of the angle marked *x*?

Fig. 17.57

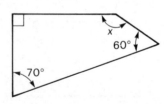

7. Fig. 17.58 shows a triangle ABC. Using your protractor measure the size of each of the three angles of this triangle. Add together your measurements and comment on your result.

Fig. 17.58

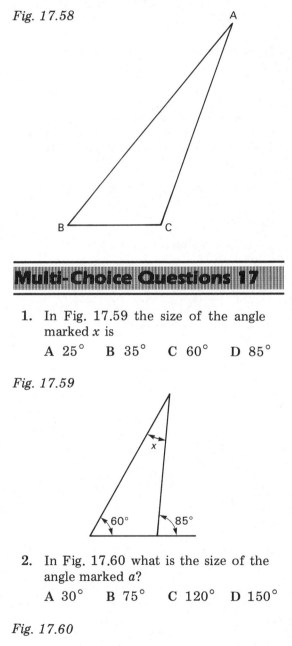

![Multi-Choice Questions 17]

1. In Fig. 17.59 the size of the angle marked x is

    A 25°   B 35°   C 60°   D 85°

Fig. 17.59

2. In Fig. 17.60 what is the size of the angle marked a?

    A 30°   B 75°   C 120°   D 150°

Fig. 17.60

3. Consider the four quadrilaterals: a square, a rectangle, a kite and a parallelogram. Which one of these has only one axis of symmetry?

    A square   B rectangle
    C kite   D parallelogram

4. The shaded part of Fig. 17.61 is called

    A an arc   B a segment
    C a sector   D a chord

Fig. 17.61

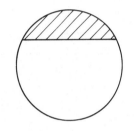

5. What is the size of the angle x in Fig. 17.62? It is

    A 60°   B 70°   C 110°   D 120°

Fig. 17.62

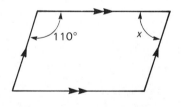

6. Consider the four plane figures: a rhombus, a pentagon, an octagon and a kite. Which one of them has five sides? The answer is

    A kite   B pentagon
    C rhombus   D octagon

7. What is the size of the interior angles of a regular octagon?

    A 45°   B 67.5°
    C 90°   D 135°

Mental Test 17

1. How many sides has a hexagon?

2. A quadrilateral has four equal sides. Which of two figures could it be?

3. How many axes of symmetry has a rhombus?

4. What is the shape shown in Fig. 17.63 called?

Fig. 17.63

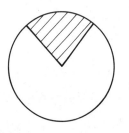

5. What kind of triangle has been drawn in Fig. 17.64?

Fig. 17.64

6. What is the size of the angle y in Fig. 17.65?

Fig. 17.65

7. The diameter of a circle is 8 cm. What is its radius?

8. What is the line AB (Fig. 17.66) called?

Fig. 17.66

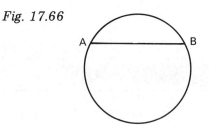

Perimeters and Areas

Perimeters

The distance all the way round a plane shape is called its **perimeter**.

Example 1

(a) The sides of an equilateral triangle (Fig. 18.1) are each 5 cm long. Calculate the perimeter of the triangle.

Fig. 18.1

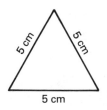

$$\text{Perimeter} = (5 + 5 + 5)\,\text{cm}$$
$$= 15\,\text{cm}$$

(b) Work out the perimeter of the rectangle shown in Fig. 18.2.

Fig. 18.2

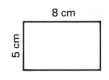

$$\text{Perimeter} = (8 + 5 + 8 + 5)\,\text{cm}$$
$$= 26\,\text{cm}$$

(c) Find the perimeter of the shape shown in Fig. 18.3.

Fig. 18.3

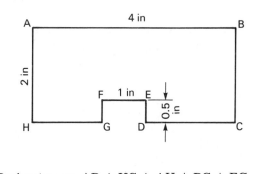

$$\text{Perimeter} = AB + HC + AH + BC + FG$$
$$\qquad\qquad + ED$$
$$= (4 + 4 + 2 + 2 + 0.5 + 0.5)\,\text{in}$$
$$= 13\,\text{in}$$

Circumference of a Circle

The perimeter of a circle is called the **circumference**.

For any circle you care to draw it will be found that its circumference is just over three times its diameter. That is

Circumference ÷ Diameter is approximately 3

The exact value of circumference ÷ diameter can never be worked out, but for most problems a value of 3.14 is accurate enough when working in decimals. When working in fractions a value of $\frac{22}{7}$ may be used. When working mentally a value of $\frac{25}{8}$ is often used.

The value of circumference ÷ diameter is so important it has been given the special symbol π (the Greek letter pi). We take the value of π as being 3.14 or $\frac{22}{7}$. We say

$$\text{Circumference} = \pi \times \text{Diameter}$$

or as an algebraic formula

$$C = \pi \times d$$

where C is the circumference, π is 3.14 or $\frac{22}{7}$ and d is the diameter.

Since the diameter of a circle is twice its radius we can write

$$C = 2 \times \pi \times r$$

where C is the circumference, π is 3.14 or $\frac{22}{7}$ and r is the radius.

Example 2

(a) A circle has a diameter of 4 inches. Calculate its circumference.

$$C = \pi \times d$$
$$= 3.14 \times 4 \text{ in}$$
$$= 12.56 \text{ in}$$

(b) The radius of a circle is 14 cm. Calculate its circumference.

$$C = 2 \times \pi \times r$$
$$= \tfrac{2}{1} \times \tfrac{22}{7} \times \tfrac{14}{1} \text{ cm}$$
$$= 2 \times 22 \times 2 \text{ cm}$$
$$= 88 \text{ cm}$$

(When the radius or the circumference is a multiple of 7 it is best to use $\pi = \frac{22}{7}$ because the numbers will then cancel.)

Exercise 18.1

1. Calculate the perimeters of each of the triangles shown in Fig. 18.4.

Fig. 18.4

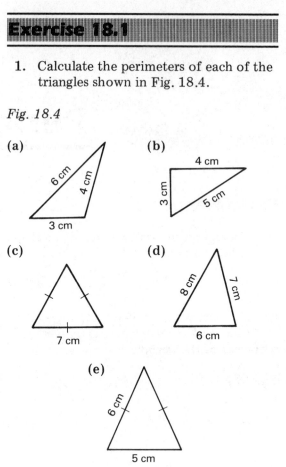

Find the perimeters of each of the following triangles:

2. a scalene triangle with sides 4 cm, 6 cm and 8 cm long

3. an equilateral triangle with each side 8 m long

4. an isosceles triangle with sides 7 in, 7 in and 5 in long.

5. Fig. 18.5 shows a number of rectangles and squares. Find the perimeters of each of them.

Fig. 18.5

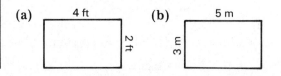

(c)

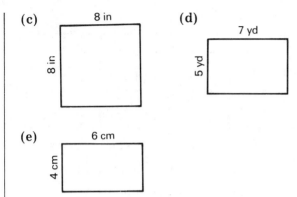

8 in

8 in

(d)

7 yd

5 yd

(e)

6 cm

4 cm

(d)

(e) **(f)**

Find the perimeters of the following rectangles:

6. length 6 cm, width 3 cm

7. length 9 in, width 5 in

8. length 10 m, width 6 m.

9. A square has sides 4 cm long. What is its perimeter?

10. Calculate the perimeter of a square if its sides are 5 ft long.

11. If the dots in Fig. 18.6 are spaced 1 cm apart, what is the length of the perimeter of each of the rectangles shown?

Fig. 18.6

(a)

(b) **(c)**

Calculate the circumference for each of the following circles:

12. diameter 6 cm (take $\pi = 3.14$)

13. diameter 2 ft (take $\pi = 3.14$)

14. diameter 21 in (take $\pi = \frac{22}{7}$)

15. diameter 14 ft (take $\pi = \frac{22}{7}$)

16. radius 5 cm (take $\pi = 3.14$)

17. radius 8 m (take $\pi = 3.14$)

18. radius 56 cm (take $\pi = \frac{22}{7}$)

19. radius 28 in (take $\pi = \frac{22}{7}$).

20. A circular flower bed has a diameter of 5 m. How far is it round the edge? (Take π 3.14.)

21. A circular garden pool has a radius of 7 ft. What is the circumference of the pool? (Take $\pi = \frac{22}{7}$.)

22. A wheel has a diameter of 70 cm. Work out the circumference of the wheel. (Take $\pi = \frac{22}{7}$.)

23. A flywheel for a motor car has a radius of 12 cm. What is the circumference of the flywheel? (Take $\pi = 3.14$.)

24. Work out the perimeters of each of the plane shapes shown in Fig. 18.7, if the dots are spaced 1 cm apart.

Fig. 18.7

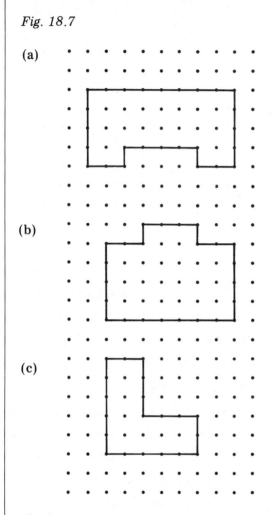

(a)

(b)

(c)

Fig. 18.8

Shape A contains 12 equal squares

Shape B contains 13 equal squares

Shape C contains 12 equal squares

We see that shape B has the greatest area because it contains the largest number of equal squares.

In practice, the squares used for measuring area have sides of 1 m, 1 cm or 1 mm if we are using the metric system, and 1 in, 1 ft or 1 yd if we are using the imperial system.

The standard abbreviations for units of area in the metric system are:

$$\text{square metre} = \text{m}^2$$
$$\text{square centimetre} = \text{cm}^2$$
$$\text{square millimetre} = \text{mm}^2$$

In the imperial system the abbreviations are:

$$\text{square inch} = \text{in}^2$$
$$\text{square foot} = \text{ft}^2$$
$$\text{square yard} = \text{yd}^2$$

Area

Area is the space taken up by a flat shape such as a sheet of metal or a table-top.

The area of a plane shape is measured by counting the number of equal squares contained in the shape.

Example 3

Look at the shapes shown in Fig. 18.8. By counting up the number of equal squares contained in each of the shapes, find out which shape has the greatest area.

Areas of Rectangles

The rectangle shown in Fig. 18.9 contains 8 equal squares each of side 1 cm. We say that the area of the rectangle is 8 cm^2.

Fig. 18.9

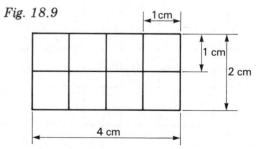

Looking at the diagram again we see that we have two rows of equal squares each containing four of these squares. So

$$\text{Area of rectangle} = 4 \times 2\,\text{cm}^2$$
$$= 8\,\text{cm}^2$$

All we have done is to multiply the length of the rectangle by its breadth. No matter how many rectangles we measure we will always find that:

$$\text{Area of rectangle} = \text{Length} \times \text{Breadth}$$

or as a formula

$$A = l \times b$$

where A is the area, l is the length and b is the breadth. Note that when using this formula l and b must be in the same units, i.e. they must be both in centimetres, or both in inches, etc.

Example 4

A rectangular carpet measures 6 metres by 4 metres. What is its area?

We are given that $l = 6$ and $b = 4$. So

$$A = l \times b$$
$$= 6 \times 4$$
$$= 24$$

The area of the carpet is $24\,\text{m}^2$.

Area of a Square

Since a square is a rectangle with all its sides equal in length

$$A = a^2$$

where A is the area and a is the length of the sides.

Example 5

A ceramic tile has sides 6 inches long. What is its area?

We are given that $a = 6$, hence

$$A = a^2$$
$$= 6^2$$
$$= 6 \times 6$$
$$= 36$$

The area of the tile is $36\,\text{in}^2$.

Exercise 18.2

1. Find the area of each of the rectangles shown in Fig. 18.10.

Fig. 18.10

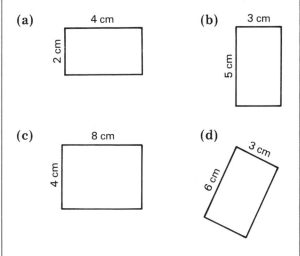

Find the areas of each of the following rectangles:

2. length 20 mm, breadth 11 mm
3. length 35 ft, breadth 8 ft
4. length 8.3 cm, breadth 5.2 cm
5. length 19 in, breadth 7 in
6. length 8.2 ft, breadth 7.6 ft
7. length 26 yd, breadth 18 yd.

8. A rectangular piece of wood is 3.7 m long and 1.3 m wide. Calculate its area.

9. A rectangular steel plate is 120 inches long by 80 inches wide. Calculate the area of the plate.

10. A rectangular floor is 5.8 yards long by 4.9 yards wide. What area of carpet is needed to completely cover it?

11. A ceramic tile is in the form of a square with a side of 8 inches. Work out its area.

12. A carpet tile is 30 cm square. What is its area?

13. Which rectangles have the same area (Fig. 18.11)?

Fig. 18.11

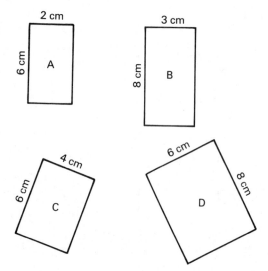

14. Four rectangles are shown in Fig. 18.12. Which two rectangles are equal in area?

Fig. 18.12

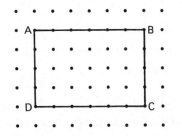

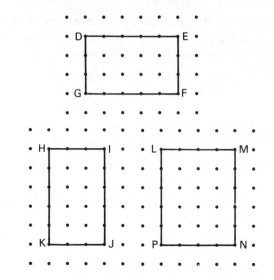

15. Fig. 18.13 shows a square and three rectangles. Which of the rectangles is equal in area to the square?

Fig. 18.13

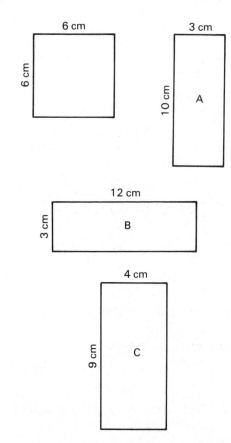

16. Calculate the areas of each of the rectangles shown in Fig. 18.14, if the spots are 1 cm apart each way.

Fig. 18.14

(a)

(b)

(c)

(d)

17. The floor of a kitchen is 12 ft by 8 ft. How many 6 in square tiles are needed to tile it?

18. To tile one wall of a bathroom 150 square tiles of 10 cm side are used. What is the area of the wall?

19. A rectangular pane of glass is 1.3 m long by 0.8 m wide. What is the area of the pane?

20. Sheet metal is sold at £8.50 per square foot. How much does a rectangular plate 3 ft by 6 ft cost?

Areas of Shapes Made from Rectangles

The areas of many shapes can be found by splitting them up into rectangles.

Example 6

Fig. 18.15 shows the cross-section of a steel girder. Work out its area.

Fig. 18.15

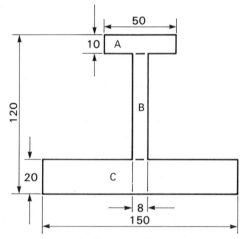

All dimensions in millimetres

The shape can be split up into three rectangles as shown in the diagram.

Area of shape = Area of rectangle A
+ Area of rectangle B
+ Area of rectangle C

$$= (50 \times 10) + (90 \times 8) + (150 \times 20)$$

$$= 500 + 720 + 3000$$

$$= 4220$$

So the area of the girder is 4220 mm^2.

Exercise 18.3

By splitting up the shapes shown in Fig. 18.16 find their total areas.

Fig. 18.16

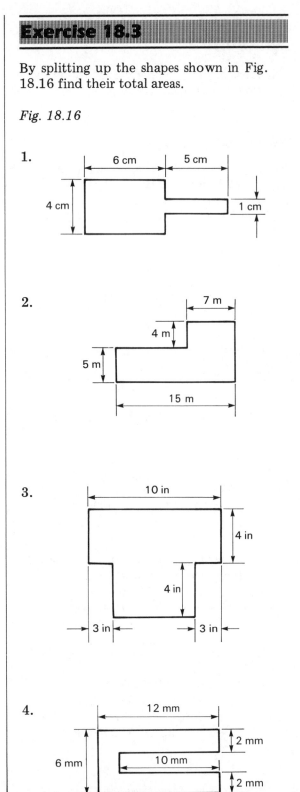

1.

2.

3.

4.

5.

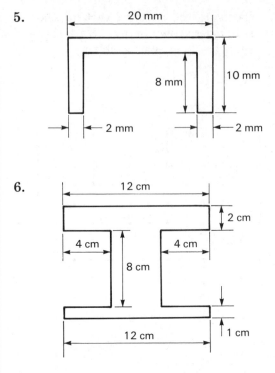

6.

Areas of Borders

In Fig. 18.17 the shaded squares form a border around the edge of the white squares.

Fig. 18.17

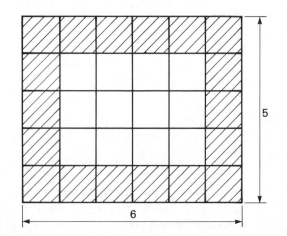

Total number of squares $= 5 \times 6$

$\qquad\qquad\qquad\qquad = 30$

Number of white squares $= 4 \times 3$

$\qquad\qquad\qquad\qquad = 12$

Number of shaded squares $=$ Area of border

$\qquad\qquad\qquad\qquad = 30 - 12$

$\qquad\qquad\qquad\qquad = 18$

To find the area of a border, work out the area of the outer rectangle and take away the area of the inner rectangle.

Example 7

Fig. 18.18 shows a room fitted with a carpet so as to leave a surround. Work out the area of the surround.

Fig. 18.18

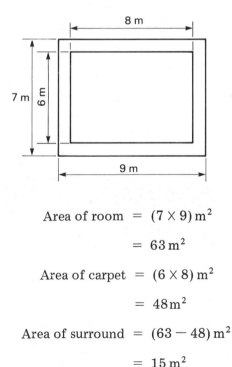

Area of room $= (7 \times 9)\,\mathrm{m}^2$

$\qquad\qquad\quad = 63\,\mathrm{m}^2$

Area of carpet $= (6 \times 8)\,\mathrm{m}^2$

$\qquad\qquad\quad = 48\,\mathrm{m}^2$

Area of surround $= (63 - 48)\,\mathrm{m}^2$

$\qquad\qquad\qquad = 15\,\mathrm{m}^2$

So the area of the surround is $15\,\mathrm{m}^2$.

Exercise 18.4

Find the areas of the shaded borders drawn in Fig. 18.19.

Fig. 18.19

1.

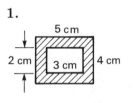

2.

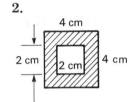

3.

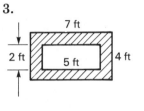

4.

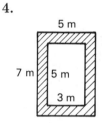

5.

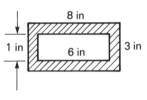

6.

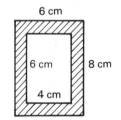

7.

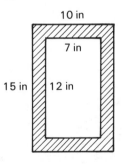

8. A rectangular lawn is 32 yards long and 23 yards wide. It has a path 2 yards wide all round it. What is the area of the path?

9. A room 8 m long and 7 m wide is to be carpeted so as to leave a surround 1 m wide round the carpet. Work out:

   (a) the area of the room

   (b) the area of the carpet

   (c) the area of the surround.

Areas of Triangles

Looking at Fig. 18.20 we see that the right-angled triangle ABC is one-half of the area of the rectangle ABCD. It is true that the area of any triangle (Fig. 18.21) is one-half of the base times the height.

Fig. 18.20

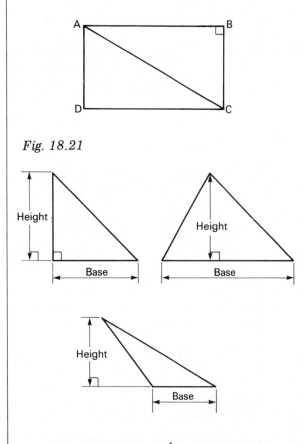

Fig. 18.21

Area of triangle $= \frac{1}{2} \times$ Base $\times$ Height

Note carefully that the height is the vertical height of the triangle.

Example 8

Work out the areas of the triangles shown in Fig. 18.22.

Fig. 18.22

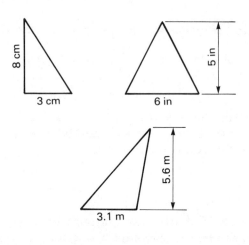

(a)   Area $= \frac{1}{2} \times 8 \times 3$ cm^2

         $= 12$ cm^2

(b)   Area $= \frac{1}{2} \times 6 \times 5$ in^2

         $= 15$ in^2

(c)   Area $= \frac{1}{2} \times 5.6 \times 3.1$ m^2

         $= 8.68$ m^2

Exercise 18.5

Find the areas of the triangles shown in Fig. 18.23:

Fig. 18.23

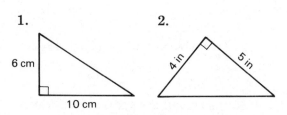

1.   2.

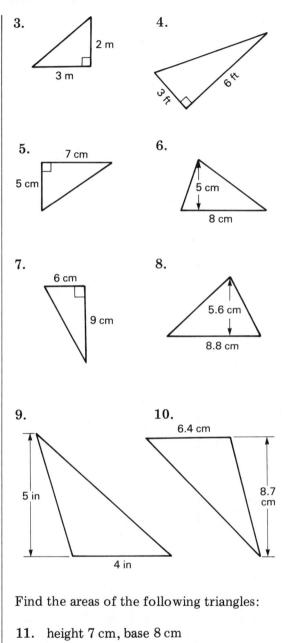

3. 2 m 3 m

4. 6 ft 3 ft

5. 7 cm 5 cm

6. 5 cm 8 cm

7. 6 cm 9 cm

8. 5.6 cm 8.8 cm

9. 5 in 4 in

10. 6.4 cm 8.7 cm

Find the areas of the following triangles:

11. height 7 cm, base 8 cm

12. height 11 in, base 12 in

13. height 8.9 cm, base 3.8 cm

14. height 50 mm, base 180 mm.

15. Which of the triangles in Fig. 18.24 have the same area as the triangle ABC?

Fig. 18.24

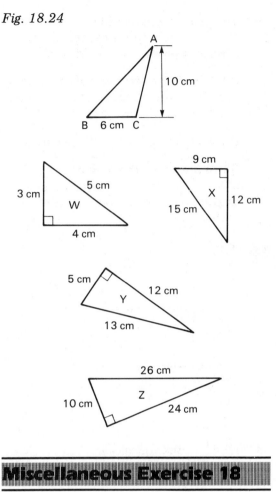

A 10 cm B 6 cm C

3 cm 5 cm W 4 cm

9 cm X 12 cm 15 cm

5 cm Y 12 cm 13 cm

26 cm Z 10 cm 24 cm

Miscellaneous Exercise 18

Section A

1. Fig. 18.25 shows a rectangle. Work out:
    (a) its perimeter
    (b) its area.

Fig. 18.25

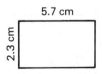

5.7 cm 2.3 cm

2. A circular pond has a radius of 14 metres:
    (a) What is its diameter?
    (b) Taking $\pi = \frac{22}{7}$, work out its circumference.

3. The pins on a pinboard are 1 cm apart. The pinboards and a rubber band are used to show a variety of rectangles:

(a) What is the perimeter of the rectangle ABCD shown in Fig. 18.26, if the pins are 1 cm apart?

(b) What is the area of this rectangle?

(c) Write down the length and width of a rectangle with the same perimeter as ABCD which can be shown on the pinboard.

Fig. 18.26

4. Which of the triangles in Fig. 18.27 have the same area as triangle ABC?

Fig. 18.27

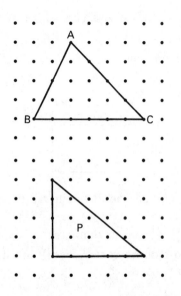

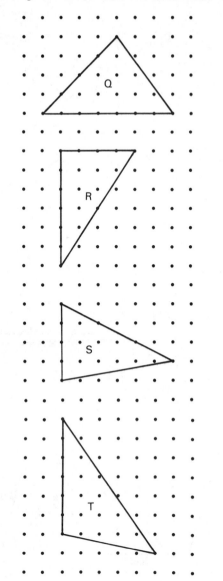

5. Work out the area and perimeter of the shape shown in Fig. 18.28, if the spots are 1 cm apart.

Fig. 18.28

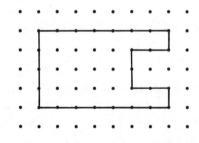

Section B

1. Work out the area of the shaded part of Fig. 18.29.

Fig. 18.29

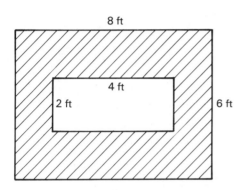

2. Fig. 18.30 shows a triangular plot of land. What is its area?

Fig. 18.30

3. A wheel has a radius of 35 cm. What is the distance around its rim? (Take $\pi = \frac{22}{7}$.)

4. Find the area of each of the shapes shown in Fig. 18.31, if a 1 cm grid has been used.

Fig. 18.31

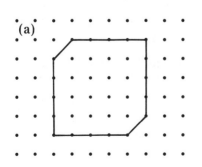

(b)

(c)

5. Work out the area and the perimeter of the shape shown in Fig. 18.32, if a 1 cm grid has been used.

Fig. 18.32

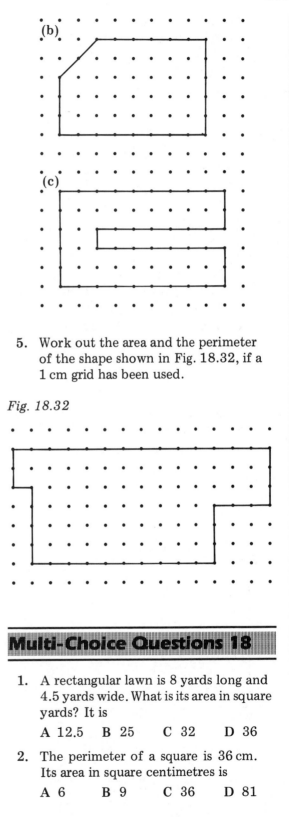

Multi-Choice Questions 18

1. A rectangular lawn is 8 yards long and 4.5 yards wide. What is its area in square yards? It is

    A 12.5   B 25   C 32   D 36

2. The perimeter of a square is 36 cm. Its area in square centimetres is

    A 6   B 9   C 36   D 81

3. Triangle ABC (Fig. 18.33) is right-angled at B. Its area in square centimetres is

   A 8.5   B 30   C 32.5   D 78

Fig. 18.33

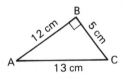

4. A swimming pool has a diameter of 7 m. Taking $\pi = \frac{22}{7}$, its circumference is

   A 11 m   B 22 m
   C 44 m   D 88 m

5. What is the area, in square centimetres, of the shape shown in Fig. 18.34? It is

   A 20   B 15   C 13   D 12.5

Fig. 18.34

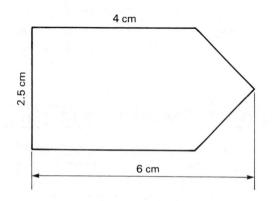

Mental Test 18

1. If the rectangle in Fig. 18.35 is drawn on a 1 cm grid, what is its perimeter?

Fig. 18.35

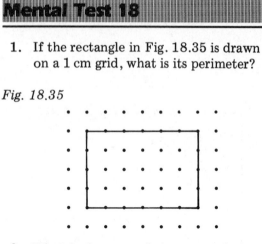

2. What is the area of the rectangle in Fig. 18.35?

3. A square has sides 9 cm long. What is its perimeter?

4. What is the area of the square in question 3?

5. What is the area of the triangle shown in Fig. 18.36?

Fig. 18.36

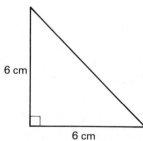

6. What is the perimeter of an equilateral triangle if one of its sides is 4 cm long?

7. Taking $\pi = \frac{25}{8}$, work out the circumference of a circle whose diameter is 8 cm long.

8. A triangle has sides 4 cm, 6 cm and 7 cm long. Work out its perimeter.

Solid Figures

Introduction

In Chapter 18 we discovered that plane shapes like triangles and rectangles had length and width but no height or thickness. Another way of describing these plane shapes is to call them two-dimensional figures.

Solid figures have three dimensions which are length, width and height or thickness. You can see that this is so by looking at a biscuit tin.

Types of Solid Figure

(1) A **sphere** (Fig. 19.1) is a circular solid. Examples are a football and a ball bearing.

Fig. 19.1

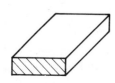

(2) A **cuboid** (Fig. 19.2) is a rectangular solid. It has a cross-section which is a rectangle. An example is a plank of wood.

Fig. 19.2

(3) A **cube** (Fig. 19.3) has all of its edges equal in length. Each of its **faces** is a square. An example is a sugar lump.

Fig. 19.3

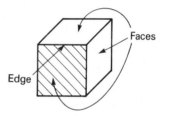

(4) A **triangular prism** (Fig. 19.4) has a constant cross-section which is a triangle. A ridge tent is an example.

Fig. 19.4

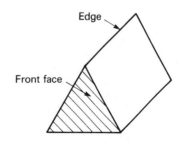

(5) Some prisms have a constant cross-section which is a regular polygon. Fig. 19.5 (overleaf) shows a **hexagonal prism**. Steel bars used in the engineering industry are sometimes made in this shape.

Fig. 19.5

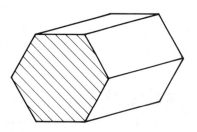

Fig. 19.6 shows an I-section which has a constant cross-section in the form of an I. Some steel bars used in the construction industry are like this.

Fig. 19.6

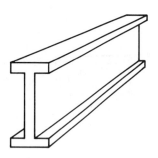

(6) A **cylinder** (Fig. 19.7) has a constant cross-section which is a circle. Many tins are cylindrical.

Fig. 19.7

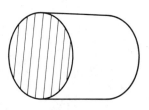

(7) A **pyramid** is a solid figure which stands upon a flat base which may be a triangle, a square, a rectangle or a polygon. As shown in Fig. 19.8 a pyramid tapers to a point. This means that each of its sides is a triangle.

Fig. 19.8

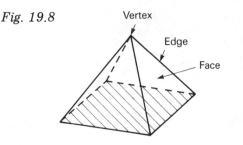

(8) A **regular tetrahedron** is a pyramid with all of its faces equilateral triangles (Fig. 19.9). Therefore a tetrahedron has all its edges equal in length.

Fig. 19.9

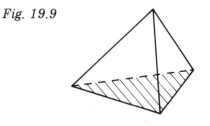

(9) A **cone** (Fig. 19.10) has a circular base and tapers to a point. An example is an ice-cream cone.

Fig. 19.10

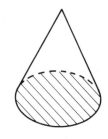

Example 1

Fig. 19.11 shows a triangular prism. How many edges and how many faces does it possess? How many vertices does it have?

Fig. 19.11

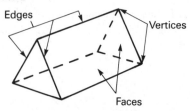

As shown in the diagram the prism has 9 edges and 5 faces. There are 6 vertices.

Exercise 19.1

For each of the solid figures named below write down

(a) the number of faces

(b) the number of edges

(c) the number of vertices

possessed by each of them:

1. cuboid

2. hexagonal prism

3. tetrahedron

4. cube

5. square pyramid

6. rectangular pyramid.

Drawing Solid Figures

We often need to draw solid figures which have three dimensions on paper which has only two dimensions. The method which follows allows us to do this.

We start off by drawing the three lines OX, OY and OZ shown in Fig. 19.12. These three lines, which are called **axes**, allow the drawing of a solid figure to be made.

Fig. 19.12

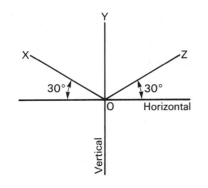

Example 2

Make a drawing of a cuboid which is 5 cm long, 3 cm wide and 2 cm high.

Start off by drawing the three axes OX, OY and OZ (Fig. 19.13). In the three-dimensional drawing the vertical edges of the cuboid are represented by vertical lines. For a rectangular solid like a cube, or a cuboid, all the lines making up the drawing either lie along the three axes or are parallel to them.

Fig. 19.13.

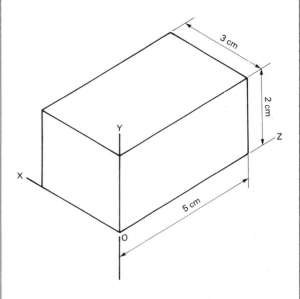

Example 3

Draw a three-dimensional picture of a pyramid with a rectangular base 3 cm by 4 cm and a vertical height of 5 cm.

Start off by drawing the axes OX, OY and OZ (Fig. 19.14). Now draw the base OABC and spot the point D which is at the intersection of the diagonals BO and AC. The vertex V lies directly above D. Join VA, VB, VC and VO to complete the diagram.

Fig. 19.14

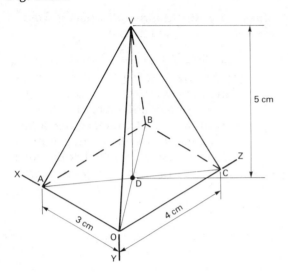

Drawing circles on a three-dimensional sketch sometimes causes difficulty. Fig. 19.15 gives the clue necessary for partly overcoming this difficulty. Diagram (a) shows a true circle drawn in a square. At the points marked X the circumference of the circle and the sides of the square touch. This must be the same in the view shown in diagram (b).

Fig. 19.15

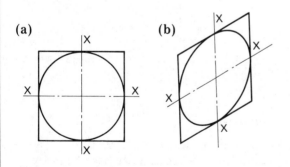

(a)   (b)

Example 4

Draw a three-dimensional picture of a cylinder which has a diameter of 4 cm and a height of 7 cm.

Start by drawing the three axes OX, OY and OZ. Next draw the cuboid ABCODEFG which is needed to

obtain the shape of the circular ends of the cylinder. The finished picture is shown in Fig. 19.16.

Fig. 19.16

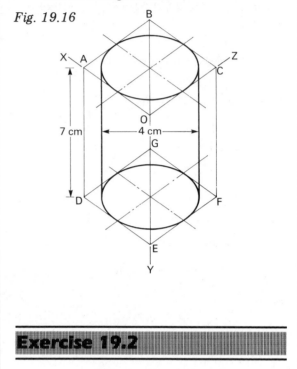

Exercise 19.2

Draw three-dimensional pictures of each of the following:

1. A cuboid with a length of 6 cm, a width of 3 cm and a height of 4 cm.

2. A cube having an edge of 5 cm.

3. A cylinder having a diameter of 4 cm and a height of 6 cm.

4. A cone having a base diameter of 6 cm and a vertical height of 5 cm.

5. A pyramid with a square base of side 4 cm and a vertical height of 6 cm.

6. The triangular prism with measurements as shown in Fig. 19.17.

Fig. 19.17

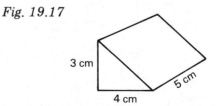

Nets

Suppose that we want to make a cube out of cardboard. We need a pattern giving us the shape of the cardboard needed to make the cube. As shown in Fig. 19.18 the pattern is six squares. This shape, which is called **a net of a cube,** can be folded to make a cube.

Fig. 19.18

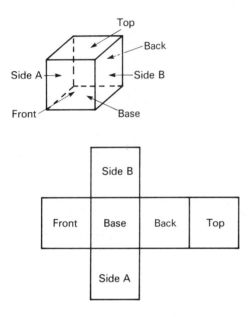

It is possible for there to be more than one net for a solid object. For instance, a cube can also be made by folding the shape shown in Fig. 19.19.

Fig. 19.19

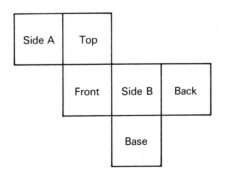

Example 5

Sketch a net of the triangular prism shown in Fig. 19.20.

Fig. 19.20

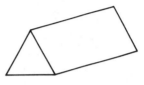

One net is sketched in Fig. 19.21 and it consists of three rectangles representing the base and the two sides together with two triangles representing the two ends.

Fig. 19.21

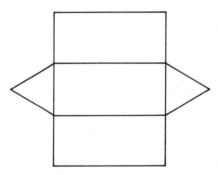

Nets of circular objects like cylinders and cones can also be drawn. Fig. 19.22 shows the net of the curved part of a cylinder. It is a rectangle whose length is equal to the circumference of the base of the cylinder and whose breadth is equal to its height.

Fig. 19.22

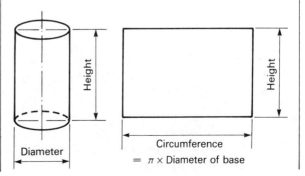

The net of a cone is shown in Fig. 19.23. It is a sector of a circle whose arms are equal to the slant height of the cone. The length of the arc is equal to the circumference of the base of the cone.

Fig. 19.23

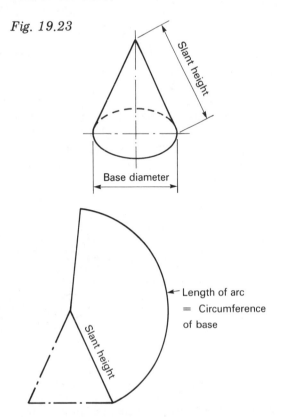

Surface Areas of Solids

We often need to find the **surface area** of solid shapes such as cubes and cuboids. The surface area of the solid can be found by drawing its net and finding the area of this net.

Example 6

A cuboid is 8 cm long, 5 cm wide and 4 cm high. Draw its net and hence find its surface area.

    The net is shown in Fig. 19.24. The total surface area is found by adding together the areas of the six rectangles making up the net.

Fig. 19.24

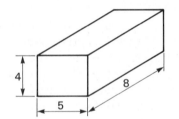

All dimensions in centimetres

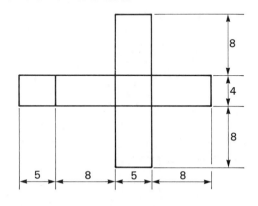

Total surface area $= 2 \times 5 \times 4 + 2 \times 4 \times 8$
$$+ 2 \times 5 \times 8$$
$$= 40 + 64 + 80$$
$$= 184 \, \text{cm}^2$$

We could write instead:

Total surface area $=$ Perimeter of end
$$\times \text{Length}$$
$$+ \text{Area of end}$$

Example 7

Calculate the total surface area of a cuboid which is 8 cm long, 3 cm wide and 2 cm high (Fig. 19.25).

Fig. 19.25

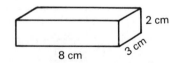

Perimeter of end $= 3 + 2 + 3 + 2$

$\qquad = 10\,\text{cm}$

Total surface area $= 10 \times 8 + 2 \times 2 \times 3$

$\qquad\qquad = 80 + 12$

$\qquad\qquad = 92\,\text{cm}^2$

Exercise 19.3

Draw nets for:

1. a cuboid

2. a cylinder

3. a triangular prism

4. a cone

5. a pyramid with a square base.

6. Fig. 19.26 shows the nets of various solid shapes. Name the solids.

Fig. 19.26

(a)

(b)

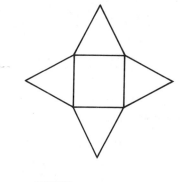

(c)

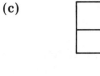

7. Which of the following diagrams (Fig. 19.27) represent the net of a cuboid?

Fig. 19.27

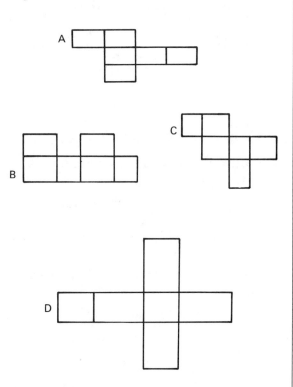

Work out the total surface areas of the cuboids shown in Fig. 19.28.

Fig. 19.28

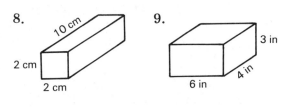

8.

9.

10.

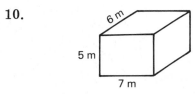

Units of Volume

The **volume** of a solid object is measured by seeing how many cubic units it contains.

Fig. 19.29 shows a unit cube whose edge is 1 cm long.

Fig. 19.29

Its volume is 1 cubic centimetre (cm³).

In the metric system the following units of volume are used:

cubic millimetres   mm³

cubic centimetres   cm³

cubic metres   m³

In the imperial system the following are used:

cubic inches   in³

cubic feet   ft³

cubic yards   yd³

Volume of a cuboid

Fig. 19.30 shows a cuboid. Three layers of 1 cm cubes fit into the shape. Each layer consists of 5 × 4 cubes. Therefore the total number of 1 cm cubes fitted in the shape is 5 × 4 × 3 = 60. The cuboid has a volume of 60 cubic centimetres.

Fig. 19.30

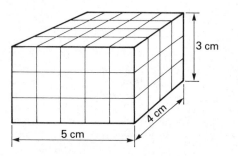

To find the volume of the cuboid we have multiplied its length by its breadth by its height. That is

Volume of cuboid = Length × Breadth
× Height

This formula is true for any cuboid even when its length, breadth and height are not whole numbers.

Example 8

Find the volume of a small tank which is 15.3 inches long, 9.8 inches wide and 5.6 inches high.

Volume of tank = 15.3 × 9.8 × 5.6 in³

= 840 in³ (to the
nearest in³)

Many shapes can be divided up into cuboids so making it easy to find their volumes.

Example 9

Find the volume of the solid shown in Fig. 19.31.

Fig. 19.31

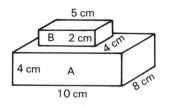

The solid can be split up into the two cuboids A and B.

Volume of A = 10 × 8 × 4 cm³

= 320 cm³

Volume of B = 5 × 4 × 2 cm³

= 40 cm³

Volume of solid = (320 + 40) cm³

= 360 cm³

Exercise 19.4

Calculate the volumes of the solid shapes shown in Fig. 19.32.

Fig. 19.32

1.

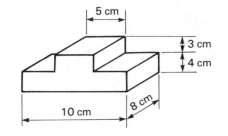

2.

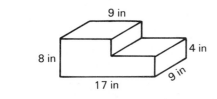

3.

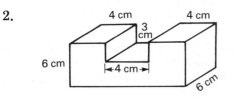

4.

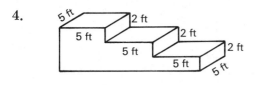

5.

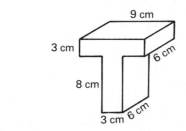

Similar Solids

Similar solids have exactly the same shape. That is the larger one must be longer, wider and taller in the same proportion.

Fig. 19.33 shows two child's building bricks.

Looking at the diagram you might say that the larger brick is three times bigger than the smaller brick. Indeed all the lengths are three times bigger but 9 small bricks could be placed on top of the large brick. Also it would take 27 small bricks to make a cube as big as the larger brick.

Fig. 19.33

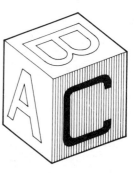

Notice carefully that:

The length is 3 times bigger.

The surface area is $3 \times 3 = 9$ times bigger.

The volume and weight (if they are both made from the same material) are $3 \times 3 \times 3 = 27$ times bigger.

From this example we can say:

(1) The surface areas of similar solids vary as the square of their lengths, widths and heights.

(2) The volumes (and weights) of similar solids vary as the cube of their lengths, widths and heights.

Example 10

Fig. 19.34 shows two tins of beans. Tin A is 7 cm high and has a diameter of 5 cm. Tin B is 14 cm tall and has a diameter of 10 cm.

(a) Are the two tins similar in shape?

(b) If tin A contains 140 grams of beans, what weight of beans are contained in tin B?

(c) The area of metal needed to make tin A is 140 square centimetres. Calculate the area of material needed to make tin B.

Fig. 19.34

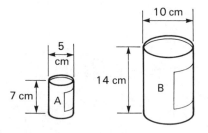

(a) For the two tins to be similar

$$\frac{\text{Diameter of tin B}}{\text{Diameter of tin A}} = \frac{\text{Height of tin B}}{\text{Height of tin A}}$$

Because $\dfrac{10}{5} = \dfrac{14}{7} = \dfrac{2}{1}$

the two tins are similar in shape.

(b) The contents of tin B will be $2 \times 2 \times 2 = 8$ times more than tin A.

Contents of tin B $= 8 \times 140$ grams

$= 1120$ grams

(c) The surface area of tin B will be $2 \times 2 = 4$ times more than that for tin A.

Surface area of tin B $= 4 \times 140 \, \text{cm}^2$

$= 560 \, \text{cm}^2$

The area of metal needed to make tin B is 440 square centimetres.

Exercise 19.5

1. A steel ball-bearing has a diameter of 1 inch and weighs 20 grams. What will be the weight of a steel ball-bearing having a diameter of 2 inches?

2. A cylindrical can has a radius of 7 cm and a height of 8 cm. It holds 1.30 litres of evaporated milk. A smaller cylindrical can has a diameter of 3.5 cm and a height of 4 cm. How much evaporated milk will the smaller can hold?

3. A bottle of paper glue holds 140 millilitres. A smaller bottle is similar in shape but is only half as high. How much glue will it contain?

4. Fig. 19.35 shows two ornamental arches which are similar in shape. The area of arch A is 4 square metres. What is the area of arch B?

Fig. 19.35

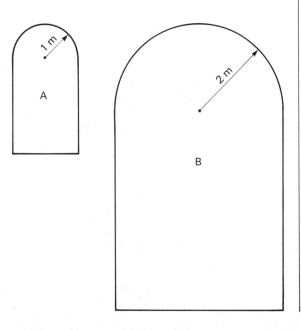

5. A rectangular box is 18 in long, 12 in wide and 6 in high:

   (a) Calculate, in cubic inches, the volume of the box.

   (b) The box is filled with soup mix which is packed in rectangular boxes 6 in long, 4 in wide and 2 in high. How many packs of soup mix are needed to fill the box?

Miscellaneous Exercise 19

Section A

1. Beef cubes are 2 cm × 2 cm × 2 cm. They are packed in cubic boxes which are 8 cm × 8 cm × 8 cm. How many beef cubes are needed to fill a box?

2. A thin cardboard sleeve is shown in Fig. 19.36:

   (a) Draw a net of the sleeve.

   (b) Work out the area of cardboard needed to make the sleeve.

Fig. 19.36

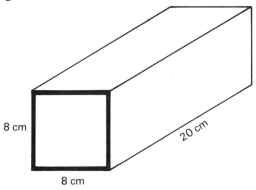

3. Draw in good proportion a pyramid with a square base of 5 cm side and a height of 6 cm.

4. (a) How many faces has a cube?

   (b) How many edges has it?

5. Work out the volume of a rectangular tin which is 8 cm long, 5 cm wide and 4 cm tall.

Section B

1. Fig. 19.37 shows two bottles which are similar in shape. Bottle A is 5 times as tall as bottle B and holds 125 litres of fluid. How many litres of fluid does bottle B hold?

Fig. 19.37

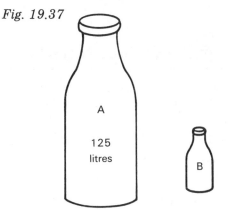

2. Fig. 19.38 shows a rectangular block 10 cm long with a square cross-section. Work out:

   (a) the total surface area of the block

   (b) the volume of the block.

Fig. 19.38

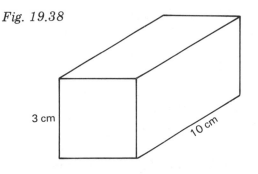

3. Name the solid shapes which can be made from the nets shown in Fig. 19.39.

Fig. 19.39

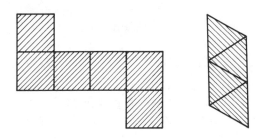

4. A cuboid 2 cm high has a volume of 24 cm³. Calculate the volume of a similar cuboid which is 4 cm high.

5. Draw and label the dimensions on the net of a cylinder which has a diameter of 7 cm and a height of 5 cm. (Take $\pi = \frac{22}{7}$.)

Multi-Choice Questions 19

Fig. 19.40 represents a solid rectangular block. Use this diagram to answer questions 1, 2 and 3.

1. What is the total length, in centimetres, of the edges of the block?

   A 13   B 16   C 36   D 52

2. What is the volume, in cubic centimetres, of the block?

   A 16   B 20   C 52   D 80

3. What is the total surface area, in square centimetres, of the block?

   A 16   B 80   C 96   D 112

Fig. 19.40

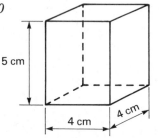

5 cm

4 cm   4 cm

4. What is the surface area of a cube having an edge of 6 cm?

   A 6 cm²   B 36 cm²
   C 144 cm²   D 216 cm²

5. A cuboid is 13 cm long, 5 cm wide and 2 cm high. Its volume in cubic centimetres is

   A 20   B 65   C 80   D 130

6. A pyramid has a square base. How many faces does it possess?

   A 4   B 5   C 6   D 7

7. How many edges does a triangular prism possess?

   A 3   B 4   C 6   D 9

8. What is the volume of the solid figure shown in Fig. 19.41?

   A 320 cm³   B 1286 cm³
   C 2560 cm³   D 2880 cm³

Fig. 19.41

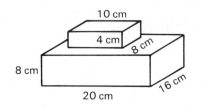

10 cm

4 cm

8 cm

8 cm

20 cm

16 cm

Mental Test 19

1. What is the volume of a cube with edges of length 2 cm?

2. Calculate the volume of a cuboid with a length of 5 cm, a width of 3 cm and a height of 2 cm.

3. What is the solid figure shown in Fig. 19.42 called?

Fig. 19.42

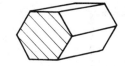

4. Calculate the volume of a box with a length of 4 cm, a width of 3 cm and a height of 5 cm.

5. How many edges does a square pyramid possess?

6. How many faces does a cuboid possess?

7. A cube has an edge 3 m long. What is its volume?

8. How many faces does an octagonal prism possess?

Maps, Bearings and Scale Drawings

Scales on Maps and Drawings

When drawings of large objects such as houses, ships and aeroplanes are to be made they are usually drawn **to scale**, for example 1 cm = 10 m. This means that a distance of 10 metres on the ground would be represented by 1 centimetre on the drawing.

On a scale drawing every measurement is in proportion to the real measurement. By scaling the drawing the real measurements of the object can be obtained.

On maps, scales are often stated as a ratio, for example, 1 : 1 000 000. This means that 1 centimetre on the map represents 1 000 000 centimetres (i.e. 10 000 metres or 10 kilometres) on the ground.

Example 1

(a) A map is drawn to a scale of 1 cm = 5 km. Express this scale as a ratio.

$$1 \text{ cm} = 5 \text{ km}$$
$$= 5 \times 1000 \text{ m}$$
$$= 5 \times 1000 \times 100 \text{ cm}$$
$$= 500\,000 \text{ cm}$$

When the units are the same on both sides of an equation they can be omitted and the scale is then 1 : 500 000.

(b) A road map is drawn to a scale of 1 : 1 000 000. Measured on the map the distance between Rheims and Luxembourg is 16 cm. What is the actual distance, in kilometres, between these two places?

The map scale is:

$$1 \text{ cm} = 1\,000\,000 \text{ cm}$$
$$= 1\,000\,000 \div 100 \text{ m}$$
$$= 10\,000 \text{ m}$$
$$= 10 \text{ km}$$

So 16 cm represents
$$16 \times 10 \text{ km} = 160 \text{ km}$$

The actual distance between Rheims and Luxembourg is 160 km.

(c) The drawing of a house is made to a scale of 1 inch = 20 feet. On the drawing one of the bedrooms measures 1.5 in by 0.8 in. What are the actual dimensions of the bedroom?

The drawing scale is:

$$1 \text{ in} = 20 \text{ ft}$$

So 1.5 in represents 1.5 × 20 ft = 30 ft

and 0.8 in represents 0.8 × 20 ft = 16 ft.

The actual size of the bedroom is 30 ft by 16 ft.

Exercise 20.1

1. A road map is drawn to a scale of 16 miles to 1 inch. On the map the distance between Coventry and Leicester is 1.4 in. Work out the actual distance, to the nearest mile, between the two places.

2. A road map of Spain is drawn to a scale of 1 : 1 000 000. On the map the distance between Barcelona and Zaragoza measures 23 cm. To the nearest kilometre, how far is it between Barcelona and Zaragoza?

3. The drawing of a house is made to a scale of 1 cm = 2 m. What is this scale as a ratio?

4. A map is made to a scale of 1 : 1 250 000. What distance, to the nearest mile, does 1 inch on the map represent?

5. The drawing of an office block is made to a scale of 1 cm = 5 m. On the drawing one of the offices measures 8 cm by 7 cm. Work out the actual size of the office.

6. Fig. 20.1 shows a map of a small island:

   (a) Measure the distance between Markton and Johnstown and so estimate the actual distance between the two places.

   (b) The map shows a road connecting Bridges and St. Thomas via Johnstown.

   (i) What is the length of the road, in centimetres, measured on the map?

   (ii) How far is it, by road, between Bridges and St. Thomas?

Fig. 20.1

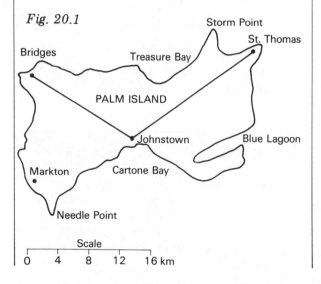

7. Fig. 20.2 shows part of a map of Devon and Cornwall drawn to a scale of 1 inch = 40 miles. Use the map to estimate the distances between

   (a) Plymouth and St. Austell

   (b) Land's End and Newquay

   (c) Exeter and Plymouth.

Fig. 20.2

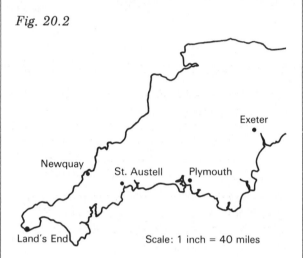

8. A map of Sweden is drawn to a scale of 1 : 1 500 000:

   (a) How many kilometres is represented by 1 cm on the map?

   (b) Measured on the map the distance between Stockholm and Gavle is 11 cm. To the nearest kilometre, how far is it between the two places?

Plans of Houses and Buildings

Architects use scale drawings when making a plan of a house or a building.

Example 2

Fig. 20.3 shows the plan of part of a school, drawn to a scale 1 : 500.

(a) How many windows are shown?

(b) How many doors are shown?

(c) Measure the drawing and write down the actual sizes of each of the two rooms.

(d) Work out the areas of each of the two rooms.

Fig. 20.3

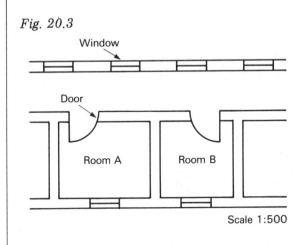

(a) There are 4 windows in the corridor and 1 window in room A and 1 window in room B, giving a total of 6 windows.

(b) There is 1 door in room A and 1 door in room B giving a total of 2 doors.

(c) Room A measures 2.5 cm by 2 cm.

2.5 cm represents

$$2.5 \times 500 \text{ cm} = 1250 \text{ cm}$$
$$= 12.5 \text{ m}$$

2 cm represents

$$2 \times 500 \text{ cm} = 1000 \text{ cm}$$
$$= 10 \text{ m}$$

Room A is 12.5 m by 10 m.

Room B measures 2 cm by 2 cm.

2 cm represents 10 m (*see above*) and so room B is 10 m by 10 m.

(d) Area of room A $= 12.5 \times 10$
$$= 125 \text{ m}^2$$

Area of room B $= 10 \times 10$
$$= 100 \text{ m}^2$$

Exercise 20.2

1. The plan of a building is drawn to a scale of 1 : 250. On the plan a rectangular room measures 6 cm by 5 cm:

    (a) Work out the actual dimensions of the room.

    (b) What is the area of the room?

2. The plan of a house and its garden is drawn to a scale of 1 cm = 2 m. The plot on which the house is to be built measures 20 cm by 8 cm. Calculate:

    (a) the actual dimensions of the plot

    (b) the area, in square metres, of the plot.

3. Fig. 20.4 shows the plan of the ground floor of a house. Taking measurements from the plan, copy and complete the following table:

Room	Length (m)	Width (m)	Area (m^2)
Lounge			
Dining room			
Kitchen			
Hall			
Bathroom			

Fig. 20.4

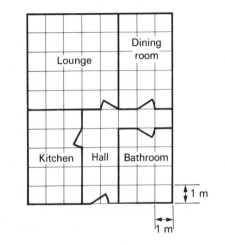

4. Fig. 20.5 shows the plan of a small flat.

    Measure the plan then copy and complete the table below:

Room	Length (m)	Width (m)	Area (m²)
Lounge			
Bedroom			
Hall			
Kitchen			
Bathroom			

Fig. 20.5

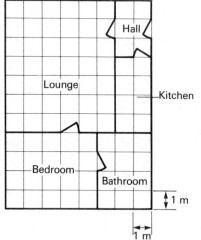

5. The diagram (Fig. 20.6) shows the plan of a garage which the householder wants building:

    (a) How many doors are there?

    (b) How many windows are there?

    (c) What is the length and width, in metres, of the garage?

    (d) What is the floor area of the garage in square metres?

Fig. 20.6

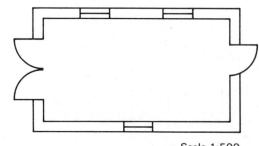

Scale 1:500

Drawing Perpendiculars

Perpendiculars may be drawn by using set-squares (Fig. 20.7) or a protractor (Fig. 20.8).

Fig. 20.7

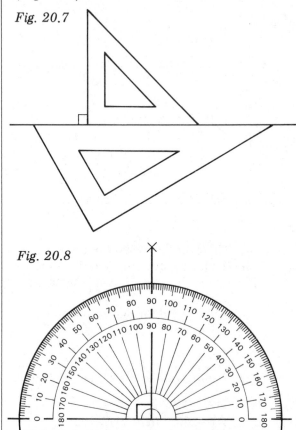

Fig. 20.8

Constructing Angles

Angles can be measured and constructed by using a protractor. The way of drawing an angle of 43° or 137° is shown in Fig. 20.9.

Fig. 20.9

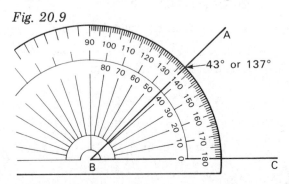

43° or 137°

When measuring an angle it is important to choose the correct number given on the protractor. Suppose that in the previous diagram we want to measure the angle ABC. The numbers on the protractor are $43°$ and $137°$. Because angle ABC is less than $90°$ we choose $43°$ rather than $137°$.

Exercise 20.3

1. Using set-squares draw lines perpendicular to the following straight lines:

    (a) AB 5 cm long at A

    (b) XY 7 cm long at Y

    (c) PQ 4 in long at P

    (d) RS 9 cm long at S.

2. Draw a line AB 5 inches long. Using set-squares draw two lines perpendicular to AB, the one passing through point A and the other passing through the point B.

3. Using a protractor draw lines perpendicular to the following straight lines:

    (a) XY 6 cm long at X

    (b) AB 8 cm long at B

    (c) CD 5 inches long at D

    (d) MN 7 cm long at M.

4. Using a protractor draw the following angles:

    (a) $54°$ (b) $78°$ (c) $22°$

    (d) $126°$ (e) $154°$

5. Measure each of the angles shown in Fig. 20.10.

Fig. 20.10

(a)

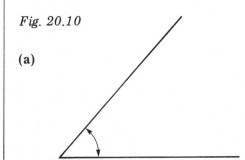

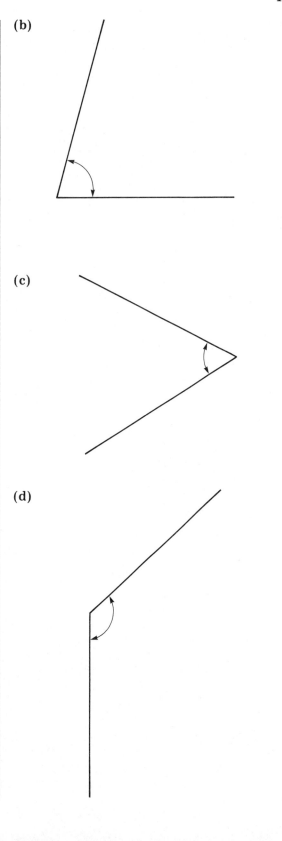

(e)

(f)

(g)

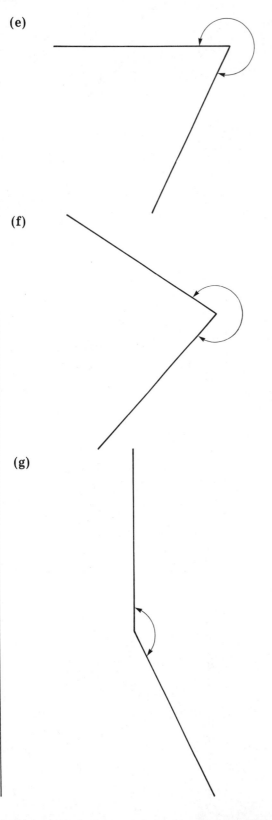

(h)

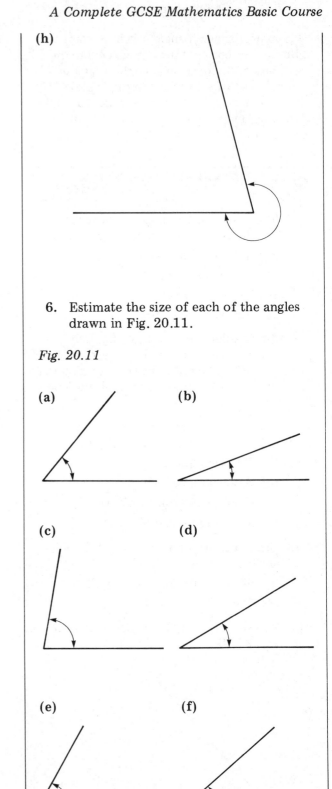

6. Estimate the size of each of the angles drawn in Fig. 20.11.

Fig. 20.11

(a)   (b)

(c)   (d)

(e)   (f)

Drawing Plane Shapes

Plane shapes such as rectangles and parallelograms can be drawn by using set-squares and protractors.

Example 3

Draw accurately a rectangle which is 6 cm long and 4 cm wide.

The drawing is made as follows (Fig. 20.12):

(1) Draw AB 6 cm long.

(2) Using a protractor or set-squares draw AX perpendicular to AB.

(3) Along AX mark off AD = 4 cm.

(4) Using set-squares draw BW parallel to AX and DY parallel to AB. C is the point where BW and DY cross.

The required rectangle is ABCD.

Fig. 20.12

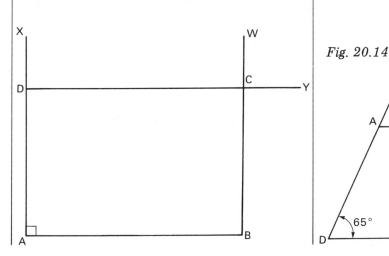

Example 4

Draw accurately the parallelogram shown in Fig. 20.13.

Fig. 20.13

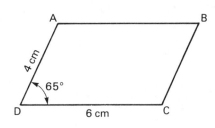

The construction is as follows (Fig. 20.14):

(1) Draw CD 6 cm long.

(2) Using a protractor draw the angle CDW = 65°.

(3) Along DW mark off DA = 4 cm.

(4) Using set-squares draw AX parallel to CD and CY parallel to AD. The point where AX and CY cross is the point B.

The required parallelogram is then ABCD.

Fig. 20.14

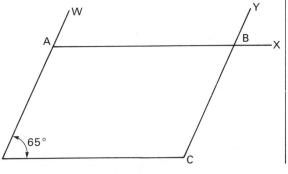

Example 5

Draw a circle whose diameter is 6 cm. In it draw a regular hexagon of side 3 cm.

The construction is as follows (Fig. 20.15):

(1) With the compasses set to the same radius as was used to draw the the circle, draw the arcs at B, C, D, E and F.

(2) Join A and B, B and C, C and D, etc. to complete the hexagon ABCDEF.

Fig. 20.15

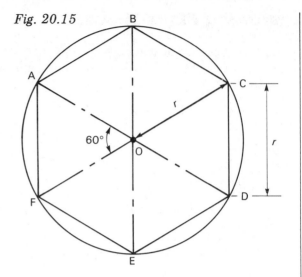

Exercise 20.4

1. Measure and write down the lengths of the lines shown in Fig. 20.16.

2. Given that the line AB is 4 cm long, estimate the lengths of the lines CD, EF, GH, KL, MN and PQ (Fig. 20.17).

Fig. 20.16

(a)

(b)

(c)

(d)

Fig. 20.17

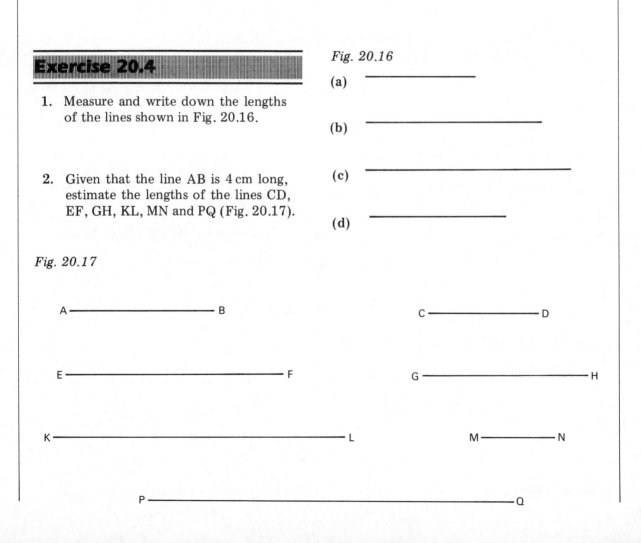

3. Draw accurately a rectangle which is 5.6 cm long and 3.8 cm wide.

4. Draw accurately a square whose sides are each 4.3 cm long.

5. Draw accurately the rhombus shown in Fig. 20.18.

Fig. 20.18

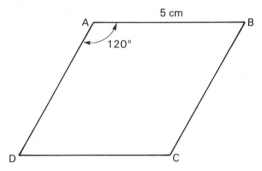

6. Fig. 20.19 shows the diagonals of a kite. Draw the kite accurately and measure the lengths of its sides.

Fig. 20.19

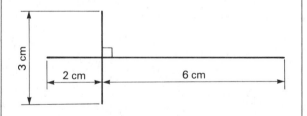

7. Draw the parallelogram shown in Fig. 20.20 accurately. Measure the diagonals AC and BD and write down their lengths.

Fig. 20.20

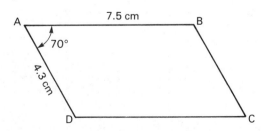

8. Draw a circle whose diameter is 5 cm. Inscribe in the circle a regular eight-sided figure.

9. Draw a regular hexagon which has sides 2.5 cm long.

10. Draw a circle whose diameter is 8 cm. Inscribe in it a regular pentagon (five-sided figure).

Compass Bearings

The four **cardinal** directions are North, South, East and West. However, the intermediate points north-east (NE), south-east (SE), south-west (SW) and north-west (NW) are often used as well (Fig. 20.21).

Fig. 20.21

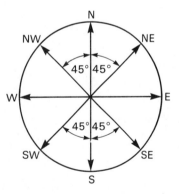

Directions between the cardinal points are called **bearings**.

One way of stating a compass bearing is to state the angle measured from the north–south line towards either west or east. A compass bearing of N20°E means an angle of 20° measured from N towards E as shown in Fig. 20.22.

Fig. 20.22

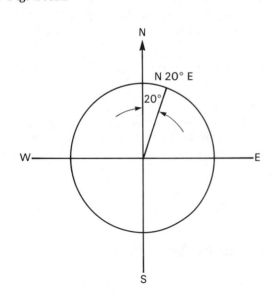

A bearing of S50°W means an angle of 50° measured from S towards W as shown in Fig. 20.23.

Fig. 20.23

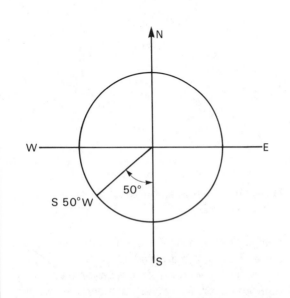

Note that bearings are always measured from N and S and never from E and W.

Exercise 20.5

Draw accurate diagrams to show each of the following compass bearings:

1. N40°W   2. N60°E   3. S45°E

4. S70°W   5. S50°E.

Using a protractor measure and write down the compass bearings shown in Fig. 20.24.

Fig. 20.24

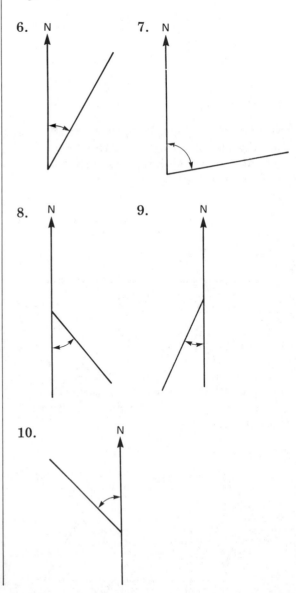

Three-digit Bearings

A second way of stating a bearing is to measure the angle from N in a clockwise direction. Three figures are always stated, N being 000°. 005° is written instead of 5° and 027° instead of 27°.

Some typical three-digit bearings are shown in Fig. 20.25.

Fig. 20.25

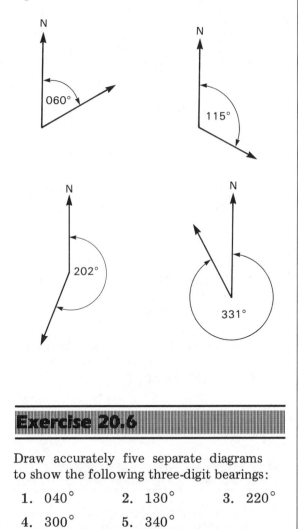

Exercise 20.6

Draw accurately five separate diagrams to show the following three-digit bearings:

1. 040°   2. 130°   3. 220°

4. 300°   5. 340°

Using a protractor measure clockwise each of the angles shown in Fig. 20.26. Then using three-digit bearings write down each of the directions.

Fig. 20.26

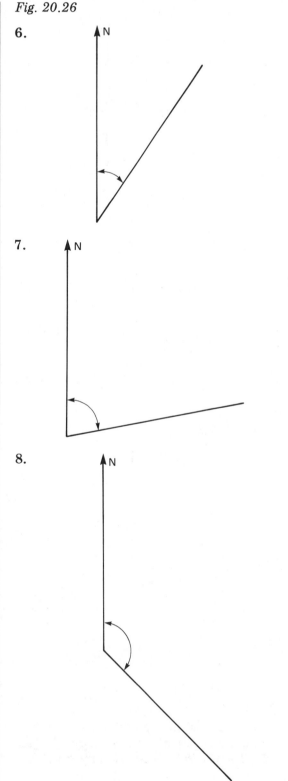

9.

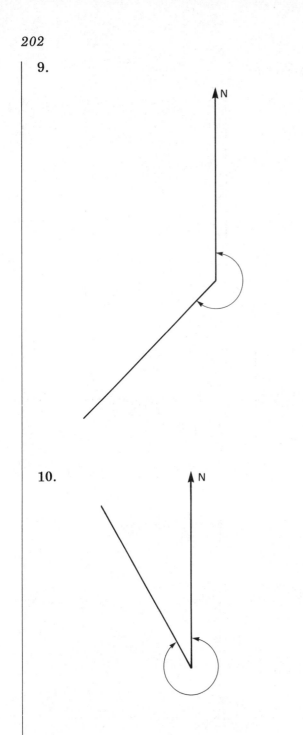

10.

Example 6

B is a point due E of a harbour A. C is a point on the coast which is 8 km due S of A. If the distance BC is 9 km, find the bearing of C from B.

Fig. 20.27

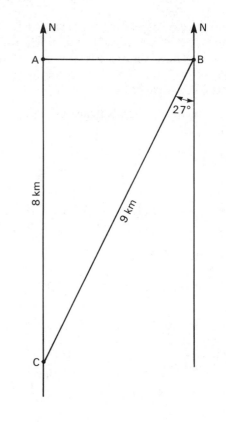

First choose a suitable scale to represent the distances AC and BC. In Fig. 20.27 a scale of 1 cm = 1 km has been chosen and a scale drawing made.

The bearing is found by using a protractor placed at B. It is S27°W or, as a three-digit bearing, 207°.

Exercise 20.7

1. Fig. 20.28 shows part of a map. Using three-digit numbers, write down the bearings of:

   (a) Beeson from Newtown

   (b) Dayton from Beeson.

Fig. 20.28

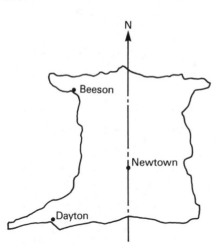

2. The bearing of A from B is 155°. What is the bearing of B from A?

3. A ship is on a bearing of 068° (N68°W) from a lighthouse. What is the bearing of the lighthouse from the ship?

4. B and C (Fig. 20.29) are both 100 km from A. C is on a bearing of 225° from B. Make a scale drawing of this information using a scale of 1 cm = 20 km. From your drawing find:

   (a) the bearing of A from B

   (b) the size of the angle ABC.

Fig. 20.29

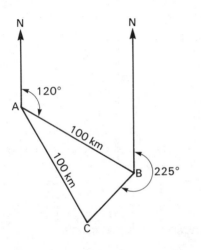

5. The map (Fig. 20.30) shows the positions of a crossroads, an electricity pylon, an oak tree, a church and a barn.

   (a) Which object is south-east of the crossroads?

   (b) What is the bearing of the barn from the crossroads?

   (c) What is the bearing of the church from the oak tree?

   Give the answers as three-digit bearings.

Fig. 20.30

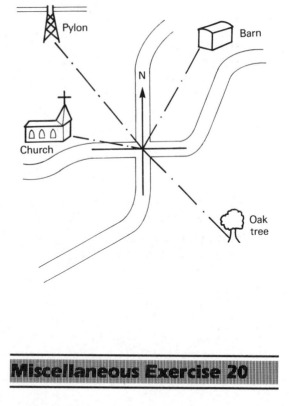

Miscellaneous Exercise 20

Section A

1. Write down three-digit bearings for the directions

   (a) south-west

   (b) north-west.

2. A boat sails from a port P to a point X (Fig. 20.31). It then sails to a second point Y. The diagram has been drawn to scale. Use it to find:

   (a) the actual distance from X to P

   (b) the three-digit bearing of X from P

   (c) the actual distance of Y from P

   (d) the three-digit bearing of Y from P

Fig. 20.31

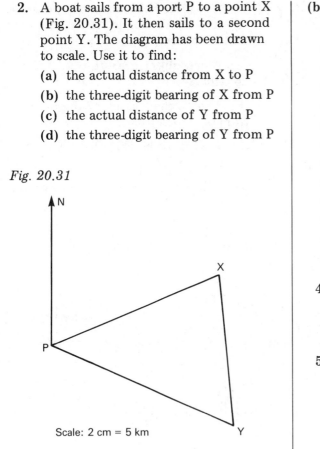

Scale: 2 cm = 5 km

3. (a) Copy the diagram and construct a square on the line AB (Fig. 20.32(a)).

   (b) Complete a rectangle using the lines WX and WY (Fig. 20.32(b)).

Fig. 20.32

(a)

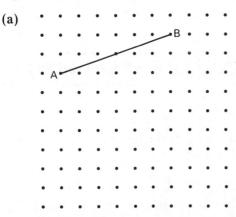

(b)

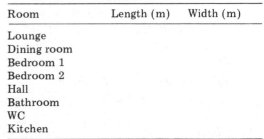

4. The scale of a map is 1 : 50 000. What distance on the ground does 6 cm on the map represent? Give your answer correct to the nearest kilometre.

5. Fig. 20.33 shows the plan of a bungalow. Measure the drawing then copy and complete the table below:

Room	Length (m)	Width (m)
Lounge		
Dining room		
Bedroom 1		
Bedroom 2		
Hall		
Bathroom		
WC		
Kitchen		

Fig. 20.33

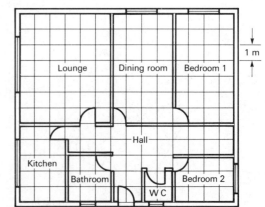

Section B

1. Draw accurately the parallelogram shown in Fig. 20.34.

Fig. 20.34

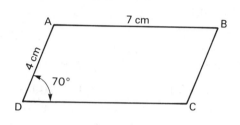

2. Fig. 20.35 shows part of a map drawn to a scale of 1 cm = 16 miles. Use this map to answer the following questions:

Fig. 20.35

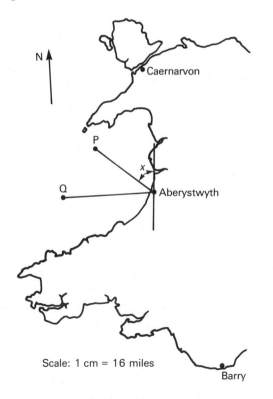

Scale: 1 cm = 16 miles

(a) An aeroplane flies from Barry to Caernarvon. Write down the distance between the two places.

(b) A fishing boat sails from Aberystwyth to a point P at sea. The bearing of P from Aberystwyth is 298°. Write down the size of the angle marked x on the map.

(c) The boat sails from the point P to a point Q which is on a bearing due west of Aberystwyth. Find the distance of P from Q.

3. Draw Fig. 20.36 accurately and measure the lengths of BW and AX.

Fig. 20.36

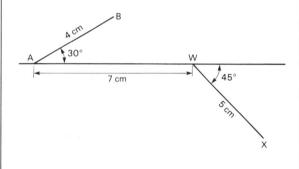

4. Draw the shape shown in Fig. 20.37 which represents a plot of land accurately. Use a scale of 1 cm = 50 ft. Use your scale drawing to find the actual length of BC.

Fig. 20.37

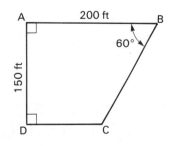

5. A ship sails from P on a bearing of 065° for a distance of 10 miles to a point Q. It then sails south for a distance of 10 miles to a point R.

   (a) Using a scale of 1 cm = 2 miles make a scale drawing to represent the ship's journey.

   (b) Use your diagram to find:

   (i) the bearing of R from P

   (ii) the distance of R from P.

Multi-Choice Questions 20

1. In Fig. 20.38, what is the bearing of P from Q?

   A 025°   B 065°
   C 245°   D 295°

Fig. 20.38

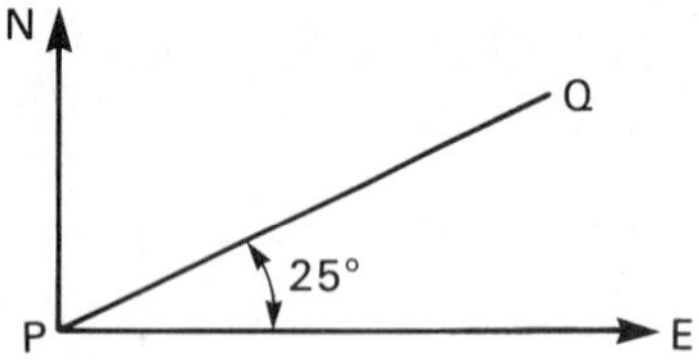

Using the information given in Fig. 20.39, answer questions 2, 3, 4 and 5.

Fig. 20.39

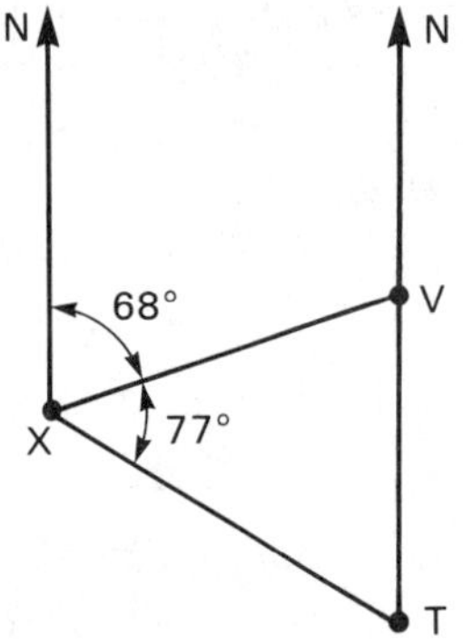

2. What is the bearing of X from V?

   A N77°E (077°)   B S68°E (112°)
   C S68°W (248°)   D N68°W (283°)

3. What is the bearing of V from T?

   A 000°   B 077°   C 180°   D 283°

4. The bearing of T from X is

   A 035°   B 077°   C 145°   D 215°

5. Determine the bearing of X from T.

   A 035°   B 145°   C 283°   D 325°

6. The scale of a map is 1 inch = 5 miles. As a ratio this scale is

   A 1:50   B 1:50 000
   C 1:316 800   D 1:500 000

7. Town A is on a bearing of 040° from town B. What is the bearing of town B from town A?

   A 130°   B 140°   C 220°   D 310°

8. The scale of a map is 1:1 000 000. The distance between two towns, measured on the map, is 8 cm. What is the actual distance between the two towns?

   A 10 km   B 80 km
   C 100 km   D 800 km

Mental Test 20

1. The scale of a map is 1:100. Find, in metres, the distances represented by

   (a) 1 cm on the map

   (b) 4 cm on the map

   (c) 0.5 cm on the map.

2. The scale of a map is 1 cm = 50 km. Find, in kilometres, the distances represented by

   (a) 3 cm on the map

   (b) 0.2 cm on the map.

3. The scale of a map is 1 cm = 5 km. What lengths on the map represent

   (a) 10 km

   (b) 30 km?

4. Fig. 20.40 shows a map of part of Kent. The scale is 1 cm = 10 km. By measuring the map, find the lengths, in centimetres, as the crow flies, between

   (a) Maidstone and Chatham

(b) High Haldon and Charing

(c) Canterbury and Margate

(d) Chatham and Faversham.

Using your measurements, work out the actual distances between the towns.

Fig. 20.40

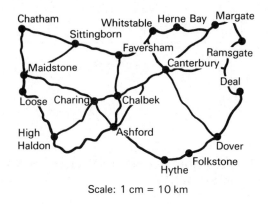

Scale: 1 cm = 10 km

5. Write down three-figure bearings for the directions

(a) south-east

(b) north-east.

6. Using a protractor and measuring the angle clockwise, express as three-figure bearings the directions shown in Fig. 20.41.

Fig. 20.41

(a)

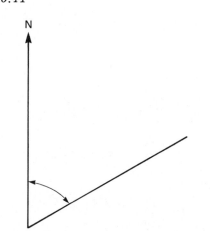

(b)

(c)

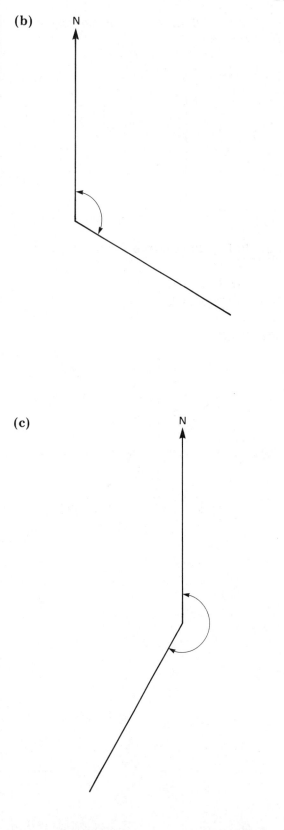

7. In Fig. 20.42, what is the bearing of A from B?

Fig. 20.42

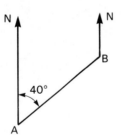

9. The rectangle shown in Fig. 20.44 is drawn to scale. Estimate its length.

Fig. 20.44

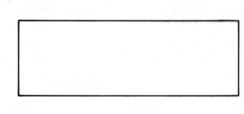

8. In Fig. 20.43, write down the bearing of
   (a) B from A
   (b) C from A.

Fig. 20.43

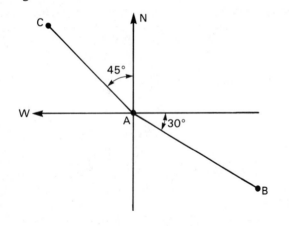

10. Estimate the size of the angle shown in Fig. 20.45.

Fig. 20.45

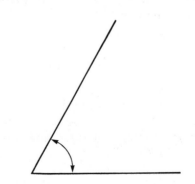

Tables, Charts and Diagrams

Tables

Table 21.1 gives the high and low water tide times for places in Cornwall in 1985. The table is used in the way shown in Example 1.

Example 1

Using Table 21.1, find:

(a) The high and low water times predicted for Newquay on Wednesday, 10th July.

(b) Using the adjustments to the tide times, find the high and low water times for Penzance on Saturday, 24th August.

   (a) Directly from the table:

High water times are 1032 and 2300 hours.

Low water times are 0435 and 1656 hours.

(b) From the table:

High water times at Newquay are 1103 and 2342 hours.

Using the adjustments to tide times:

High water times at Penzance are $1103 + 17 = 1120$ hours and $2342 + 17 = 2359$ hours.

Low water times at Newquay are 0456 and 1735 hours.

Using the adjustments to tide times:

Low water times at Penzance are $0456 + 09 = 0505$ hours and $1735 + 09 = 1744$ hours.

TABLE 21.1

HIGH AND LOW WATER TIDE TIMES 1985
For Newquay   See adjustments below for other places

JUNE

	HIGH WATER		LOW WATER	
	AM	PM	AM	PM
1 Sa	3 07	15 34	9 40	22 00
2 Su	3 57	16 23	10 26	22 47
3 M	4 47	17 09	11 13	23 34
4 Tu	5 35	17 57	11 58	....
5 W	6 21	18 45	0 22	12 44
6 Th	7 09	19 31	1 09	13 29
7 F	7 56	20 18	1 59	14 15
8 Sa	8 44	21 09	2 47	15 03
9 Su	9 34	22 03	3 39	15 55
10 M	10 32	23 03	4 37	16 59
11 Tu	11 38	....	5 42	18 10
12 W	0 10	12 47	6 48	19 18
13 Th	1 12	13 45	7 47	20 14
14 F	2 05	14 34	8 35	20 59
15 Sa	2 52	15 15	9 18	21 38
16 Su	3 34	15 54	9 56	22 16
17 M	4 12	16 30	10 31	22 52
18 Tu	4 49	17 05	11 07	23 27
19 W	5 26	17 42	11 43	....
20 Th	6 04	18 20	0 05	12 21
21 F	6 45	19 00	0 45	13 00
22 Sa	7 28	19 45	1 27	13 44
23 Su	8 12	20 33	2 14	14 29
24 M	9 04	21 27	3 05	15 24
25 Tu	10 02	22 28	4 01	16 23
26 W	11 04	23 32	5 05	17 29
27 Th	....	12 12	6 12	18 38
28 F	0 40	13 17	7 20	19 44
29 Sa	1 44	14 18	8 22	20 46
30 Su	2 44	15 14	9 18	21 42

JULY

	HIGH WATER		LOW WATER	
	AM	PM	AM	PM
1 M	3 41	16 08	10 10	22 36
2 Tu	4 35	16 59	11 00	23 24
3 W	5 25	17 47	11 47	....
4 Th	6 13	18 32	0 14	12 30
5 F	6 57	19 16	0 58	13 15
6 Sa	7 38	19 58	1 42	13 56
7 Su	8 18	20 40	2 24	14 36
8 M	8 59	21 23	3 05	15 17
9 Tu	9 42	22 09	3 47	16 03
10 W	10 32	23 00	4 35	16 56
11 Th	11 31	....	5 32	17 59
12 F	0 03	12 40	6 37	19 10
13 Sa	1 11	13 45	7 42	20 14
14 Su	2 12	14 41	8 40	21 08
15 M	3 05	15 28	9 28	21 53
16 Tu	3 51	16 10	10 10	22 34
17 W	4 33	16 49	10 49	23 13
18 Th	5 12	17 29	11 29	23 54
19 F	5 53	18 09	....	12 08
20 Sa	6 34	18 52	0 35	12 50
21 Su	7 16	19 35	1 19	13 33
22 M	8 00	20 21	2 03	14 17
23 Tu	8 47	21 08	2 48	15 05
24 W	9 35	22 00	3 36	15 55
25 Th	10 31	22 57	4 29	16 52
26 F	11 35	....	5 32	18 02
27 Sa	0 06	12 47	6 47	19 21
28 Su	1 21	14 00	8 02	20 35
29 M	2 34	15 07	9 08	21 37
30 Tu	3 38	16 04	10 03	22 31
31 W	4 31	16 54	10 53	23 19

AUGUST

	HIGH WATER		LOW WATER	
	AM	PM	AM	PM
1 Th	5 18	17 36	11 36	....
2 F	5 59	18 17	0 01	12 16
3 Sa	6 37	18 55	0 41	12 54
4 Su	7 13	19 31	1 17	13 29
5 M	7 46	20 05	1 51	14 03
6 Tu	8 21	20 39	2 25	14 36
7 W	8 54	21 16	2 58	15 10
8 Th	9 34	21 59	3 36	15 50
9 F	10 20	22 52	4 20	16 45
10 Sa	11 24	....	5 23	18 03
11 Su	0 03	12 49	6 47	19 31
12 M	1 32	14 11	8 04	20 42
13 Tu	2 42	15 07	9 05	21 34
14 W	3 33	15 52	9 51	22 18
15 Th	4 17	16 33	10 33	22 59
16 F	4 56	17 12	11 13	23 38
17 Sa	5 36	17 53	11 52	....
18 Su	6 16	18 32	0 19	12 32
19 M	6 56	19 14	1 00	13 13
20 Tu	7 38	19 57	1 41	13 56
21 W	8 21	20 40	2 22	14 37
22 Th	9 05	21 29	3 06	15 25
23 F	9 56	22 25	3 55	16 20
24 Sa	11 03	23 42	4 56	17 35
25 Su	....	12 28	6 23	19 14
26 M	1 14	13 58	8 00	20 36
27 Tu	2 39	15 07	9 08	21 38
28 W	3 38	15 58	10 01	22 26
29 Th	4 23	16 41	10 43	23 07
30 F	5 02	17 19	11 20	23 41
31 Sa	5 38	17 54	11 54	....

Approximate adjustments to Tide Times

	H. Water Time (mins)	L. Water Time (mins)
Bude	+25	+18
Falmouth	+05	+13
Fowey	+23	+15
Looe	+25	+18
Mevagissey	+23	+01
Padstow	+13	+13
Penzance	+17	+09
Porthleven	+20	+12
St. Ives	+03	+06

TABLE 21.2

CHEPSTOW · CALDICOT · NEWPORT · CARDIFF

via M4 Motorway **Service X70**

Mondays to Saturdays (except Bank Holidays)

									M-F	S	#			
CHEPSTOW (Bus Station)	0645	....	0810	0930	1030	1130	1230	1330	1430	1430	1530	....	1630	1735
Portskewett (Church)	0654	....	0821	0941	1041	1141	1241	1341	1441	1441	1541	....	1641	1746
Caldicot (Mitel)	0655	....	0822	0942	1042	1142	1242	1342	1442	1442	1542	....	1642	1747
CALDICOT (Cross)	0658	....	0825	0945	1045	1145	1245	1345	1445	1445	1545	....	1645	1750
Caldicot (Longfellow Court)	0703	....	0830	0950	1050	1150	1250	1350	1450	1450	1550	....	1650	1755
Rogiet (Pool)	0709	....	0836	0956	1056	1156	1256	1356	1456	1456	1556	....	1656	1801
Magor (Rear of Golden Lion)	0715	....	0842	1002	1102	1202	1302	1402	1502	1502	1602	....	1702	1807
NON-STOP VIA M4	↓	....	↓	↓	↓	↓	↓	↓	↓	↓	↓	....	↓	↓
NEWPORT (Bus Station) ....arr	0730	....	0857	1017	1117	1217	1317	1417	1517	1517	1617	....	1717	1822
NEWPORT (Bus Station) ....dep	....	0825	0900	1020	1120	1220	1320	1420	....	1520	1620	1625	1720	....
VIA M4	....	↓	↓	↓	↓	↓	↓	↓	....	↓	↓	↓	↓	....
CARDIFF (Bus Station)	....	0855	0930	1050	1150	1250	1350	1450	....	1550	1650	1655	1750	....

								M-F	S				
CARDIFF (Bus Station)	0900	1000	1100	1200	1300	1400	1500	....	1600	1705	1800	....	....
VIA M4	↓	↓	↓	↓	↓	↓	↓	....	↓	↓	↓	....	....
NEWPORT (Bus Station) ....arr	0930	1030	1130	1230	1330	1430	1530	....	1630	1735	1830	....	....
NEWPORT (Bus Station) ....dep	0935	1035	1135	1235	1335	1435	1535	1635	1635	1740	1835	....	....
NON-STOP VIA M4	↓	↓	↓	↓	↓	↓	↓	↓	↓	↓	↓	....	....
Magor (Rear of Golden Lion)	0950	1050	1150	1250	1350	1450	1550	1650	1650	1755	1850	....	....
Rogiet (Pool)	0956	1056	1156	1256	1356	1456	1556	1656	1656	1801	1856	....	....
Caldicot (Longfellow Court)	1002	1102	1202	1302	1402	1502	1602	1702	1702	1807	1902	....	....
CALDICOT (Cross)	1007	1107	1207	1307	1407	1507	1607	1707	1707	1812	1907	....	....
Caldicot (Mitel)	1010	1110	1210	1310	1410	1510	1610	1710	1710	1815	1910	....	....
Portskewett (Church)	1011	1111	1211	1311	1411	1511	1611	1711	1711	1816	1911	....	....
CHEPSTOW (Bus Station)	1022	1122	1222	1322	1422	1522	1622	1722	1722	1827	1922	....	....

CODE: **M-F** – Mondays to Fridays
S – Saturdays
– Saturdays and Gwent School Holidays

Table 21.2 shows a typical bus timetable. The way in which it is used is shown in Example 2.

Example 2

Use the bus timetable (Table 21.2) to answer the following questions:

(a) At what time does the 1130 bus from Chepstow arrive at Newport?

(b) How long does the journey take?

(c) How long does the bus wait at Newport before going on to Cardiff?

    (a) From the timetable, the 1130 bus from Chepstow arrives at Newport at 1217.

    (b) The time taken for the journey is $1217 - 1130 = 47 \text{ min}$.

    (c) From the timetable the bus arrives at Newport at 1217 and leaves at 1220. The bus waits 3 min at Newport.

Route maps are often associated with bus timetables. Fig. 21.1 shows a typical route map for the bus services of Gloucester.

Example 3

Use the route map (Fig. 21.1) to answer the following questions:

(a) What is the number of the bus service from Clarence Street to Caledonian Road?

(b) I wish to travel from Clarence Street to the Wheatway. Which bus service should I use?

(c) Write down the number of the service from the Bus Station to Brockworth.

    (a) bus service number 1

    (b) bus service number 3

    (c) bus service number 50.

Fig. 21.1

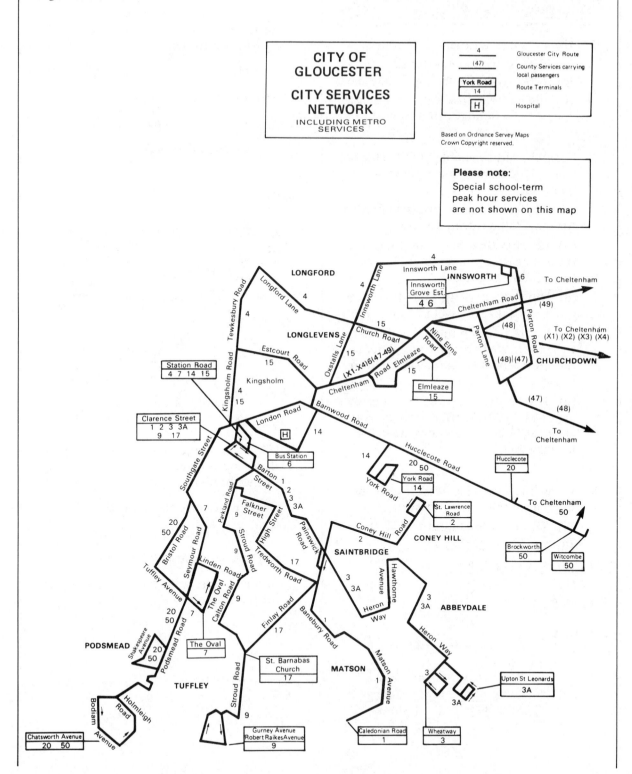

Exercise 21.1

1. Using Table 21.1 (page 209), find:

   (a) the high and low water tide times for Newquay on Thursday, 27th June

   (b) the high and low water tide times for Looe on Sunday, 18th August.

2. Using the bus timetable (Table 21.2 on page 210), answer the following questions:

   (a) What is the time of arrival at Cardiff for the 1330 bus from Chepstow?

   (b) At what time does the 0810 bus from Chepstow arrive at Caldicot (Longfellow Court)?

   (c) How long does the 0645 bus from Chepstow take to travel to Newport?

   (d) I want to catch the bus at Rogiet to arrive in Newport at 1420. What is the latest time that I should arrive at the bus stop?

   (e) The bus from Cardiff to Chepstow leaves at 1100. At what time does it arrive at Magor? How long does the journey take?

3. Using the map (Fig. 21.2), give the number of the bus services from

   (a) Thornton to Munson

   (b) Thornton to Newton

   (c) Townley to Patton.

Fig. 21.2

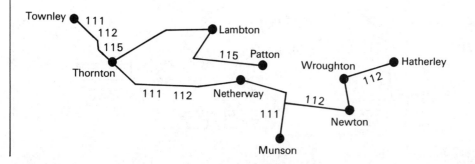

4. Fig. 21.3 shows part of an airway map. If the map shows the various airports in the correct order, write down the correct routes from

   (a) Montreal to Bangkok

   (b) London to Nairobi

   (c) Rome to Sydney

   (d) Chicago to Tokyo.

Fig. 21.3

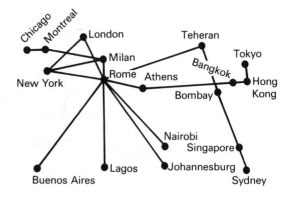

5. Fig. 21.4 shows a route map for a city. It shows the various places in correct order. Write down the routes from

   (a) City Centre to Churchdown

   (b) City Centre to Hardwicke

   (c) City Centre to Witcombe.

   In each case write down the number of the bus service.

Fig. 21.4

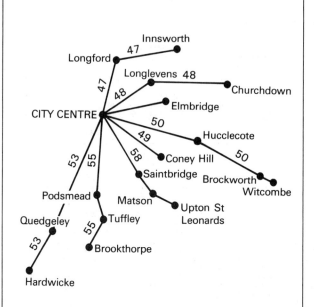

6. Fig. 21.5 shows a road map of Cleveland. The distances between the various towns and villages are given in miles.

(a) There are three possible routes from West Auckland to Houghton via Darlington:

(i) Write down the three routes.

(ii) Which is the shortest route?

(iii) Which is the longest route?

(b) There are extensive road repairs on the Thorpe to Billingham Road. Motorists are advised to use the Stockton, Middlesbrough Road. How much farther is this route?

(c) Name the two possible routes from Rushyford to Stockton. Write down the two routes and find the difference in length between the two.

Fig. 21.5

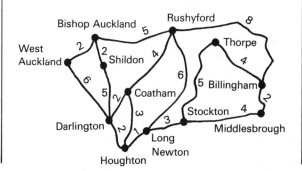

7. Using Fig. 21.6 and the associated bus timetables, answer the following questions:

(a) I want to get to Field Court by 1310 hours. Which bus should I catch from Gloucester Bus Station? In your answer give the departure time and the service number.

(b) On a Friday what is the time of the first bus to Dursley? What time does it arrive in Dursley? How long does the journey take?

(c) I want to travel to Stonehouse from Quedgeley. What is the service number of the bus I should use?

(d) How long does the 6.11 bus from Gloucester take to reach Sharpness? How much slower is the 5.15 bus?

(e) Is there a bus from Gloucester to Whitminster at 1415 hours on a Monday?

GLOUCESTER · WHITMINSTER via Quedgeley, Field Court, Hardwicke

Service 53: Gloucester Bus Station, Leisure Centre, St. Paul's Church, Bristol Road, Quedgeley, School Lane, Field Court Drive, Bristol Road, Hardwicke, Moreton Valence, Whitminster.

Timetable includes extracts from Services X1, X2 and 51 to show complete service, Gloucester—Whitminster.

Note: **Services X1** and **51** operate via The Cross between Gloucester (Bus Station) and Bristol Road.

    Service X2 operates via St. Paul's Church, Stroud Road, St. Barnabus Church and Cole Avenue between Gloucester (Bus station) and Bristol Road.

Mondays to Saturdays

Service No.	X1 NS	53 NS	X1 S	X1 NS	X1 S	53	53	X1	53	X1	53	X1	53	X1	51	X1	X1	X1 NS	X1 S	51 NS
GLOUCESTER, Bus Station	6.11	6.36‡	6.48	7.25	8.05	8.25	8.45	9.15	9.40	10.15	10.40	11.15	11.40	12.15	12.40	1.15	2.15	3.15	3.15	3.10
Quedgeley, Field Court, Holly Grove ..	†	†	†	7.40	†	8.40	†	9.30	9.55	10.30	10.55	11.30	11.55	12.30	12.55	1.30	2.30	3.30	3.30	†
Hardwicke, Morning Star	6.26	6.56	7.03	7.45	8.20	8.45	....	9.35	10.00	10.35	11.00	11.35	12.00	12.35	1.00	1.35	2.35	3.35	....	3.25
HARDWICKE, Springfield	....	....	....	....	....	....	9.00	....	10.01	....	11.01	....	12.01	....	....	....	....	3.36	....	....
WHITMINSTER, Hotel	6.36	7.06	7.13	7.55	8.30	8.55	....	9.45	....	10.45	....	11.45	....	12.45	....	1.45	2.45	....	3.45	....

GLOUCESTER · DURSLEY or SHARPNESS
(with connections at Dursley for Wotton-under-Edge)

Mondays to Saturdays

Service No.	X1 NS	53 NS	X1 S	X1 NS	X1 S	X1	X1	X1	X1	X1	X1	X1 NS	X1 S	X1	X1 F	X2 NFS	X1	X1	X1
GLOUCESTER, Bus Station, Bay D	6.11	6.36	6.48	7.25	8.05	9.15	10.15	11.15	12.15	1.15	2.15	3.15	3.15	3.30	4.00	4.15	5.15	5.40	8.40
Stroud Rd. St. Barnabas Church															4.22				
Field Court, Holly Grove	†	†	†	7.40	†	9.30	10.30	11.30	12.30	1.30	2.30	3.30	3.30	3.45	4.15	4.30	5.30	5.55	8.55
Hardwicke, Morning Star	6.26	6.56	7.03	7.45	8.20	9.35	10.35	11.35	12.35	1.35	2.35	3.35	3.35	....	....	4.35	5.35	6.00	9.00
Whitminster Hotel	6.36	7.06	7.13	7.55	8.30	9.45	10.45	11.45	12.45	1.45	2.45	■		3.45	....	4.45	5.45	6.10	9.10
Eastington, Cross																	5.51		
Claypits	6.40	7.10	7.17	7.59	8.34	9.49	10.49	11.49	12.49	1.49	2.49	....		3.49	....	4.49	5.53	6.14	9.14
Cambridge, Stores	6.44	7.14	7.21	8.03	8.38	9.53	10.53	11.53	12.53	1.53	2.53	....		3.53	....	4.53	5.57	6.18	9.18
Berkeley Road	6.51																6.04		
Berkley, Salter Street	6.58																6.11		
SHARPNESS, Hinton Turn	7.06																6.19		
Lower Cam, Police Station	....	7.22	7.29	8.11	8.46	10.01	11.01	12.01	1.01	2.01	3.01	....		4.01	....	5.01	....	6.26	9.26
Tilsdown	....	7.24	7.31	8.13	8.48	10.03	11.03	12.03	1.03	....	3.03	....		4.05	....	5.03	....	6.28	9.28
Upper Cam, Church	....	□								2.05		....		4.05					
DURSLEY, May Lane	....	7.34	7.36	8.18	8.53	10.08	11.08	12.08	1.08	2.10	3.08	....		4.10	....	5.08	....	6.33	9.33

	NS	S														NS			
Dursley, May Lane	8.13	8.38	....	....	10.10	....	12.20	....	2.20	....	4.20	....				5.13	....	6.42	....
Wotton-under-Edge, Memorial	8.38	9.02	....	....	10.34	....	12.44	....	2.44	....	4.42	....				5.35	....	7.04	....

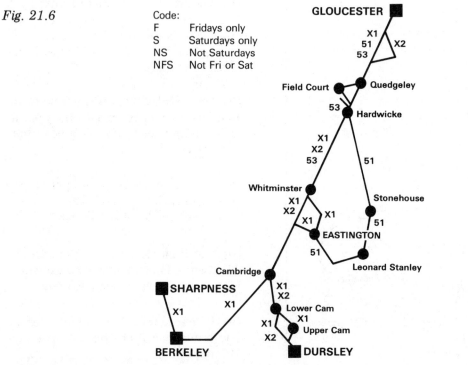

Fig. 21.6

Code:
F	Fridays only
S	Saturdays only
NS	Not Saturdays
NFS	Not Fri or Sat

Conversion Tables and Charts

Conversion graphs will be discussed on page 235 but conversion charts and tables are extensively used by scientists and engineers. Table 21.3 shows part of a temperature conversion chart.

TABLE 21.3

TEMPERATURE			Degrees Fahrenheit to degrees Celsius						
°F	0	1	2	3	4	5	6	7	8	9
0	−17.8	−17.2	−16.7	−16.1	−15.6	−15.0	−14.4	−13.9	−13.3	−12.8
10	−12.2	−11.7	−11.1	−10.6	−10.0	−9.4	−8.9	−8.3	−7.8	−7.2
20	−6.7	−6.1	−5.6	−5.0	−4.4	−3.9	−3.3	−2.8	−2.2	−1.7
30	−1.1	−0.6	0	0.6	1.1	1.7	2.2	2.8	3.3	3.9
40	4.4	5.0	5.6	6.1	6.7	7.2	7.8	8.3	8.9	9.4
50	10.0	10.6	11.1	11.7	12.2	12.8	13.3	13.9	14.4	15.0
60	15.6	16.1	16.7	17.2	17.8	18.3	18.9	19.4	20.0	20.6
70	21.1	21.7	22.2	22.8	23.3	23.9	24.4	25.0	25.6	26.1
80	26.7	27.2	27.8	28.3	28.9	29.4	30.0	30.6	31.1	31.7
90	32.2	32.8	33.3	33.9	34.4	35.0	35.6	36.1	36.7	37.2
100	37.8	38.3	38.9	39.4	40.0	40.6	41.1	41.7	42.2	42.8
110	43.3	43.9	44.4	45.0	45.6	46.1	46.7	47.2	47.8	48.3
120	48.9	49.4	50.0	50.6	51.1	51.7	52.2	52.8	53.8	53.9
130	54.4	55.0	55.6	56.1	56.7	57.2	57.8	58.3	58.9	59.4
140	60.0	60.6	61.1	61.7	62.2	62.8	63.3	63.9	64.4	65.0
150	65.6	66.1	66.7	67.2	67.8	68.3	68.9	69.4	70.0	70.6
160	71.1	71.7	72.2	72.8	73.3	73.9	74.4	75.0	75.6	76.1
170	76.7	77.2	77.8	78.3	78.9	79.4	80.0	80.6	81.1	81.7
180	82.2	82.8	83.3	83.9	84.4	85.0	85.6	86.1	86.7	87.2
190	87.8	88.3	88.9	89.4	90.0	90.6	91.1	91.7	92.2	92.8
Interpolation: deg F:	0.1	0.2	0.3	0.4	0.5	0.6	0.7	0.8	0.9	
deg C:	0.1	0.1	0.2	0.2	0.3	0.3	0.4	0.4	0.5	

Example 4

Convert:

(a) 54°F to degrees Celsius

(b) 122.7°F to degrees Celsius

(c) 35°C to degrees Fahrenheit

(d) 62.7°C to degrees Fahrenheit.

(a) We first find 50 in the first column and move along this row until we find the column headed 4. We find the number 12.2 and hence 54°F = 12.2°C.

(b) To convert 122.7°F into degrees Celsius we make use of the figures given under the heading 'interpolation'. From the table we find 122°F = 50°C.

Looking at the figures given at the foot of the table we can find 0.7°F and this is equivalent to 0.4°C which is shown immediately below. Thus

$$122.7°F = 50°C + 0.4°C$$
$$= 50.4°C$$

(c) To convert degrees Celsius into degrees Fahrenheit we use the table in reverse. To convert 35°C into degrees Fahrenheit we search in the body of the table until we find the figure 35.

This occurs in the column headed 5 in the row starting with 90. Hence 35°C is equivalent to 95°F.

(d) To convert 62.7°C into degrees Fahrenheit we look in the body of the table for a number as close to 62.7 as possible, but less than 62.7. This number is 62.2 corresponding to 144°F. Now 62.2°C is 0.5°C less than 62.7°C.

Using the interpolation figures we see that 0.5°C corresponds to 0.9°F. Therefore

$$62.7°C = 62.2°C + 0.5°C$$
$$= 144°F + 0.9°F$$
$$= 144.9°F$$

Mileage Charts

Mileage charts show the distance between various towns and cities. They are usually laid out like the one shown in Table 21.4.

TABLE 21.4

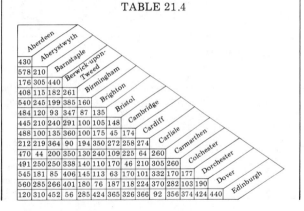

Example 5

Use Table 21.4 to find the distance between

(a) Aberdeen and Carlisle

(b) Barnstaple and Cambridge.

(a) To find the distance from Aberdeen to Carlisle, find Aberdeen and Carlisle in the sloping list. Move vertically down the column for Aberdeen and horizontally along the row for Carlisle until the two movements coincide. The figure given in the table at this point is 212 and thus the distance between Aberdeen and Carlisle is 212 miles.

(b) Find Barnstaple in the sloping list and move down the third column. The row for Cambridge is the seventh row from the top. In the third column of the seventh row we find the figure 240 and hence the distance between Barnstaple and Cambridge is 240 miles.

Postal rates

Table 21.5 gives details of the inland letter rates (Sept. 1986).

Example 6

Use Table 21.5 to answer the following questions:

(a) A letter weighs exactly 150 g. Write down the cost of sending the letter by

   (i) first-class post
   (ii) second-class post.

(b) A postal package weighs 425 g. Find the cost of sending the package by

   (i) first-class post
   (ii) second-class post.

(c) A postal package weighs 1600 g. How much will it cost to post it?

   (a) Directly from Table 21.5,

      (i) Cost of first-class post is 31 p.
      (ii) Cost of second-class post is 22 p.

   (b) Since the package weighs 425 g we must pay the rate for a 450 g package and hence

      (i) Cost of first-class post is 78 p.
      (ii) Cost of second-class post is 59 p.

TABLE 21.5

Inland postal rates

Letter Post

**Rates for letters within the UK and from the UK to
the Isle of Man, the Channel Islands and the Irish Republic**

Weight not over	First Class	Second Class	Weight not over	First Class	Second Class
60g	17p	12p	400g	69p	52p
100g	24p	18p	450g	78p	59p
150g	31p	22p	500g	87p	66p
200g	38p	28p	750g	£1.28	98p
250g	45p	34p	1000g	£1.70	Not admissible over 750g
300g	53p	40p	Each extra 250g or part thereof	42p	
350g	61p	46p			

(c) Since the package weighs 1600 g:

1600 g = 1000 g + 600 g

= 1000 g + (600 ÷ 250) g

= 1000 g + 2.4 units of 250 g

Therefore we must pay for an extra 3 units.

Postal charge = £1.70 + 3 × £0.42

= £1.70 + £1.26

= £2.96

Parallel Scale Conversion Charts

A system of **parallel scales** may be used when we wish to convert from one set of units to another related set. Fig. 21.7 is a chart relating degrees Fahrenheit and degrees Celsius.

Fig. 21.7

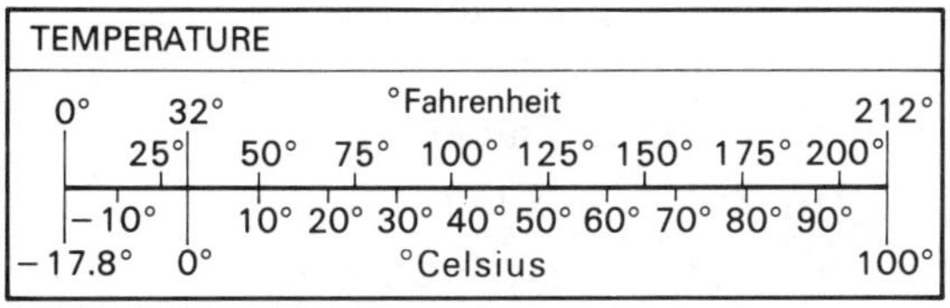

Example 7

Using the chart in Fig. 21.7 convert

(a) 50°F to degrees Celsius

(b) 65°F to degrees Celsius

(c) 20°C to degrees Fahrenheit.

(a) From the chart we see that 50°F corresponds with 10°C and therefore 50°F is equivalent to 10°C.

(b) From the chart we see that 75°F corresponds roughly with 24°C. Note carefully that when using the chart we cannot expect accuracy greater than 1°.

(c) From the chart we see that 20°C is about 68°F.

Exercise 21.2

1. Using the route map and the bus time-table on page 214, answer the following questions:

   (a) Only one bus per day picks up at Stroud Road St. Barnabas Church. At what time does it leave Gloucester Bus Station and what number is the service?

   (b) Three buses run on Saturdays only. At what time do they leave Gloucester Bus Station and at what time do they arrive in Dursley?

   (c) How long does the 11.15 bus from Gloucester wait at Dursley before proceeding to Wotton-under-Edge?

2. Use the temperature conversion table on page 215 to convert

   (a) 83°F to degrees Celsius

   (b) 114.6°F to degrees Celsius

   (c) 55°C to degrees Fahrenheit

   (d) 68.7°C to degrees Fahrenheit.

3. Use the mileage chart on page 215 to find the distances between

   (a) Aberdeen and Cardiff

   (b) Birmingham and Carmarthen

   (c) Bristol and Dorchester.

4. Table 21.6 shows a comparison between gradients expressed as a ratio and gradients expressed as a percentage. Use the table to convert

   (a) a gradient of 1 : 7 to a percentage

(b) a gradient of 7.1% to a ratio.

TABLE 21.6

Gradients	
Ratio	%
1:3	33.3
1:4	25
1:5	20
1:6	16.7
1:7	14.3
1:8	12.5
1:9	11.1
1:10	10
1:11	9.1
1:12	8.3
1:13	7.7
1:14	7.1
1:15	6.7
1:16	6.3
1:17	5.9
1:18	5.6
1:19	5.3
1:20	5

5. Using Table 21.5 (page 216) find the first-class charges for the following postal packages:

   (a) 300 g   (b) 550 g   (c) 1850 g

6. Using Table 21.5 (page 216) find the second-class rates for the following letters:

   (a) 200 g   (b) 320 g   (c) 475 g

7. Table 21.7 allows conversion from miles to kilometres (and also miles per hour and kilometres per hour) to be made. Use the table to convert

   (a) 15 miles to kilometres

   (b) 48.27 km/h to miles per hour

   (c) 58 miles to kilometres

   (d) 568 miles per hour to kilometres per hour.

TABLE 21.7

Distance and Speed				
Miles to kilometres				
Miles per hour to kilometres per hour				
1	1.60	20	32.18	75	120.7
2	3.21	25	40.23	80	128.7
3	4.82	30	48.27	85	136.8
4	6.43	35	56.32	90	144.8
5	8.04	40	64.37	95	152.9
6	9.65	45	72.41	100	160.9
7	11.26	50	80.46	200	321.9
8	12.87	55	88.51	300	482.8
9	14.48	60	96.55	400	643.7
10	16.09	65	104.60	500	804.7
15	24.13	70	112.70	1000	1609.3

1 mile = 1.609 344 km
1 kilometre = 0.621 371 miles

8. Use the chart (Fig. 21.7 page 217) relating temperatures in degrees Fahrenheit to degrees Celsius to convert

   (a) 100°F to degrees Celsius

   (b) 80°C to degrees Fahrenheit

   (c) 64°F to degrees Celsius

   (d) 48°C to degrees Fahrenheit.

9. Fig. 21.8 is a chart relating pounds avoirdupois to kilograms. Use the chart to convert

   (a) 3.4 lb to kilograms

   (b) 568 lb to kilograms

   (c) 1.8 kg to pounds.

Fig. 21.8

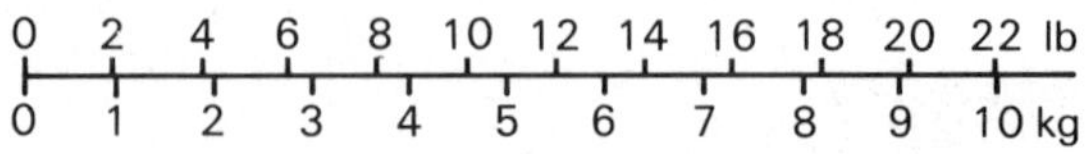

10. Table 21.8 shows a comparison of tyre pressures. Convert:

   (a) 22 lb/in^2 into kilograms per square centimetre

   (b) 2.10 kg/cm^2 into pounds per square inch.

(c) Estimate the tyre pressure in pounds per square inch corresponding to 3 kg/cm^2.

TABLE 21.8

Tyre Pressures					
Pounds per sq in to kg per sq cm					
16	1.12	26	1.83	40	2.80
18	1.26	28	1.96	50	3.50
20	1.40	30	2.10	55	3.85
22	1.54	32	2.24	60	4.20
24	1.68	36	2.52	65	4.55

Simple Bar Charts

The information in a **simple bar chart** is represented by a series of bars all of the same width. The bars may be drawn vertically or horizontally. The height (or length) of the bars represents the magnitude of the figures given. A simple bar chart shows clearly the size of each item of information but it is not easy to obtain the total of all the items from the diagram.

Example 8

A family spends it weekly income of £300 as follows:

Food	£100
Clothes	£50
Fuel	£40
Mortgage	£90
Other expenses	£20
Total	£300

Draw (a) a vertical bar chart, (b) a horizontal bar chart to represent this information.

(a) Fig. 21.9 shows the information in the form of a vertical bar chart whilst

(b) Fig. 21.10 shows the information in the form of a horizontal bar chart.

Fig. 21.9

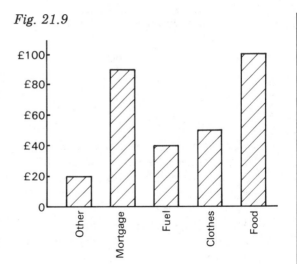

Fig. 21.10

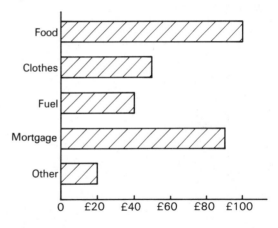

Chronological Bar Charts

A **chronological bar chart** compares quantities over periods of time. It is similar to the vertical bar chart, except that vertical lines replace the bars.

Example 9

The table overleaf gives the world population in millions of people from 1750 to 1950.

Draw a chronological bar chart to represent this information.

Year	Population (millions)
1750	728
1800	906
1850	1171
1900	1608
1950	2504

When drawing a chronological bar chart, time is always marked off along the horizontal axis, as shown in Fig. 21.11.

The chart clearly shows how the population of the world has increased over the last 200 years.

Fig. 21.11

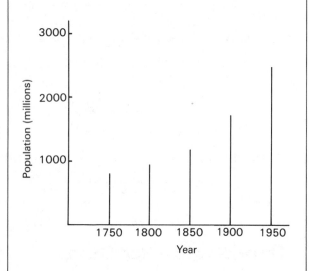

Proportionate Bar Charts

The **proportionate bar chart** relies on heights (if drawn vertically) or lengths (if drawn horizontally) to convey proportions of the whole. It should be the same width throughout its height or length.

The proportionate bar chart shows clearly the total of all the items, but it is rather difficult to obtain the proportion of each item accurately.

Example 10

The table below shows the number of people employed on various kinds of work in a factory. Represent this information in the form of a proportionate bar chart.

Type of personnel	Number employed
Unskilled workers	45
Craftsmen	27
Draughtsmen	5
Clerical workers	8
Total	85

Suppose that the total height of the proportionate bar chart is to be 6 cm.

The heights of the component parts must first be calculated and then drawn accurately using a rule.

45 unskilled workers are

represented by $\dfrac{45}{85} \times 6 \,\text{cm} = 3.18 \,\text{cm}$

27 craftsmen are

represented by $\dfrac{27}{85} \times 6 \,\text{cm} = 1.91 \,\text{cm}$

5 draughtsmen are

represented by $\dfrac{5}{85} \times 6 \,\text{cm} = 0.35 \,\text{cm}$

8 clerical workers are

represented by $\dfrac{8}{85} \times 6 \,\text{cm} = 0.56 \,\text{cm}$

The proportionate bar chart is shown in Fig. 21.12.

Fig. 21.12

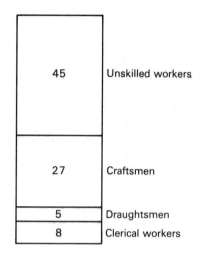

Pie Charts

A **pie chart** displays the proportion of a whole as a **sector angle** or **sector area**. The circle as a whole represents the total of the component parts.

Pie charts are very useful when component parts of a whole are to be represented, but it is not easy to discover the total quantity represented. Up to eight component parts can be represented, but above this number the chart loses its clarity and effectiveness.

Example 11

24 sixth-form students were asked what sort of career they would like. Represent this information in the form of a pie chart.

        8 chose work with animals

        3 chose office work

        2 chose teaching

        5 chose outdoor work

        6 said they did not know

The first step is to calculate the angles at the centre of the pie chart. Since there are $360°$ in the circle we divide $360°$ by the total number of students

quizzed. This gives the angle for one pupil. To find the angle for each section, we multiply the angle for one pupil by the number of pupils in the section and correct the product to the nearest degree. The work may be set out as follows:

$$360° \div 24 = 15°$$

Sector	Number of pupils	Sector angle
Work with animals	8	$8 \times 15° = 120°$
Office work	3	$3 \times 15° = 45°$
Teaching	2	$2 \times 15° = 30°$
Outdoor work	5	$5 \times 15° = 75°$
Don't know	6	$6 \times 15° = 90°$

The pie chart is shown in Fig. 21.13.

Fig. 21.13

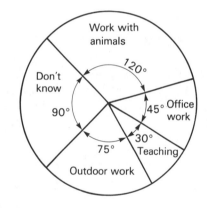

Line Graphs

Line graphs are sometimes used instead of chronological bar charts.

Example 12

Represent the information given in Example 9 in the form of a line graph.

    The line graph is shown in Fig. 21.14. As with chronological bar charts the year is always taken along the horizontal axis. The points are joined by straight lines because no information is given about the world population between the years given in the table.

Fig. 21.14

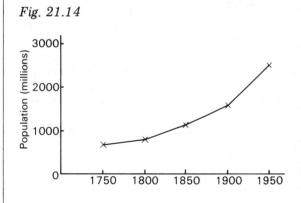

Pictograms

Pictograms are diagrams in the form of pictures which are used to present information to those who are unskilled in dealing with figures or to those who have only a limited interest in the topic depicted.

Example 13

The table below shows the output of bicycles for the years 1980 to 1984.

Year	1980	1981	1982	1983	1984
Output	2000	4000	7000	8500	9000

Represent this information in the form of a pictogram.

    The pictogram is shown in Fig. 21.15. It will be seen that each bicycle in the diagram represents an output of 2000 bicycles. Part of a symbol is shown in 1982, 1983 and 1984 to represent a fraction of 2000 but clearly this is not a precise way of representing the output.

Fig. 21.15

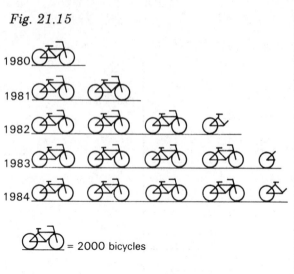

A method not recommended is shown in Fig. 21.16. Comparison is difficult because the reader is not sure whether to compare heights, areas or volumes.

Fig. 21.16

Sales of milk in 1960 and 1980 (millions of litres)

Exercise 21.3

1. Fig. 21.17 is a pictogram showing the method by which first-year boys come to school. How many

(a) come by bus

(b) come by car?

Fig. 21.17

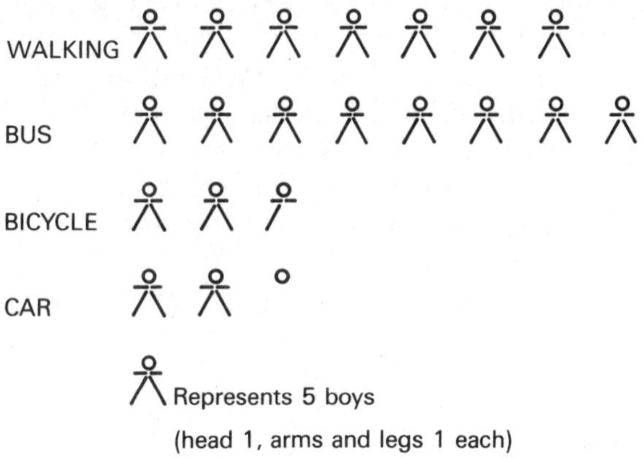

WALKING

BUS

BICYCLE

CAR

Represents 5 boys

(head 1, arms and legs 1 each)

2. Draw a proportionate bar chart 5 cm long for the figures shown below which represent the way in which people travel to work in Gloucester.

Type of transport	Numbers using
Bus	780
Private motoring	420
Other (foot, bicycle, motor cycle, etc.)	160
Total	1360

3. Draw a proportionate bar chart 8 cm long to represent the following information, which relates to the way in which a family spends its income.

Item	Amount spent (£)
Food and drink	95
Housing	42
Transport	29
Clothing	33
Other	51
Total	250

4. The table below shows the results of a survey of the colours of doors on a housing estate. Draw a vertical bar chart to represent this information.

Colour of door	Number of houses
White	85
Red	17
Green	43
Brown	70
Blue	15

5. The figures below show the result of a survey to find the favourite sports of a group of boys. Draw a pie chart to represent this information.

Sport	Number of boys
Athletics	20
Cricket	58
Football	32
Hockey	10

6. The table below shows the output of bicycles from a certain factory for the years 1980–84. Represent this information in the form of

    (a) a chronological bar chart

    (b) a line graph.

Year	1980	1981	1982	1983	1984
Number of bicycles	2000	4000	7000	8500	9000

7. The table below shows the number of houses completed in a certain town for the years 1978–83. Represent this information by means of

    (a) a chronological bar chart

    (b) a line graph.

Year	1978	1979	1980	1981	1982	1983
Number of houses	81	69	73	84	80	120

8. The pie chart (Fig. 21.18) shows a local election result. There were three candidates, White, Green and Brown. The angle representing the votes cast for White is $140°$ and that for Green is $90°$.

    (a) Work out the angle representing the votes cast for Brown.

    (b) Calculate the votes cast for each candidate if the total number voting was 18 000.

Fig. 21.18

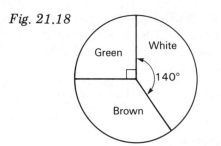

9. A pie chart was drawn to show the number of tonnes of various crops grown by a group of farmers. The sector angles corresponding to each crop was as follows:

Crop	Sector angle (degrees)
Vegetables	137
Grain	93
Potatoes	80
Fruit	50

If the total amount grown was 500 tonnes, calculate the amount of each crop grown.

10. The information shown below gives the output of motor tyres made by the Treadwell Tyre Company for the first six months of 1985. Draw

    (a) a chronological bar chart

    (b) a line graph, to represent this information.

Month	Jan	Feb	Mar	Apr	May	June
Output (thousands)	40	43	39	38	37	45

11. Fig. 21.19 shows how Mrs Lever spends her housekeeping money. If she receives £50 per week, how much does she spend on each item shown in the diagram?

Fig. 21.19

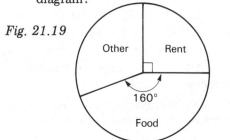

1. The bar chart in Fig. 21.20 shows the distribution of children in years 1 to 6. Each child is given three exercise books for the term. If each book costs 20p, how much was spent altogether on exercise books?

    A £180 **B** £360 **C** £600 **D** £1800

Fig. 21.20

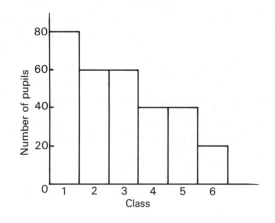

2. The pie chart (Fig. 21.21) illustrates the sports preferred by a group of male six-formers. What percentage preferred tennis?

    A 20% **B** 25% **C** 30% **D** 50%

Fig. 21.21

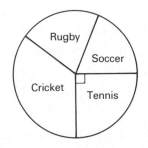

3. The bar chart (Fig. 21.22) shows the ages of the children in a school. How many children took part in the survey?

A 5   B 17   C 65   D 222

Fig. 21.22

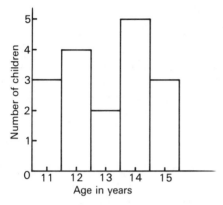

4. 2000 students were interviewed to find out which of four sports they preferred. The results were as follows:

Sport	Cricket	Tennis	Swimming	Athletics
Number of students	800	480	400	320

A pie chart is to be drawn to represent this information. What will be the sector angle for swimming?

A 95°   B 72°   C 65°   D 60°

5. The main leisure interests of 150 fifth-year pupils are shown as percentages below:

Football   40%   Television   30%
Reading   10%   Scouts   8%
Other   12%

If a pie chart is drawn of this information, what size of angle will represent those pupils whose main interest is television?

A 30°   B 72°   C 108°   D 120°

6. The table below shows the results of a survey to find the colour that the external woodwork was painted on the houses of a housing estate.

Colour	White	Red	Blue	Green
Number of houses	90	70	60	80

The information is to be illustrated in a proportionate bar chart which is to have a total height of 15 cm. What height must be used to represent the houses with green external paintwork?

A 4.5 cm   B 4.0 cm
C 3.5 cm   D 3.0 cm

7. There are 300 pupils in a school. The pie chart (Fig. 21.23) shows the proportion of children travelling to school by private bus, public transport, bicycle and on foot. How many travel to school by bicycle?

A 30   B 36   C 120   D 360

Fig. 21.23

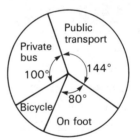

8. Fig. 21.24 shows the frequency distribution of the marks obtained in a test. How many candidates took the test?

A 15   B 24   C 50   D 62

Fig. 21.24

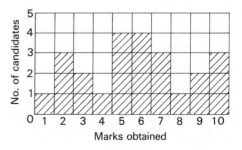

9. Fig. 21.25 is a pie chart which shows the sales for a departmental store during one day. The total sales were £4000. What were the sales of clothing?

   A £1200   B £1000

   C £900   D £600

Fig. 21.25

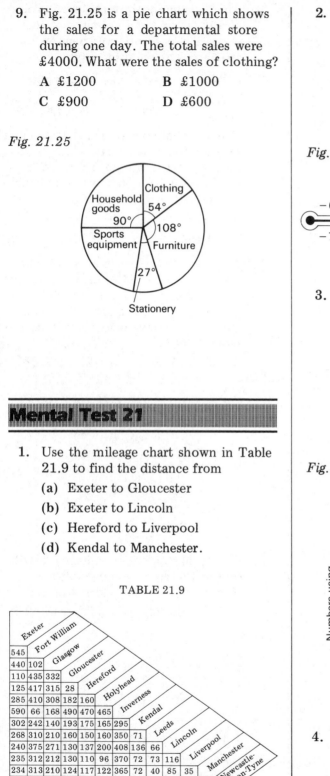

Mental Test 21

1. Use the mileage chart shown in Table 21.9 to find the distance from

   (a) Exeter to Gloucester

   (b) Exeter to Lincoln

   (c) Hereford to Liverpool

   (d) Kendal to Manchester.

TABLE 21.9

2. Fig. 21.26 shows a diagram which can be used to change degrees Celsius to degrees Fahrenheit and vice versa. Use the diagram to change

   (a) −40°C to degrees Fahrenheit

   (b) 20°C to degrees Fahrenheit

   (c) 100°F to degrees Celsius.

Fig. 21.26

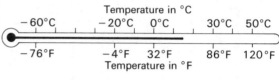

3. Fig. 21.27 is a simple bar chart which shows the way that commuters in the south-east region travel to work. Use the chart to find the number of commuters in the sample who travelled to work by

   (a) private motoring

   (b) bus and underground

   (c) British Rail.

Fig. 21.27

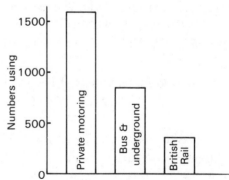

4. The chronological bar chart in Fig. 21.28 shows the number of television sets sold in southern England during the period 1970–75. Use the

diagram to estimate the numbers of television sets sold in the years 1972 and 1975.

Fig. 21.28

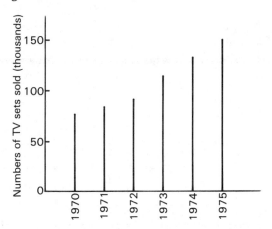

5. The pie chart shown in Fig. 21.29 depicts the sales of the various departments of a large store. If the sales for the week depicted were £600 000, find the value of

    (a) the clothing sold

    (b) the household goods sold.

Fig. 21.29

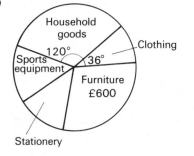

6. The pictogram (Fig. 21.30) shows the sales of cars in the years 1975–79:

    (a) In which year were the most cars sold?

    (b) How many cars were sold in 1978?

    (c) How many cars were sold in 1976?

    (d) Estimate the total number of cars sold in the years represented on the pictogram.

Fig. 21.30

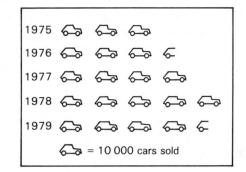

7. Mrs Wood has £100 per week for housekeeping. A pie chart is drawn which illustrates how she spends the money. If she spends £20 on food, work out the angle of the sector representing this amount.

22

Graphs

Introduction

In newspapers, business reports and government publications great use is made of pictorial illustrations to help the reader understand the report. Some of the charts and diagrams used were discussed in Chapter 21. Graphs which are pictures of numerical information are also commonly used as illustrations.

Axes

The first step in drawing a graph is to draw two lines, one horizontal and the other vertical (Fig. 22.1). These lines are called the **axes** and the point where they cross is called the **origin.** The vertical axis is often called the y-axis and the horizontal axis is then called the x-axis. Both axes should be clearly labelled as shown, for instance, in Fig. 22.2.

Fig. 22.1

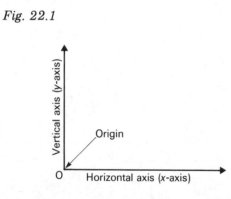

Fig. 22.2

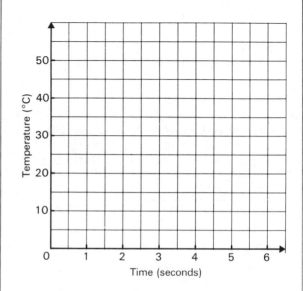

Scales

The number of units represented by a unit length along an axis is called the **scale.** For instance, in Fig. 22.2 the scale on the vertical axis is 1 cm = 10 degrees. The scales need not be the same on both axes.

The most useful scales are 1 cm to 1, 2, 5 and 10 units. Some multiples of these are also satisfactory, for instance, 1 cm to 20, 50 and 100 units.

No matter which scale is chosen it is important that it is easy to read. When graph or squared paper is used, the scale will depend upon the type and size of the paper used.

Cartesian Coordinates

Cartesian **coordinates** are used to mark the points on a graph. In Fig. 22.3 the point P has the coordinates $(4, 6)$. The first number, it is 4, gives the horizontal distance from the origin. The second number, it is 6, gives the vertical distance from the origin.

Fig. 22.3

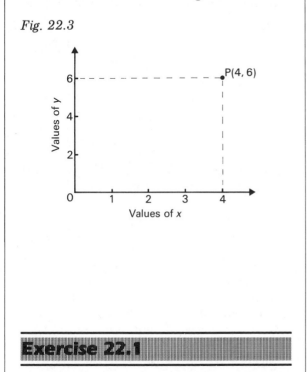

Exercise 22.1

1. Write down the coordinates of the points A, B, C, D, E and F (Fig. 22.4).

Fig. 22.4

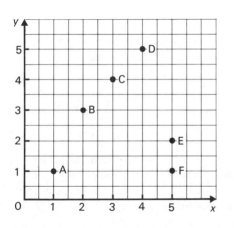

2. Write down the coordinates of the points R, S, T, U and V shown in Fig. 22.5.

Fig. 22.5

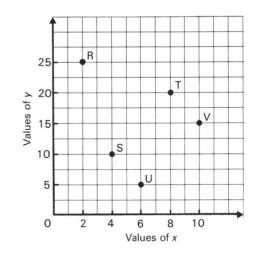

3. On graph or squared paper draw a pair of axes. Along the horizontal axis use a scale of 1 cm to represent 2 units and along the vertical axis use a scale of 1 cm = 10 units. Now plot the following points:

   (a) $(4, 20)$ (b) $(6, 25)$

   (c) $(5, 35)$ (d) $(7, 60)$

   (e) $(3, 45)$

4. On graph paper plot the points $A(2, 2)$, $B(6, 2)$, $C(6, 4)$ and $D(2, 4)$. Label each of the points and join them to form the shape ABCD. Use scales of 1 cm to 1 unit on each axis and cover the range from 0 to 8 units:

   (a) What is the length of the sides AB and BC?

   (b) What is the length of the diagonal AC?

   (c) What is the area of ABCD?

5. Plot the points A(2, 6), B(7, 6), C(7, 3) and D(3, 3). Use scales of 1 cm to 1 unit on both axes. Join the points in alphabetical order and answer the following questions:

(a) What is the name of the shape ABCD?

(b) Write down the length of the side BC.

Drawing a Graph

Every graph shows the relationship between two sets of numbers.

Example 1

The table below shows the speed of a racing car at various times during its first lap.

Time (seconds)	0	10	20	30	40
Speed (mile/hour)	0	10	40	90	160

Plot these points on graph paper using a scale of 1 cm = 5 seconds on the horizontal axis and 1 cm = 20 mile/hour on the vertical axis:

(a) Use the graph to find the speed of the car after 25 seconds.

(b) How many seconds did it take the car to attain a speed of 100 mile/hour?

The graph is shown in Fig. 22.6 and the points can be joined by a smooth curve. When a graph is a smooth curve or a straight line it can be used to find corresponding values which were not given in the original table.

(a) To find the speed after a time of 25 seconds we draw the vertical and horizontal lines shown in the diagram

and we find that the speed of the car after 25 seconds is 62 mile/hour.

(b) Working in a similar way we find that it takes 32 seconds for the car to reach a speed of 100 mile/hour.

Fig. 22.6

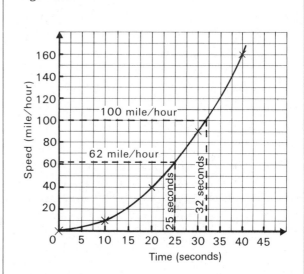

The numbers given in a table of values do not have to be whole numbers. As shown in Example 2, they can also be decimal numbers.

Example 2

The table below gives the average diameter of ash trees of varying ages:

Age (years)	10	20	30	40	50
Diameter (cm)	9.3	16.2	27.7	43.8	64.5

Plot these values on graph paper and join them to form a smooth curve. Use your graph to find the diameter of a tree which is 35 years old. How old is a tree with a diameter of 20 cm?

The graph is shown in Fig. 22.7. The diameter of a tree 35 years old is 35 cm and a tree with a diameter of 20 cm is 23 years old.

Fig. 22.7

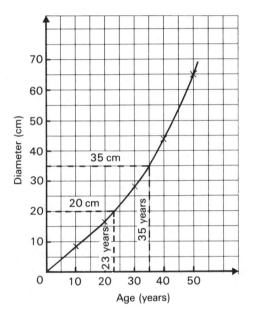

Diameter (cm) / Age (years)

35 cm

20 cm

23 years

35 years

Example 3

The table below gives the temperature at 12.00 noon on seven successive days.

Day	June	1	2	3	4	5	6	7
Temp. (°C)		16	20	16	18	22	15	17

Plot a graph to illustrate this information with the date taken along the horizontal axis.

Since the temperatures range from 15 to 22 degrees we can make 14 degrees our starting point. This will allow us to use a larger scale on the vertical axis which makes for greater accuracy in plotting the graph. Note that zero is shown by breaking the vertical axis. It is bad practice not to show zero. The reasons are given later.

On plotting the points (Fig. 22.8) we see that it is impossible to join them with a straight line or a smooth curve. The best we can do is to join them with a series of straight lines. The graph shows in pictorial form the variations in temperature over the seven days. We can see at a glance that the 1st, 3rd and 6th June were cool days whilst the 2nd and 5th were warm days.

Fig. 22.8

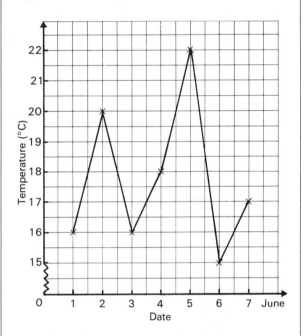

Temperature (°C) / Date

Misleading Graphs

Graphs are sometimes drawn so that they are misleading. Look at Fig. 22.9. The top graph looks as if wages have risen a great deal during the period 1982-85. However the bottom graph seems to show that wages have risen by only a small amount.

Fig. 22.9

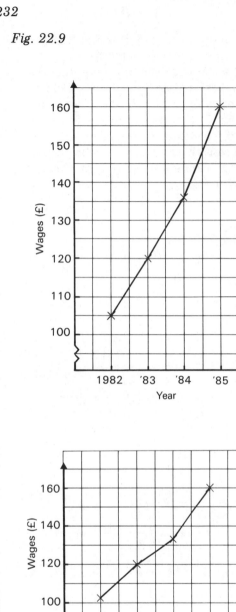

Fig. 22.10

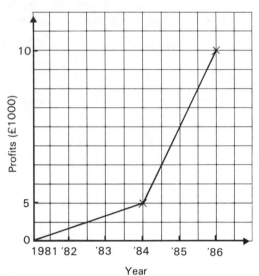

Fig. 22.11 shows no origin (i.e. no zero value). It is therefore impossible to see exactly what the graph is trying to depict because we cannot work out the scale on the vertical axis.

Fig. 22.11

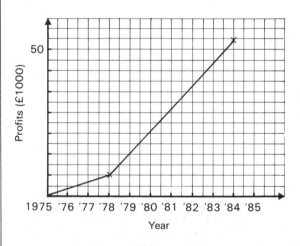

Fig. 22.10 shows a graph with an inconsistent scale which gives an entirely false impression of the speed at which the profits have risen.

Fig. 22.12 shows two graphs with no scales. It is impossible to decide if both are drawn with the same or different scales. Has the cost of living risen slowly (graph A) or quickly (graph B)? Graphs like these are often used for political purposes.

Fig. 22.12

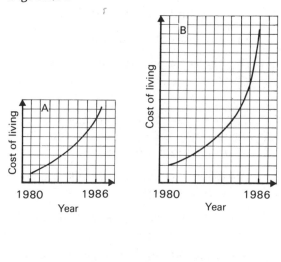

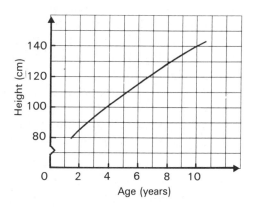

2. Fig. 22.14 shows how the height of a girl alters with her age. Use the graph to find:

   (a) her age when she attained a height of 108 cm

   (b) her height when she was 7 years old.

Fig. 22.14

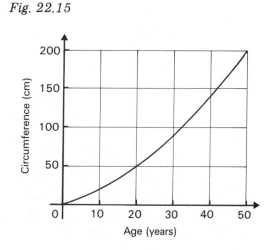

Exercise 22.2

1. The distance a car travels (in metres) after various times (in seconds) starting from rest is shown in Fig. 22.13. Find:

   (a) the distance travelled by the car at the end of 3 seconds

   (b) the time taken to travel 40 metres.

Fig. 22.13

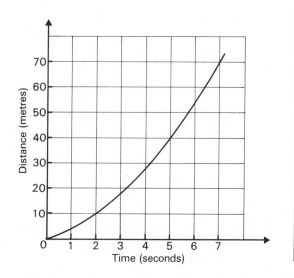

3. Fig. 22.15 shows how the circumference of the trunk of a tree increased with its age. Use the graph to find:

   (a) the age of the tree when its circumference was 100 cm

   (b) the circumference of the trunk when it was 40 years old.

Fig. 22.15

4. The table below shows how the area of a square increases as the length of the side increases. Plot these points on graph paper and draw a smooth curve to represent the information given in the table.

Length of side	0	2	4	6	8	10
Area		0	4	16	36	64	100

Use scales of 1 cm = 1 unit along the horizontal axis to represent length of side and 1 cm = 10 units along the vertical axis to represent area. Use your graph to find:

(a) the area of a square whose length of side is 7 units

(b) the length of side of a square with an area of 25 square units.

5. The table below gives the sales (in millions of pounds) for the years from 1982 to 1986.

Year	1982	1983	1984	1985	1986
Sales	1.2	1.8	2.2	3.4	5.6

Taking the year along the horizontal axis to a scale of 2 cm = 1 year and sales along the vertical axis to a scale of 1 cm = 1 million pounds, plot the above values. Join the points up with a series of straight lines.

6. The table below gives the amount of grain (in tonnes) grown by A & S Farms Ltd. during six successive years.

Year	1	2	3	4	5	6
Amount	200	176	233	258	243	197

Plot the year along the horizontal axis to a scale of 1 cm = 1 year and the amount grown along the vertical axis to a scale of 1 cm = 10 tonnes. You may start the graph at 160 tonnes on the vertical axis. Join the points up with a series of straight lines. Use your graph to find the years in which

(a) the most grain was grown

(b) the least amount of grain was grown.

7. Fig. 22.16 shows how the amount of an electricity bill increases as the number of units used increases:

(a) When the number of units used is 0, what is the amount of the bill?

(b) Why is there a charge when no electricity has been used?

(c) When 800 units are used what is the amount of the bill?

(d) What is the cost per unit used?

Fig. 22.16

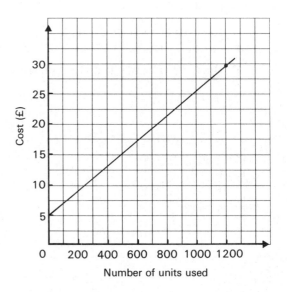

8. Graphs can be used to deceive. Briefly explain why the graphs in Fig. 22.17 are misleading.

Fig. 22.17

(a)

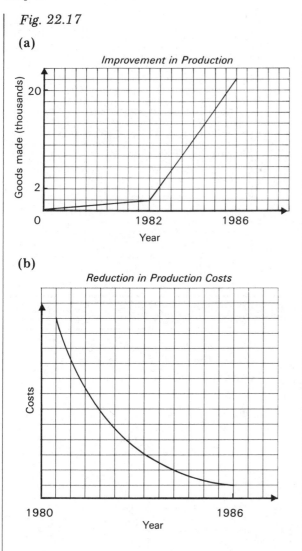

Improvement in Production

(b)

Reduction in Production Costs

Conversion Graphs

Conversion graphs are used to convert one set of units into another set of units e.g. inches into centimetres, German marks into pounds sterling, degrees Celsius into degrees Fahrenheit, etc.

Example 4

If the rate of exchange between the pound and the French franc is £1 = 11 francs, construct a graph to show the value of the

franc in pounds. Find from the graph the value of

(a) 60 francs in pounds

(b) £9 in francs.

The first step is to draw up a table of corresponding values as follows:

Pounds	0	2	4	6	8	10	12
Francs	0	22	44	66	88	110	132

The graph is shown plotted in Fig. 22.18 where it will be seen to be a straight line passing through the origin. If one quantity is directly proportional to another (as they are in this case) a straight line passing through the origin always results.

From the graph

(a) 60 francs = £5.50

(b) £9 = 99 francs

Fig. 22.18

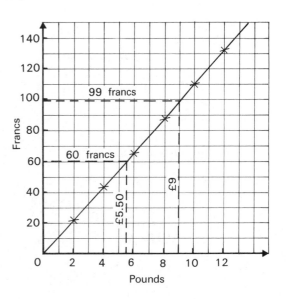

Not all graphs pass through the origin. For instance the cost of electricity consists of a fixed charge plus so much per unit used.

This means that even if no units are used the fixed charge must be paid and if we draw a graph of cost of electricity against number of units used, the straight line will start at a point above the origin on the vertical axis.

Example 5

In a certain area electricity costs 6 p per unit used plus a fixed charge of £7. Draw a graph of units used and the cost of electricity with cost on the vertical axis. The graph should cater for up to 250 units used. Use your graph to find the cost of 180 units.

The first step is to draw up a table of corresponding values as follows:

Units used	0	50	100	150	200	250
Cost (£)	7	10	13	16	19	22

The values in the table have been worked out as follows:

When 0 units are used,

$$\text{Cost} = (700 + 6 \times 0) \text{ pence}$$

$$= (700 + 0) \text{ pence}$$

$$= 700 \text{ pence}$$

$$= £7$$

When 50 units are used,

$$\text{Cost} = (700 + 6 \times 50) \text{ pence}$$

$$= (700 + 300) \text{ pence}$$

$$= 1000 \text{ pence}$$

$$= £10$$

and so on.

The graph is shown in Fig. 22.19. From the graph the cost of 180 units is £17.80.

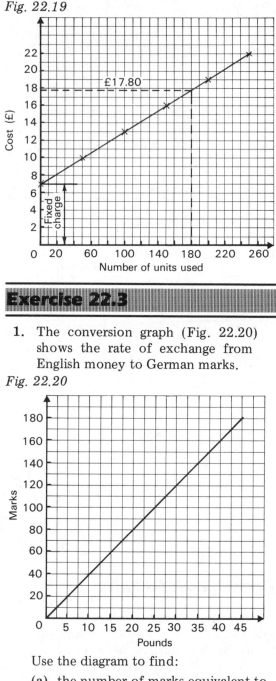

Fig. 22.19

Fig. 22.20

Exercise 22.3

1. The conversion graph (Fig. 22.20) shows the rate of exchange from English money to German marks.

Use the diagram to find:

(a) the number of marks equivalent to £20

(b) the cost in pounds for a gift bought in Germany for 140 marks.

(c) If the new exchange rate gives £1 = 2.5 marks, draw another graph to represent this.

2. The graph shown in Fig. 22.21 shows the increase in the cost of a gas bill as the number of therms increases. Use the diagram to find:

(a) the standing charge

(b) the cost of gas per therm

(c) the cost of the gas bill when 250 therms are used

(d) the number of therms used when the gas bill is £90.

Fig. 22.21

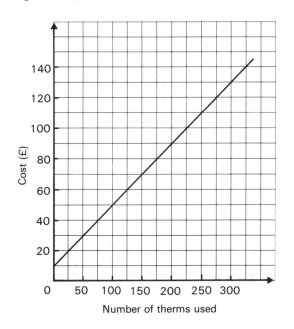

3. To convert inches into millimetres we multiply the number of inches by 25:

(a) Make a table giving corresponding values of inches and millimetres. Your table should range from 0 to 20 inches.

(b) From your table draw a graph taking inches along the horizontal axis. Suitable scales are 1 cm to represent 1 inch and 1 cm to represent 20 millimetres.

(c) Use your graph to convert (i) 8 inches into millimetres, (ii) 400 millimetres into inches.

4. The table below gives corresponding values of degrees Celsius and degrees Fahrenheit:

°C	0	5	10	15	20	25	30
°F	32	41	50	59	68	77	86

(a) Draw a graph of this information taking values of degrees Celsius along the horizontal axis. Suitable scales are 1 large square to 5° along the horizontal axis and 1 large square to 10° along the vertical axis.

(b) Use your graph to convert (i) 18°C to °F, (ii) 80°F to °C.

5. The rate of exchange between the Swiss franc and the pound is £1 = 3 francs:

(a) Draw up a table giving corresponding values of the pound and the franc. Your table should cover a range between £0 and £80.

(b) Construct a graph from your table taking pounds on the horizontal axis. Suitable scales are 1 large square to £5 on the horizontal axis and 1 large square to 20 francs on the vertical axis.

(c) Use your graph to find (i) the number of francs equivalent to £25, (ii) the number of pounds equivalent to 135 francs.

6. In calculating the amount of a telephone bill, dialled calls are charged at 4p for each time unit. In addition a rental charge of £7 is charged:

(a) A user estimates that he will dial up to 1000 calls. Draw up a table giving corresponding values of cost against the number of dialled calls covering the range from 0 calls to 1000 calls.

(b) Draw a graph from your table with the number of calls along the horizontal axis. Suitable scales are 1 large square to 10 calls along the horizontal axis and 1 large square to 500 pence along the vertical axis.

(c) Use your graph to find (i) the cost when 800 calls are made, (ii) the number of calls made when the cost was £23.

Distance-Time Graphs

When the speed of a vehicle is constant the distance–time graph is a straight line.

Example 6

Fig. 22.22 shows a distance–time graph for a car. Use the graph to find:

(a) the distance travelled after 2 hours

(b) the average speed of the car.

Fig. 22.22

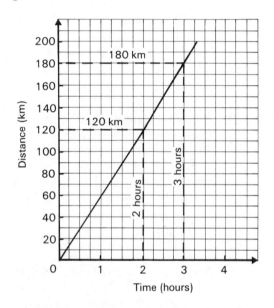

(a) From the graph we see that the distance travelled after 2 hours is 120 km.

(b) To find the average speed of the car we recall that

$$\text{Average speed} = \frac{\text{Distance travelled}}{\text{Time taken}}$$

From the graph we see that after 3 hours the distance travelled is 180 km. So

$$\text{Average speed} = \frac{180}{3}\,\text{km/h}$$

$$= 60\,\text{km/h}$$

Example 7

The graph (Fig. 22.23) shows details of a girl's cycle journey. She leaves her home X at 0900 hours and after cycling 20 km she stops for a rest. She then cycles on until she arrived at her destination Y at 1300 hours. After a stop for lunch she cycled back home arriving at 1700 hours. Find:

(a) the distance between X and Y

(b) the length of time she stopped for lunch.

(c) How long did she stop on the outward journey?

(d) What was her average speed on the return journey?

Fig. 22.23

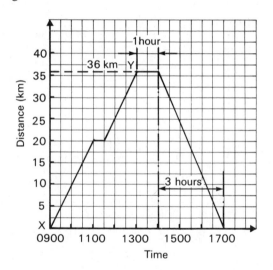

(a) From the graph the distance between X and Y is 36 km.

(b) She stopped for lunch between 1300 hours and 1400 hours, a length of time of 1 hour.

(c) On the outward journey she rested for half-an-hour.

(d) $\text{Average speed} = \dfrac{36}{3}\,\text{km/h}$

$$= 12\,\text{km/h}$$

Example 8

The graph (Fig. 22.24) shows the journey made by a man. The first part was made by car, the second part by cycle and the third part by walking. Find the average speed for the entire journey.

Fig. 22.24

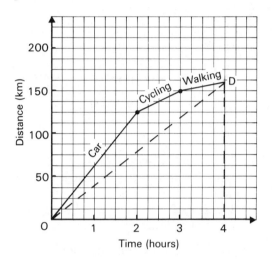

To find the average speed we draw the line OD shown dotted on the graph. This line shows that the man travelled 160 km in 4 hours.

$$\text{Average speed} = \frac{160}{4}$$

$$= 40 \text{ km/h}$$

Exercise 22.4

1. Fig. 22.25 shows a distance–time graph for a cyclist. Find:

   (a) the distance travelled in the first 2 hours

   (b) the time taken to cycle 30 miles

   (c) the average speed attained by the cyclist.

Fig. 22.25

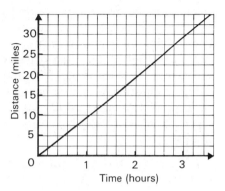

2. The graph in Fig. 22.26 shows the distance travelled by a motorist between 1000 hours and 1400 hours:

   (a) How far did the motorist drive in the four hours?

   (b) For how long did the motorist stop?

   (c) Calculate the average speed between 1000 hours and 1200 hours.

   (d) What was the average speed for the complete journey?

Fig. 22.26

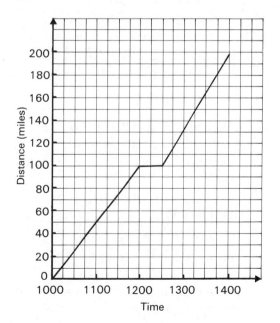

3. The graph (Fig. 22.27) shows the journeys made by a motorist. He leaves his home town A at 0800 hours and after driving 40 km he stops at a service station for petrol. He then drives on and arrives at town B at 1100 hours. After conducting his business he drives back home arriving at 1300 hours:

(a) How far is it between towns A and B?

(b) What was the motorist's average speed between town A and the service station?

(c) How long did it take the motorist to conduct his business in town B?

(d) What was the average speed on the return journey?

Fig. 22.27

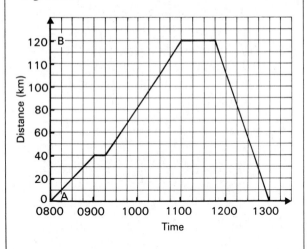

4. The graph (Fig. 22.28) shows a car journey from Leeds to Hull and back:

(a) How long does the journey to Hull take?

(b) What is the distance between Hull and Leeds?

(c) How long was the stop in Hull?

(d) What was the average speed of the car on the journey to Hull?

(e) What was the average speed on the journey from Hull back to Leeds?

Fig. 22.28

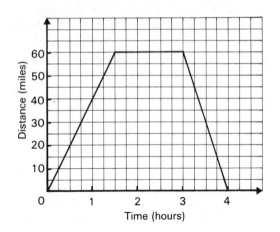

5. Fig. 22.29 shows a journey made by a lady. She travels the first 90 km by car and she then cycles for 2 hours. She then rests for a short time before completing her journey on foot:

(a) How far did she travel altogether?

(b) What was the average speed during the car drive?

(c) How many kilometres did she cycle?

(d) What was her walking speed?

(e) What was the average speed for the entire journey?

Fig. 22.29

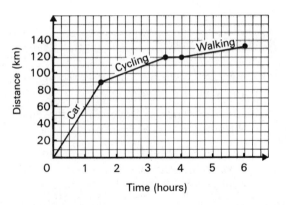

Miscellaneous Exercise 22

Section A

1. Fig. 22.30 shows part of a seating plan for a cinema. Tom's seating position is written as B5. Write down the seat positions of

   (a) Sue   (b) Dick   (c) Madge

Fig. 22.30

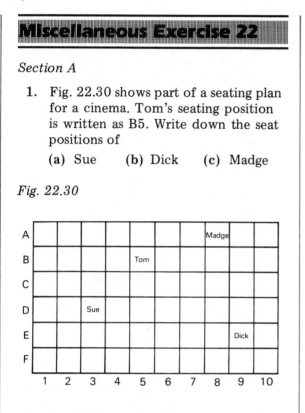

2. Fig. 22.31 represents the screen of a computer game. W shows the position of a spaceship:

   (a) Write down the coordinates of W.

   (b) A second spaceship S is at a point with the coordinates (4, 3). Copy the diagram and mark on it the position of S.

   (c) Mark on your diagram the position of a third spaceship which is the same distance as S from W and label it T.

Fig. 22.31

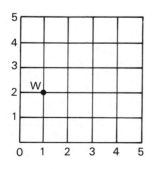

3. Fig. 22.32 shows a conversion graph which is used to convert ounces into grams:

   (a) How many ounces approximately are equivalent to 150 grams?

   (b) How many grams are there in 4 ounces?

Fig. 22.32

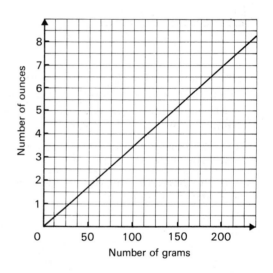

4. Fig. 22.33 represents a journey by car from Liverpool to Birmingham. Use the graph to find:

   (a) the time the car arrives at Birmingham

   (b) how long the journey took

   (c) how long the car stopped at Macclesfield

   (d) the distance from Macclesfield to Birmingham.

Fig. 22.33

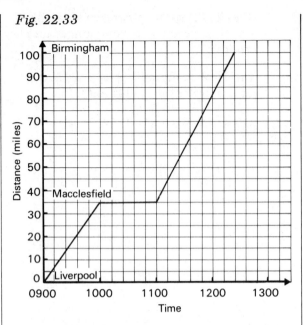

5. The graph in Fig. 22.34 shows the fusing current (in amperes) for various wire diameters (in millimetres). Use the graph to find:

(a) the fusing current for a wire which has a diameter of 17 mm.

(b) the diameter of wire needed to withstand a current of 25 amp.

Fig. 22.34

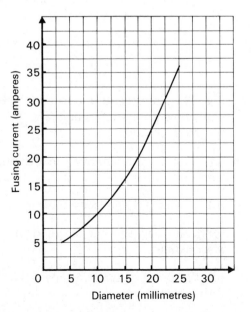

Section B

1. An electric train starts from A and travels to its next stop 5 km from B. The following readings were taken of the time since leaving A (in minutes) and the distance from A (in kilometres).

Time	1.0	1.5	2.0	2.5	3.0	3.5
Distance	0.20	0.45	0.80	1.25	1.80	2.45

Time	4.0	4.5	5.0
Distance	3.20	4.05	5.00

Draw a graph of these values taking time horizontally. Suitable scales are 2 cm to 1 unit on both axes. Use your graph to estimate the time taken to travel 2 km from A.

2. The travel graph (Fig. 22.35) shows a journey by car from London to Fishguard via Cardiff:

(a) How far is it from Cardiff to Fishguard?

(b) How long was the stop in Cardiff?

(c) What was the average speed for the entire journey?

Fig. 22.35

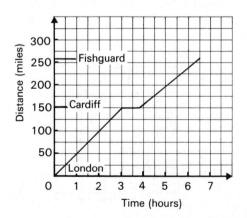

3. Using a 1 cm grid plot the following points: A(2, 2), B(4, 2) and C(2, 4). Join the points to form the triangle ABC. Use your diagram to find:

   (a) the length of the side BC

   (b) the area of triangle ABC.

4. The exchange rate is $2 = £1. Draw a conversion graph to represent this information. Take pounds on the horizontal axis to cover the range £0 to £50.

5. The graph in Fig. 22.36 shows the flight of a cricket ball after a hit by a batsman. Use the graph to find:

   (a) the height after a distance of 30 yards

   (b) the highest height reached by the ball

   (c) the height after a distance of 70 yards.

Fig. 22.36

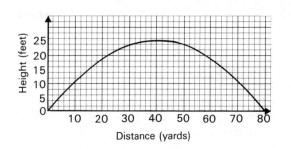

Transformations and Tessellations

Translation

We can move a shape from one position to another by sliding the shape in one direction. For instance we can slide the parallelogram (Fig. 23.1) from position A to position A′. We say that the **parallelogram has been translated** from position A to position A′.

Fig. 23.1

In Fig. 23.2 the point P has been displaced by 3 units to the right and 5 units upwards. We say that P′ is the image of P after translation under 3 units to the right and 5 units upwards.

Fig. 23.2

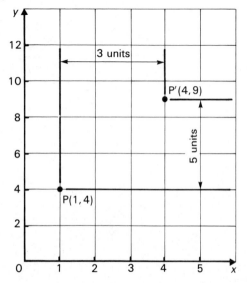

We can describe a translation by using a **column vector.** Thus the translation in Fig. 23.2 would be described by the column vector $\begin{pmatrix} 3 \\ 5 \end{pmatrix}$.

To obtain the coordinates of P′ we add 3 to the x-coordinate and 5 to the y-coordinate. The coordinates of P′ are

$(1 + 3, 4 + 5) = (4, 9)$.

Example 1

The triangle ABC (Fig. 23.3) is formed by joining the points A(1, 1), B(3, 1) and C(1, 4). Write down the coordinates of the image of ABC under the translation $\begin{pmatrix} 3 \\ 5 \end{pmatrix}$.

Fig. 23.3

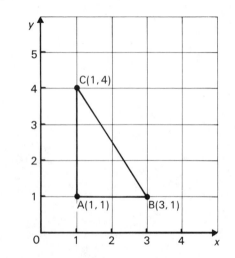

The image of A is
A′(1 + 3, 1 + 5) = A′(4, 6).

The image of B is
B′(3 + 3, 1 + 5) = B′(6, 6).

The image of C is
C′(1 + 3, 4 + 5) = C′(4, 9).

The triangle ABC and its image A′B′C′ are shown in Fig. 23.4.

Fig. 23.4

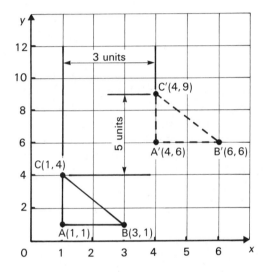

When two plane shapes are related by a translation the lengths of their corresponding sides are equal and they are also parallel to each other. We say that the two shapes are **congruent** meaning that they are exactly the same shape and size. Thus in Fig. 23.4, triangles ABC and A′B′C′ are congruent.

Reflection

The idea of **reflection** is familiar to all of us. If you look in a mirror you see a reflection of yourself. Reflection is a second transformation.

In Fig. 23.5 the point P has been reflected in the mirror M. The image of P is P′. The line PP′ is perpendicular to M and P′ is as far behind the mirror as P is in front of it. We say that M is the perpendicular bisector of PP′.

Fig. 23.5

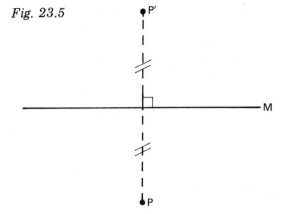

Example 2

Fig. 23.6 shows the triangle ABC.

Fig. 23.6

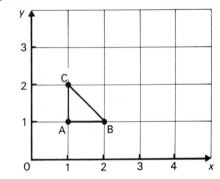

(a) Draw the image of ABC after reflection in the x-axis and mark it A′B′C′.

(b) Draw the image of ABC after reflection in the y-axis and label it A″B″C″.

The two reflections are shown dashed in Fig. 23.7.

Fig. 23.7

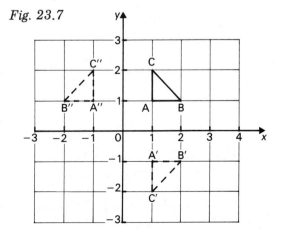

After reflection the original shape and its image are congruent shapes, i.e. they have the same shape and size. In Fig. 23.7 the triangles ABC, A′B′C′ and A″B″C″ are all congruent triangles.

Symmetry

If you look at a leaf one half looks the same as the other half. We say that the leaf is **symmetrical** about its central vein (Fig. 23.8(a)). The butterfly (Fig. 23.8(b)) is symmetrical about the line AB.

Fig. 23.8

(a)

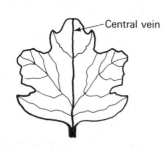

—Central vein

(b)

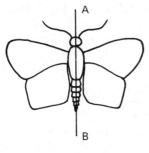

A

B

Now look at the kite drawn in Fig. 23.9. We see that it is symmetrical about the line AC. This means that triangle ABC is a reflection of triangle ADC, AC being the mirror line.

Fig. 23.9

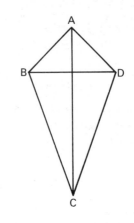

For a plane figure to be symmetrical, the shape on one side of the axis must be a reflection of the shape on the other side (Fig. 23.10). The line about which the shape is symmetrical is called the line of symmetry, the **axis of symmetry** or the mirror line.

Fig. 23.10

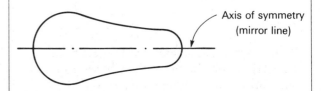

Axis of symmetry (mirror line)

Some shapes have more than one axis of symmetry. For instance, the square shown in Fig. 23.11 has four axes of symmetry.

Fig. 23.11

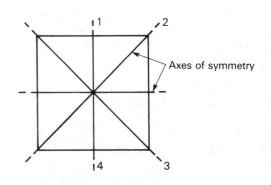

Axes of symmetry

Example 3

Complete the shape shown in Fig. 23.12 so that it has two axes of symmetry.

Fig. 23.12
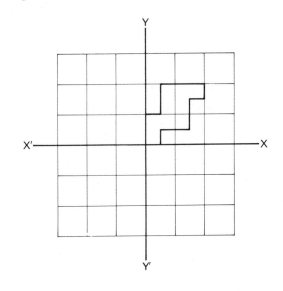

The shape is completed by first reflecting it in the line YY′ and then reflecting the double shape in the line XX′. The shape now has the two axes of symmetry, XX′ and YY′ as shown in Fig. 23.13.

Fig. 23.13
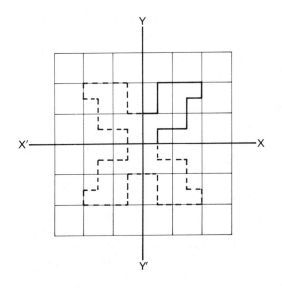

Rotation

To define a **rotation** we need to know the angle of rotation and its direction (taken to be anticlockwise if not stated). We also need to know the position of the centre of the rotation.

In Fig. 23.14, the triangle ABC has been rotated to a new position A′B′C′. The centre of rotation is the point O and the angle of rotation is $\theta°$ which has been made in a clockwise direction.

Fig. 23.14
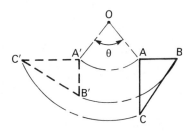

Example 4

The rectangle ABCD (Fig. 23.15) is to be rotated through one-quarter of a turn anti-clockwise with O the centre of rotation. Draw the image of ABCD under this rotation.

The image of ABCD has been drawn dashed in the diagram and labelled A′B′C′D′. Note that one-quarter of a turn anticlockwise is the same as a rotation of 90° anticlockwise.

Fig. 23.15
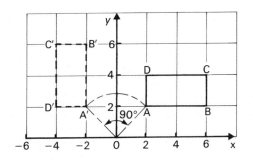

Rotational Symmetry

If the shape shown in Fig. 23.16 was cut out of cardboard and a pin put through its centre it could be rotated. It would appear to be in the same position three times during one complete turn. We say that the shape has **rotational symmetry** of order 3.

Fig. 23.16

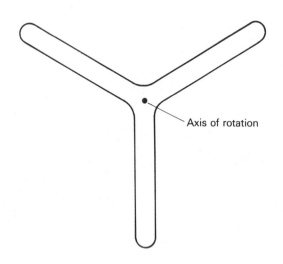

The regular pentagon (Fig. 23.17) has rotational symmetry of order 5 whilst the rectangle ABCD (Fig. 23.18) has rotational symmetry of order 2.

Fig. 23.17

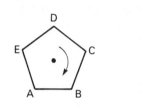

Fig. 23.18

The shape shown in Fig. 23.19 has a line of symmetry. However, when it is rotated it is in the same position only once during one complete turn. The shape is said to have rotational symmetry of order 1.

Fig. 23.19

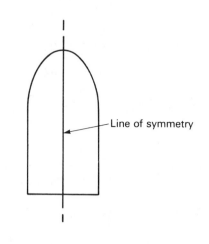

Translations, reflections and rotations are often used for wallpaper, carpet and curtain material patterns.

Example 5

Fig. 23.20 shows part of a wallpaper design. Complete the pattern.

Fig. 23.20

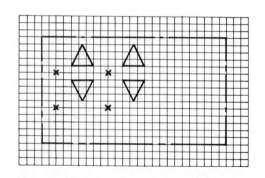

The completed pattern is shown in Fig. 23.21 where it will be seen that translations and reflections have been used to complete the design.

Fig. 23.21

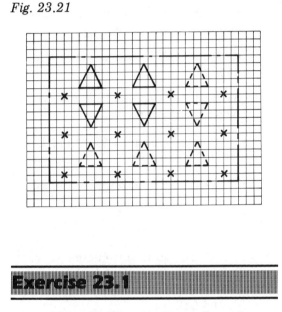

Exercise 23.1

1. In Fig. 23.22 write down the column vectors which describe the translations of

   (a) P to P′   (b) Q to Q′

   (c) R to R′

Fig. 23.22

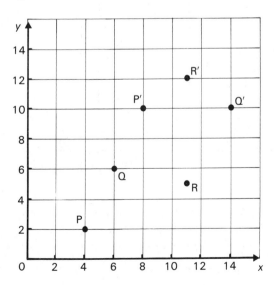

2. The image of triangle ABC under a translation is A′B′C′ (Fig. 23.23). Write down the column vector which describes this translation.

Fig. 23.23

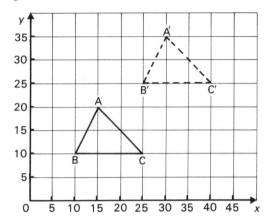

3. Copy Fig. 23.24 and draw the image of the rectangle ABCD under the translation $\begin{pmatrix} 4 \\ 6 \end{pmatrix}$.

Fig. 23.24

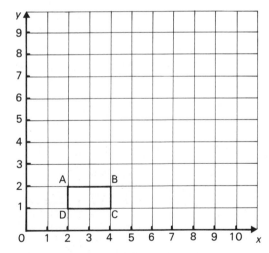

4. Draw a pair of axes to cover the range of values from 0 to 8 units on both axes using scales of 1 cm to 1 unit:

   (a) Plot the points A(1, 3) and B(5, 2). Join the points A and B.

   (b) Translate the line AB under $\begin{pmatrix} 3 \\ 5 \end{pmatrix}$.

5. Copy Fig. 23.25 and reflect the shaded shape in **(a)** the x-axis, **(b)** the y-axis.

Fig. 23.25

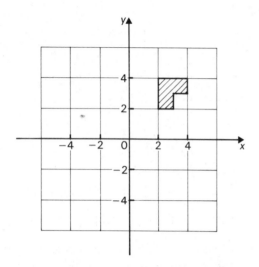

6. Copy Fig. 23.26 on to squared paper and reflect the shape ABCDEFGH

(a) in the x-axis

(b) in the y-axis.

(c) Complete the drawing so that a shape having two axes of symmetry is produced.

Fig. 23.26

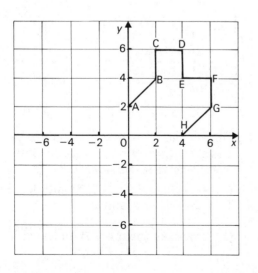

7. Draw on graph paper two axes to cover the range −6 to +6 on both axes using a scale of 1 cm to 1 unit:

(a) Plot the point A(4, 5).

(b) Reflect A in the x-axis to give point B.

(c) Reflect A in the y-axis to give point D.

(d) Reflect the point D in the x-axis to give the point C.

(e) Join the points A, B, C and D in alphabetical order to give a plane shape.

(f) What is the name given to this shape?

8. Fig. 23.27 shows one-quarter of a decorative door. Complete the door by reflecting the given shape in the lines AB and CD.

Fig. 23.27

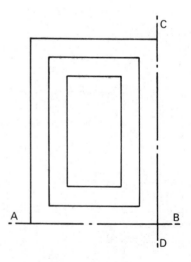

9. Copy the shape shown in Fig. 23.28 and reflect it

(a) in the x-axis

(b) in the y-axis.

Fig. 23.28

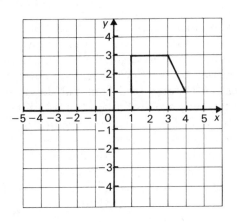

10. Copy Fig. 23.29 and reflect the letter E in the *y*-axis.

Fig. 23.29

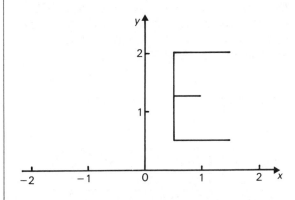

11. Fig. 23.30 shows a rhombus drawn inside a rectangle. Copy the diagram and show on it all the axes of symmetry.

Fig. 23.30

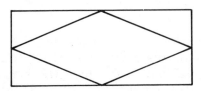

12. Patterns based upon translations, reflections and rotations of simple shapes are often used in carpet designs. Fig. 23.31 shows part of such a pattern. Copy the diagram and complete it.

Fig. 23.31

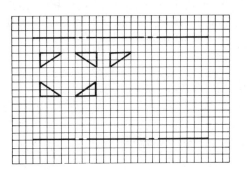

13. Using a scale of 1 cm to 1 unit on both the *x* and *y* axes draw a pair of axes on graph paper for values of *x* and *y* from -5 to $+5$ units:

   (a) Plot the point P(4, 3).

   (b) Rotate P through a one-quarter turn anticlockwise about the origin and label the image P′.

   (c) Now rotate P through half a turn anticlockwise about the origin. Label this image P″.

14. Copy Fig. 23.32 and rotate the triangle ABC through 90° clockwise about the origin. Label the image A′B′C′.

Fig. 23.32

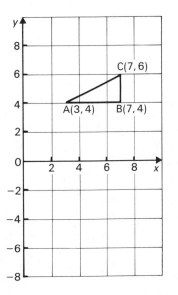

15. Copy Fig. 23.33 and then rotate the rectangle ABCD through 90° anti-clockwise about the origin. Label the resulting image A'B'C'D'.

Fig. 23.33

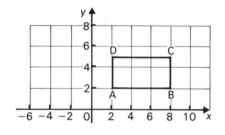

16. Plot the points A(0, 0), B(2, 0), C(2, 3), D(4, 5) and E(0, 5). Join them up in alphabetical order to form the plane shape ABCDE. Reflect ABCDE in the *x*-axis and then in the *y*-axis. Finally complete the drawing so that the resulting shape has two axes of symmetry.

17. Write down the order of rotational symmetry for each of the shapes shown in Fig. 23.34.

Fig. 23.34

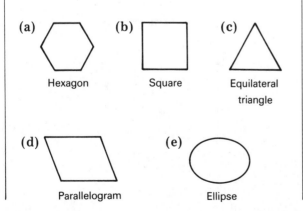

Similarities

We can enlarge a photograph to make it any size, within reason, that we wish. Suppose that the original photograph is 7 cm long by 4 cm wide. We can enlarge it so that its length is 14 cm and its width is 8 cm. We have multiplied the length and width by 2. The original photograph and its enlargement are the same shape and we call them **similar** figures. The amount of the enlargement is called the **scale factor.** In the case of the photograph the scale factor is 2.

Example 6

Draw the enlargement of rectangle ABCD (Fig. 23.35) if the scale factor is 2. Mark the enlargement A'B'C'D' and write down the lengths of A'B' and B'C'.

Fig. 23.35

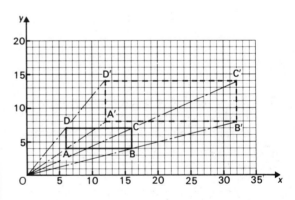

To form the enlargement of ABCD draw the radial lines OAA', OBB', OCC' and ODD' where OA' = 2 × OA, OB' = 2 × OB, OC' = 2 × OC and OD' = 2 × OD.

From the diagram we see that A'B' = 20 units and B'C' = 6 units. That is, A'B' is twice as long as AB and B'C' is twice as long as BC.

Example 7

Fig. 23.36 shows the triangles ABC and XYZ:

(a) Is XYZ an enlargement of ABC and if so, what is the scale factor?

(b) What is the length of XY?

Fig. 23.36

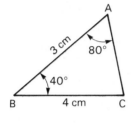

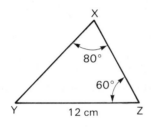

(a) The three angles of triangle ABC are 40°, 60° and 80° and so are the angles of triangle XYZ.

Therefore the two triangles are similar in shape and so XYZ is an enlargement of ABC.

The scale factor is $12 \div 4 = 3$.

(b) AB and XY are corresponding sides because they both lie opposite to the angle of 60°. Therefore

$$XY = 3 \times AB$$

$$= 3 \times 3$$

$$= 9 \, cm$$

Exercise 23.2

1. In Fig. 23.37, what is the scale factor of the enlargement?

Fig. 23.37

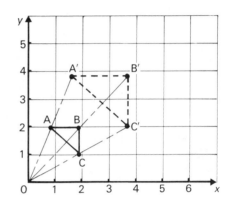

2. **(a)** In Fig. 23.38, what is the scale factor of the enlargement?

   (b) Write down the lengths of A'B' and B'C'.

Fig. 23.38

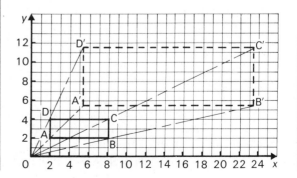

3. **(a)** Are the triangles PQR and XYZ similar (Fig. 23.39)?

   (b) If so what is the scale factor?

   (c) What is the length of YZ?

Fig. 23.39

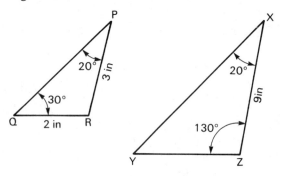

4. The rectangle ABCD (Fig. 23.40) is an enlargement of rectangle EBGF with a scale factor of 4. Work out the lengths of EF and CD.

Fig. 23.40

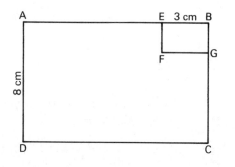

5. Two regular octagons (eight-sided figures) have sides whose lengths are 3 cm and 12 cm respectively:

   (a) What is the scale factor of the enlargement?

   (b) Work out the perimeters of both octagons.

Patterns

Patterns can often be made by using sequences of numbers. It will be recalled that a set of numbers connected by some definite law is called a **sequence** of numbers.

Each row of the stack of tins shown in Fig. 23.41 contains one more tin than the row above. The sequence of the total number of tins, row by row, is

$$1, 3, 6, 10, \ldots$$

This is the sequence (see Chapter 1) of **triangular numbers**.

Fig. 23.41

Example 8

The snooker balls in Fig. 23.42 form a definite pattern. Work out the number of snooker balls in row five.

   The sequence is:

   $$1, \ 1 + 2, \ 3 + 3, \ 6 + 4, \ 10 + 5$$

   The number of balls in row five is 5.

Fig. 23.42

Fig. 23.43 shows the pattern of **square numbers**. From the diagram we see that the sequence is

$$1 \times 1 = 1, \quad 2 \times 2 = 4, \quad 3 \times 3 = 9,$$
$$4 \times 4 = 16$$

and so on.

Fig. 23.43

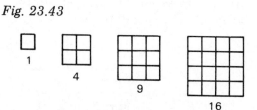

Example 9

Susan forms the pattern shown in Fig. 23.44 by using sugar lumps. Draw the fourth and fifth arrangements and work out the number of sugar lumps needed to make the seventh arrangement.

Fig. 23.44

The number of sugar lumps in each arrangement is given by each term in the sequence

$$1, 4, 9, 16, 25, \ldots$$

They are, in fact, the sequence of square numbers.

The fourth and fifth arrangements are shown in Fig. 23.45.

Fig. 23.45

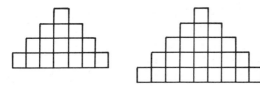

The number of sugar lumps in the seventh arrangement is $7 \times 7 = 49$.

Numbers which are not prime can be arranged in either a square or a rectangle and they are called **rectangular numbers**. Some examples are shown in Fig. 23.46.

Fig. 23.46

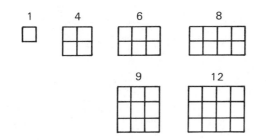

Tessellations

A **tessellation** is a pattern which could go on for ever. Tessellations are often used, for instance, in tiling and carpet patterns.

Fig. 23.47 shows a tessellation of rectangles.

Fig. 23.47

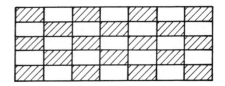

Example 10

Fig. 23.48 shows a tessellation of equilateral triangles:

(a) How many triangles fit round the point A?

(b) What do the angles round point A add up to?

Fig. 23.48

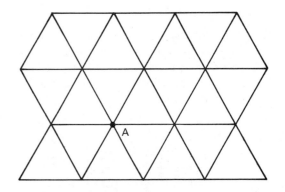

(a) 6 equilateral triangles fit round point A.

(b) Each angle of an equilateral triangle is $60°$. The angles round point A add up to $6 \times 60° = 360°$.

Exercise 23.3

1. John wants to tile a wall using the triangular-shaped tiles shown in Fig. 23.49. On a 1 cm grid make a diagram showing how John started tiling the wall using nine triangular tiles.

Fig. 23.49

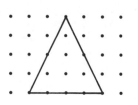

2. Maud made the arrangements shown in Fig. 23.50 by using wooden building blocks:

   (a) Make a drawing showing the fourth arrangement.

   (b) How many blocks will be needed for the sixth arrangement?

Fig. 23.50

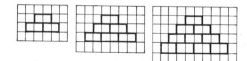

3. Fig. 23.51 shows a tiling pattern made with square black and white tiles. Copy the diagram and complete it. How many white tiles are there in the complete pattern?

Fig. 23.51

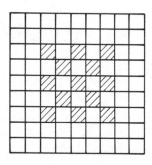

4. Fig. 23.52 shows three arrangements for a display of tiles. Draw the next two arrangements if the sequence is to be maintained. How many white tiles will there be in the seventh arrangement?

Fig. 23.52

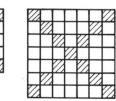

5. Fig. 23.53 shows part of a pattern made with hexagonal tiles:

   (a) Continue the pattern to show nine tiles.

   (b) How many tiles fit around the point X?

   (c) What size is each angle of a regular hexagon?

Fig. 23.53

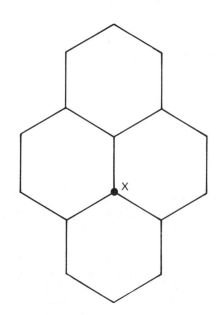

Miscellaneous Exercise 23

Section A

1. Fig. 23.54 shows a point P plotted on squared paper. Copy the diagram and answer the following questions:

   (a) What are the coordinates of P?

   (b) Reflect P in the x-axis and label the image P'. What are the coordinates of P'?

   (c) Reflect P in the y-axis and label the image P''. What are the coordinates of P''?

Fig. 23.54

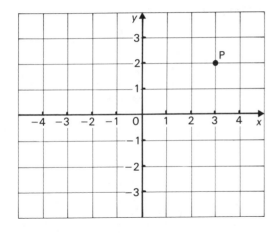

2. Fig. 23.55 shows pictures of some of the shapes found in nature and architecture. Make sketches of these shapes. On each of your sketches mark the axis of symmetry.

Fig. 23.55

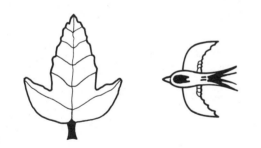

3. Fig. 23.56 shows part of a wallpaper pattern. Copy the diagram and complete the pattern.

Fig. 23.56

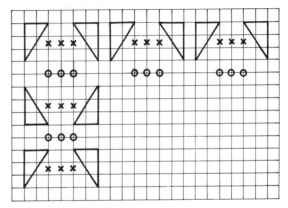

4. Copy Fig. 23.57 and show the result of giving the triangle a quarter turn anticlockwise about the point A.

Fig. 23.57

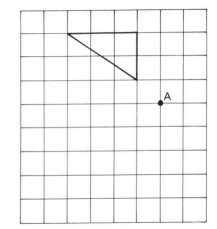

5. Two squares are drawn. Square A has a side 2 inches long whilst square B has a side which is 6 inches long:

  (a) What is the enlargement factor?

  (b) What is the perimeter of square B?

  (c) What is the area of square A?

  (d) What is the area of square B?

Section B

1. Fig. 23.58 shows part of the pattern for the black squares of a crossword puzzle. The finished pattern is to be symmetrical in both directions. Copy the diagram and complete the pattern.

Fig. 23.58

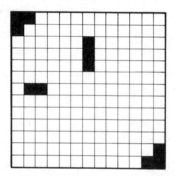

2. Fig. 23.59 shows a tile in the shape of a diamond. Peter wants to tile a wall using tiles of this shape. Make a diagram showing how the first nine tiles should be fitted. (Use dots 1 cm apart.)

Fig. 23.59

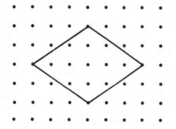

3. Fig. 23.60 shows a bathroom wall tile with part of the pattern shaded. Copy the diagram and complete the figure so that it is symmetrical about the lines AB and CD. Shade the figure where appropriate.

Fig. 23.60

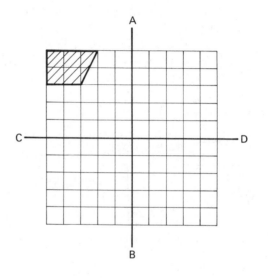

4. The equilateral triangles in Fig. 23.61 are made with matchsticks:

  (a) Draw the fourth arrangement.

  (b) How many matchsticks are needed for the sixth arrangement?

Fig. 23.61

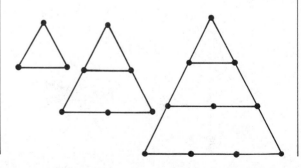

5. Fig. 23.62 shows a rectangle ABCD. Copy the diagram and enlarge it with a scale factor of 2. Label the enlarged figure A′B′C′D′:

(a) Write down the coordinates of the points A, B, C and D.

(b) What are the lengths of the sides AB and BC?

(c) What is the length of A′B′?

(d) What is the area of ABCD?

(e) What is the area of A′B′C′D′?

Fig. 23.62

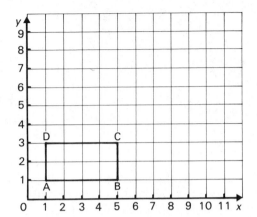

Statistics

Statistics is the name given to the science of collecting and analysing facts. Originally statistics used only facts about the state, hence its name. Nowadays, though, it is used in business, pure science, insurance, etc.

Raw Data

Raw data is collected information which is not arranged in any kind of order.

Consider the scores obtained by 50 students in a test:

```
4 3 3 5 5 6 5 8 7 6 7 8
9 5 4 1 8 7 5 6 6 7 5 2
5 2 6 9 5 7 6 5 6 2 8 6
7 3 3 8 7 6 5 5 6 4 3 4
5 7
```

This is an example of raw data and we see that the information is not arranged in any sort of order.

Frequency Distributions

The information about the scores of the students does not mean very much but if the scores are grouped and placed in a table then it is easier to see if a pattern emerges.

Example 1

Form a **frequency distribution** for the information about the scores of 50 students in a test. The easiest way to form a frequency distribution is to use a **tally chart**.

On looking at the information we see that the lowest score is 1 whilst the highest score is 9. The scores 1 to 9 are written in the first column of the tally chart. We now take each score in the raw data just as it comes and put a tally mark in the appropriate row.

The fifth tally mark for each score is made in an oblique direction, thereby tying the tally marks into bundles of five. This makes counting much easier. When all the tally marks have been entered they are counted and the total for each score placed in the column headed frequency. The frequency is simply the number of times a score appears in the raw data.

From the tally chart it will be seen that the score 2 occurs three times (a frequency of 3) whilst the score 6 occurs ten times (a frequency of 10), etc.

TALLY CHART 1

Score	Tally	Frequency
1	I	1
2	I I I	3
3	⊥⊥⊥⊤	5
4	I I I I	4
5	⊥⊥⊥⊤ ⊥⊥⊥⊤ I I	12
6	⊥⊥⊥⊤ ⊥⊥⊥⊤	10
7	⊥⊥⊥⊤ I I I	8
8	⊥⊥⊥⊤	5
9	I I	2

The frequencies can be displayed on a bar chart (Fig. 24.1) with all the rectangles of equal width. The frequencies are represented by the heights of the rectangles.

Fig. 24.1

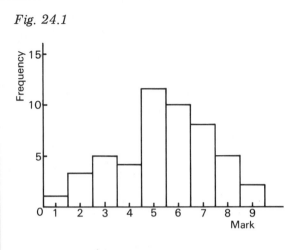

Grouped Distributions

When a large amount of information has to be organised a grouped frequency distribution is used.

Example 2

The scores obtained in a recent physics examination were as follows:

    19 37 42 19 27 42 37 27 7 33
    30 12 6 31 24 25 50 29 38 35

Use this information to obtain a grouped frequency distribution with classes 1–10, 11–20, 21–30, 31–40 and 41–50.

TALLY CHART 2

Score	Tally	Frequency
1–10	I I	2
11–20	I I I	3
21–30	⊥⊦⊦⊦ I	6
31–40	⊥⊦⊦⊦ I	6
41–50	I I I	3

As shown in Fig. 24.2 the frequencies can be displayed on a bar chart.

Fig. 24.2

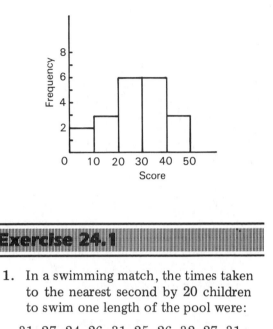

Exercise 24.1

1. In a swimming match, the times taken to the nearest second by 20 children to swim one length of the pool were:

   31 27 24 26 31 25 26 32 27 31
   26 32 30 32 29 25 29 27 26 28

   Draw up a tally chart and hence form a frequency distribution. Represent this frequency distribution on a bar chart.

2. The goals scored by a football team in 30 matches were:

   4 1 3 2 0 1 1 1 1 0 2 5 0 0 4
   1 1 0 1 1 5 1 2 1 1 0 2 1 2 0

   (a) Arrange this information in a frequency table using classes of 0 goals, 1 goal, 2 goals, 3 goals, 4 goals and 5 goals.

   (b) Draw a bar chart to represent this information.

3. The bar chart (Fig. 24.3) gives the heights of a group of 54 fourteen-year-old boys. Copy and complete the following table:

Height (cm)	150	151	152	153	154	155
Frequency						

Height (cm)	156	157	158	159	160	161
Frequency						

Fig. 24.3

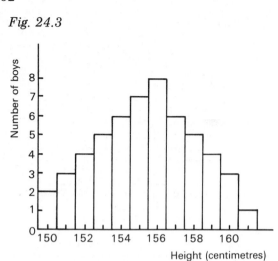

Height (centimetres)

4. A survey one evening was made of the ages of 30 members of a youth club with the following results:

    14 16 16 15 15 14 14 17 18 15
    14 16 16 14 14 15 17 15 14 15
    14 14 18 16 15 14 15 17 17 15

    Tabulate these results to form a frequency distribution using classes 14, 15, 16, 17 and 18 years. Draw a bar chart to represent this information.

5. The runs scored on one Saturday afternoon by 50 batsmen belonging to St. John's Cricket Club were:

    16   1 17   8   9 11 19 13 10 18
    21   2 13 10 14 15 10   4 11 20
    12 10 28   9 11 12   9 11   9   5
     6 16   8 24   7   9   8   7   6   0
     3   7 12 10 10 13   5 15 14   8

    Using these scores copy and complete the following table:

Score	Tally	Frequency
0–3		
4–7		
8–11		
12–15		
16–19		
20–23		
24–27		
28–31		

6. The bar chart (Fig. 24.4) relates to the ages of children in a small school:

    (a) How many children were aged 13 years?

    (b) What was the total number of children attending the school?

Fig. 24.4

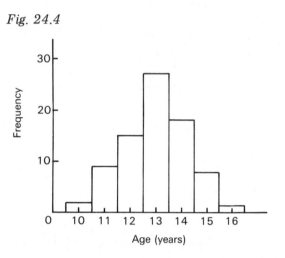

Age (years)

7. The bar chart of Fig. 24.5 shows the shoe sizes of pupils attending a comprehensive school. Complete the following table:

Shoe size	2	3	4	5	6	7	8	9
Frequency								

Fig. 24.5

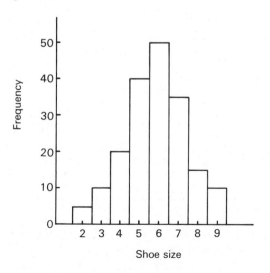

Shoe size

8. Draw a bar chart to represent the information given in this table:

Length (nearest cm)	Frequency
11–15	4
16–20	7
21–25	12
26–30	8
31–35	3

Statistical Averages

Three kinds of statistical average are used. They are the **arithmetic mean** (usually called the mean), the **median** and the **mode**.

The arithmetic mean is found by adding all the values in the set and dividing by the number of values making up the set.

$$\text{Mean} = \frac{\text{Total of all the values}}{\text{The number of values}}$$

Example 3

Find the mean of the numbers 3, 5, 8, 9 and 10.

$$\text{Mean} = \frac{3 + 5 + 8 + 9 + 10}{5}$$

$$= \frac{35}{5}$$

$$= 7$$

When a set of values is arranged in ascending (or descending) order the median is the middle value.

Example 4

Find the median of the numbers 5, 4, 2, 8, 7, 2, 9, 7 and 3.

Arranging the numbers in ascending order we have

2, 2, 3, 4, 5, 7, 7, 8, 9

The middle number is 5 and so the median is 5. When there is an even number of values the median is found by taking the mean of the two middle values.

Example 5

Find the median of the numbers 8, 5, 4, 7, 9, 10, 2 and 5.

Arranging the numbers in descending order we have

10, 9, 8, 7, 5, 5, 4, 2

The middle two values are 7 and 5 and so

$$\text{Median} = \frac{7 + 5}{2}$$

$$= 6$$

The mode of a set of values is the value which occurs most frequently.

Example 6

Find the mode of the numbers 5, 5, 5, 6, 6, 7, 7, 8, 9 and 9.

The mode is 5 because there are more fives than any other number.

Uses of Statistical Averages

The mean is the most commonly used average. Indeed when most people talk of an average they are talking about the mean, for example, batting averages in cricket.

Example 7

The hourly rate of pay for four office workers are £4.06, £4.32, £9.40 and £2.22. Calculate their average rate of pay.

$$\text{Mean} = \frac{4.06 + 4.32 + 9.40 + 2.22}{4}$$

$$= \frac{20.00}{4}$$

$$= 5$$

So the mean hourly rate of pay is £5.

The mean value in Example 7 is greatly distorted by the low hourly rate of £2.22 and the high rate of £9.40 per hour — it is not really a reliable guide to the amounts paid to the four office workers. In fact, the median rate will give a better guide.

Example 8

Find the median for the rates of pay given in Example 7.

Arranging the values in ascending order we have

2.22, 4.06, 4.32, 9.4

The two middle values are 4.06 and 4.32.

$$\text{Median} = \frac{4.06 + 4.32}{2}$$

$$= \frac{8.38}{2}$$

$$= 4.19$$

So the median wage is £4.19 and this is a better average to use in this case.

Example 9

A shop sells 2 pairs of shoes of size 4, 3 pairs of size 5, 2 pairs of size 6, 4 pairs of size 7 and 3 pairs of size 8. What is the mean, median and modal size?

Putting the sizes in order we have

4, 4, 5, 5, 5, 6, 6, 7, 7, 7, 7, 8, 8, 8

The mean size = 6.21

The median size = 6.5

The modal size = 7

The mean and median sizes tell us something about the size of feet but neither average is a stock size and is not of much use to the shopkeeper. The modal value tells us that more shoes of size 7 were sold than any other size. This is what the shopkeeper wants to know when he is ordering more shoes.

The Range

As shown in Fig. 24.6 the range is a measure of the spread of the data. The larger the range the greater the spread of the data.

The range of a set of numbers is the largest number minus the smallest number.

Fig. 24.6

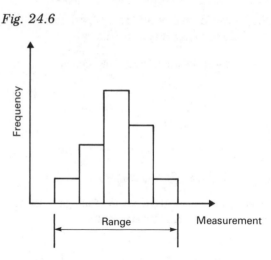

Example 10

The wages paid in an office are £114, £124, £168, £136 and £160 per week. Work out the range of these wages.

Highest wage = £168

Lowest wage = £114

Range of wages = £168 − £114 = £54

Exercise 24.2

1. Work out the mean of the numbers 5, 8, 9 and 10.

2. Find the mean of £23, £27, £30, £28 and £32.

3. The heights of some men are 172, 170, 168, 181, 175, 179 and 173 cm. What is the mean height of these men?

4. Find the median of the numbers 2, 3, 4, 5, 6, 8 and 8.

5. Find the median of the number 2, 4, 4, 6, 6, 6, 7 and 9.

6. Find the median of 8, 9, 6, 5, 5, 2 and 9.

7. Find the median of 2, 5, 7, 6, 9, 7, 5 and 8.

8. What is the mode of the numbers 3, 5, 4, 8, 3, 6, 5, 9, 5, 4 and 7?

9. Ten people were asked to guess the weight of a cake. Their estimates, to the nearest half kilogram, were as follows: $3\frac{1}{2}$, $2\frac{1}{2}$, 2, 1, $3\frac{1}{2}$, 2, $3\frac{1}{2}$, 3, 3, 1 kg. What was:

   (a) the estimated mean weight

   (b) the estimated modal weight

   (c) the estimated median weight?

10. Six people occupy a lift. Their weights, to the nearest kilogram, are 84, 67, 73, 76, 80 and 82 kg. Work out:

    (a) their median weight

    (b) their mean weight.

11. The weights of five boys were 112, 120, 106, 126 and 121 lb. Find the range of the weights.

12. The wages of six workers in a factory were £127, £122, £136, £142, £114 and £117. Find the range of these wages.

Probability

Probability has great importance in insurance (the risks have to be carefully calculated before the premiums can be set), investment in business, medical research and so on.

If a coin is tossed the result is equally likely to be heads or tails. We say that the events of tossing heads or tossing tails are **equiprobable events.**

Similarly if we throw a normal die (plural dice) with six faces the events of throwing a 1, 2, 3, 4, 5 or 6 are all equally likely to happen, i.e. they are equiprobable events. The probability of an event occurring is defined as

$$\frac{\text{Number of ways the event can occur}}{\text{Total number of possible outcomes}}$$

A probability can be written either as a fraction or as a decimal.

Example 11

(a) A fair coin is tossed once. What is the probability that a head will result?

Number of ways in which a head can occur = 1. Total number of possible outcomes = 2.

Probability of a head $= \frac{1}{2}$

$= 0.5$

(b) What is the probability of throwing a 5 in a single roll of a fair die?

Number of ways in which a 5 can occur = 1. Total number of possible outcomes = 6.

Probability of a 5 $= \frac{1}{6}$

(c) A card is dealt from the top of a well shuffled pack of playing cards. Work out the probability that it will be an ace.

Number of ways in which an ace can occur = 4 (because there are 4 aces in the pack).

Number of possible outcomes = 52 (because there are 52 cards in the pack each having the same chance of being the top card).

Probability of dealing an ace $= \frac{4}{52}$

$= \frac{1}{13}$

All probabilities have a value between 0 and 1. A probability of 0 means absolute impossibility (for example, the probability that you will swim the Atlantic Ocean). A probability of 1 means absolute certainty (for example, the probability of picking the winner of a one-horse race).

Example 12

A die has six faces numbered 1, 2, 3, 4, 5 and 6. Find the probability of

(a) throwing a 7

(b) throwing a 1, 2, 3, 4, 5 or 6.

(a) Throwing a 7 on a die with six faces is impossible, so the probability of this happening is 0.

(b) When the die is thrown one of the numbers 1, 2, 3, 4, 5 or 6 must turn up. So we are certain to throw a 1, 2, 3, 4, 5 or 6. Hence the probability of this happening is 1.

Total Probability

If we toss a fair coin it will come down heads or tails. That is

$$\text{Probability of a head } = \tfrac{1}{2}$$

$$\text{Probability of a tail } = \tfrac{1}{2}$$

Total probability covering all possible outcomes is $\tfrac{1}{2} + \tfrac{1}{2} = 1$.

Another way of saying this is

Probability of success

+ Probability of failure $= 1$

Example 13

A bag contains 5 blue balls, 3 red balls and 2 black balls. A ball is drawn at random from the bag. Calculate the probability that it will not be black.

Probability of a black ball $= \tfrac{2}{10} = 0.2$

(because there are 2 black balls and 10 balls in all, each of which has an equal chance of being selected).

Probability of not drawing a black ball
$= 1 - 0.2 = 0.8$.

Experimental Probability

Although it is possible to calculate many probabilities in the way previously shown, in a great many cases we have to rely on an experiment or an enquiry to establish the probability of an event happening.

Example 14

100 ball bearings are examined and 4 are found not to be round. What is the probability of selecting a non-round ball bearing out of the 100 examined?

Total number of trials conducted $= 100$.

Number of ways in which a non-round ball bearing can be selected $= 4$.

Probability of selecting a non-round ball bearing $= \tfrac{4}{100} = \tfrac{1}{25}$.

In industry probabilities are worked out just like this, and these experimental probabilities can usually be relied upon. In the test on the ball bearings the calculated probability $\tfrac{1}{25}$ would be used to estimate the number of defective (i.e. non-round) ball bearings that will be produced. The sample size is very important and, generally speaking, the larger the sample the more accurate the probability will be.

When we say that the probability of an event happening is $\tfrac{1}{3}$ we do not mean that if we repeat the experiment three times the event will happen once. Even if we repeat the experiment 30 times it is unlikely that the event will happen exactly ten times.

Probability tells us what to expect in the long run. If the experiment is repeated 300 times then we would expect the event to happen about 100 times.

Exercise 24.3

1. A die has six faces numbered 1, 2, 3, 4, 5 and 6. It is thrown once. Find the probability that:

   (a) a 3 will turn up

   (b) a number less than 4 will turn up

   (c) an even number will turn up.

2. A letter is chosen at random from the word TERRIFIC. Find the probability that it will be:
   (a) an F   (b) an R.

3. A card is dealt from the top of a well shuffled pack of 52 playing cards. Calculate the probability that it will be:
   (a) the jack of hearts
   (b) a king
   (c) a red card
   (d) a spade.

4. A bag contains 3 red balls, 5 blue balls and 2 green balls. A ball is drawn at random from the bag. Calculate the probability that it will be:
   (a) red   (b) blue
   (c) green   (d) not green.

5. The probability that John will be selected for the school football team is 0.6. What is the probability that he will *not* be selected?

6. There are seven tomatoes in a bag. Four of them are red and the remainder are green. A tomato is selected at random. What is the probability that it will be green?

7. In a game it is my turn to throw the die. I need a six or a one to win the game. What is the probability that I will win?

8. The probability of a particular couple having a child with brown eyes is 0.4. What is the probability that the child will not have brown eyes?

9. Tests showed that out of 5000 electric light bulbs produced in a factory, 50 were defective. If a light bulb is chosen at random, what is the probability of choosing a defective bulb?

10. 50 invoices were checked and 3 of them were found to contain errors. Find the probability that an invoice chosen at random would contain errors.

Miscellaneous Exercise 24

Section A

1. The information below relates to the scores of a class of 30 students in a mathematics test. The marks are out of 10:

   4, 3, 8, 8, 9, 7, 7, 6, 5, 6, 7, 8, 4, 6, 4,
   8, 7, 6, 7, 8, 5, 5, 7, 9, 6, 9, 5, 7, 6, 9

   (a) Using classes 3, 4, 5, 6, 7, 8 and 9 draw up a tally chart and so obtain a frequency distribution.

   (b) Draw a bar chart to represent this information.

2. The lengths of 50 pieces of wood were measured, with the following results:

Length (mm)	Frequency
295	1
296	2
297	6
298	9
299	15
300	10
301	4
302	2
303	1

   Draw a bar chart of this information.

3. Fig. 24.7 is a bar chart which illustrates the frequency distribution for the heights (in centimetres) of a group of boys all aged 14 years. Copy and complete the following table:

Height (cm)	Frequency
150	
151	
152	
153	
154	
155	
156	
157	
158	
159	
160	
161	

Fig. 24.7

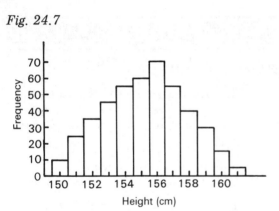

4. The temperatures, in degrees Celsius, at noon on five days in January were as follows:

8, 12, 10, 9, 11

(a) What was the mean noon temperature over the five days?

(b) What was the range of temperatures?

5. The figures which follow give the population of seven villages taken after a census:

1864 2467 1392 1459 2134 9803 5072

Find the median population.

6. The probability of Susan being selected for the school hockey team is 0.3. What is the probability that she will not be selected?

7. The shoe sizes of 9 girl pupils were:

2, 4, 4, 6, 5, 7, 4, 2, 6

What is the modal shoe size?

Section B

1. The scores of a student in five different examinations were:

54, 63, 49, 78, 61

What was the candidate's mean score?

2. The pointer shown in Fig. 24.8 is spun once:

(a) Find the probability that the pointer stops in section B.

(b) Find the probability that the pointer stops in R or B sections.

Fig. 24.8

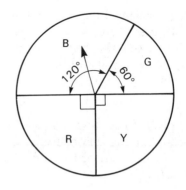

3. As part of an experiment in botany the lengths of 40 laurel leaves were measured, to the nearest millimetre, with the following results:

161 146 168 146 138 164 158
126 173 145 135 142 138 140
150 132 147 176 147 142 150
135 163 136 144 156 153 120
148 125 145 140 154 152 149
128 135 165 144 157

(a) Make a tally chart to form a grouped distribution using the classes 120-129, 130-139, 140-149, 150-159, 160-169, 170-179.

(b) Draw a bar chart to represent this information.

4. In a certain town it has been calculated that the probability of a child catching measles is 0.14. What is the probability that a child in that town will not catch measles?

5. Ten pupils were asked to estimate the length, to the nearest centimetre, of their teacher's table. Their estimates were as follows:

148, 134, 146, 129, 133, 140, 132, 140, 132, 140

(a) Find the median estimate.

(b) What is the modal estimate?

Multi-Choice Questions 24

1. The mean of 12, 18, 26, 33 and 36 is
   A 13   B 25   C 26   D 36

2. Find the median of 13, 15, 14, 11, 12, 15, 17, 13, 14 and 13.
   A 12.5   B 13   C 13.5   D 14

3. A car park contains 100 cars of which 28 are blue and 34 are red. What is the probability that if a car is selected at random it is neither blue nor red?
   A 0.28   B 0.34   C 0.38   D 0.62

4. The tally chart below shows the distribution of a sample of mixed nuts. What fraction of the nuts were almonds?

   Brazils   ⊦⊦⊦⊦
   Walnuts   ⊦⊦⊦⊦ I I
   Almonds   ⊦⊦⊦⊦ I I I
   Chestnuts   ⊦⊦⊦⊦ I I I
   Cobnuts   ⊦⊦⊦⊦ ⊦⊦⊦⊦ I I

   A $\frac{1}{8}$ B $\frac{7}{40}$ C $\frac{1}{5}$ D $\frac{3}{10}$

5. Calculate the mean of 1, 2, 5, 7 and 15.
   A 6   B 7   C 15   D 30

6. The bar chart (Fig. 24.9) shows the ages of children in a school. How many children took part in the survey?
   A 5   B 20   C 65   D 263

Fig. 24.9

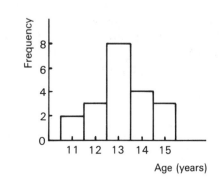

7. The marks of ten students in a test were:

   17, 13, 14, 19, 18, 17, 15, 14, 17, 15

   What is the modal mark?
   A 19   B 17.5   C 17   D 15.9

8. In a ten-horse race, the probability that one or other of two particular horses will win is $\frac{2}{5}$. The probability that the race will be won by one of the remaining eight horses is
   A $\frac{1}{8}$ B $\frac{1}{5}$ C $\frac{3}{5}$ D $\frac{4}{5}$

Mental Test 24

1. The tally marks show the distribution of sports preferred by a sample of schoolboys. How many preferred rugby?

   Hockey   I I I I
   Rugby   ⊦⊦⊦⊦ I I
   Soccer   ⊦⊦⊦⊦ ⊦⊦⊦⊦ I I
   Cricket   ⊦⊦⊦⊦ I I I
   Tennis   ⊦⊦⊦⊦

2. What is the mean of £14 and £16?

3. What is the median of 2, 4, 6, 7, 8, 9 and 10?

4. Find the mode of 3, 5, 2, 7, 5, 8, 2, 7 and 2.

5. The following numbers give the probability of an event occurring: 0, 0.6, 1 and 2. One of these cannot be correct. Which one?

6. A girl's handbag contains 7 five-pence pieces and 3 ten-pence pieces. She chooses one coin at random. Find the probability that it will be a ten-pence piece.

7. One letter is chosen from the word EMERGENCY. What is the probability that it will be an E?

8. Find the range of the following measurements: 9, 7, 11, 5, 8, 12 and 6 mm.

Coursework

Coursework tasks which are separately assessed at the examination centre may be set by the Examiner or the choice may be left to the teaching staff and the individual candidate. The content may or may not be directly related to the main core topics and the details connected with this aspect of the examination must be read through carefully in the chosen syllabuses.

You will be asked to work on the various coursework assignments during the two years (or one year) before the final examinations. Coursework may include mathematical investigations as well as practical work, historical research, problem solving or perhaps a real life application of mathematics based on statistical surveys carried out locally.

A written report on each coursework study will be required to explain how the various results were obtained and in some cases the the candidate may be asked for a verbal explanation to clarify some particular aspect of the work. The presentation should include all the relevant calculations and observations with diagrams, tables, graphs and constructions where necessary. The final package should contain sufficient explanation and detail for any outside reader to understand the development and conclusion. The percentage of marks allocated to coursework varies from one Examining Group to another, and an assessment will be made according to instructions received by the centre. However, the following points will certainly receive consideration:

(1) The candidate's comprehension of the task
(2) The clarity of the plan of attack
(3) Recognition and reaction to results at various stages
(4) Mathematical accuracy where appropriate
(5) Employment of equipment, graphics and suitable materials
(6) Final evaluation and presentation.

An investigation begins with some facts and ideas concerning a given situation. The way in which the clues are applied is then left to the individual who will decide which particular direction the development should take. Two students may well follow quite different lines of enquiry and could arrive at different — but equally valid — conclusions for a given set of information. Even if no final solution is obtained, the reasoning processes involved are always valuable and should receive recognition in the final presentation. Topics requiring such an approach are included under the title 'Investigations'.

The syllabuses may require some form of practical work. This could involve the construction of models or a sequence of practical tasks not normally associated with everyday mathematics lessons. Coursework may include investigations of real life situations; statistical surveys could be carried out, or you may be asked to undertake some historical research involving mathematical ideas and concepts. This element of the work is covered in the 'Further Studies' section.

Mathematical problem solving requiring a logical approach or even some trial-and-error tactics in search of a finite solution, may also be included in certain coursework assignments. Examples of a variety of mathematical puzzles are offered in the 'Problems' part of this chapter.

At various times during the two-year (or one-year) GCSE preparation period you may like to attempt a selection of these different tasks. The investigations are varied and your choice will depend on the particular Examining Group's instructions that you are following. They should provide solid practice and may even stimulate a range of interesting ideas for inclusion in your own coursework.

Investigations

The first two are presented as examples and a logical sequence in pursuit of conclusions is outlined in some detail. Suggestions are made and questions are posed. In addition, you may find other useful avenues of enquiry which are not mentioned in the text.

Investigation 1

A gardener has decided to cover parts of his garden with concrete blocks and is looking for useful suggestions. The first area for investigation is the square shown in the first diagram. The size of block he has chosen is drawn to the right and broken lines are included to assist with the experiments.

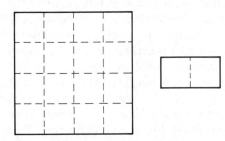

Investigate the pattern of blocks in this and other squares.

The gardener has already worked out the most obvious ways of using the stones to cover his first square area:

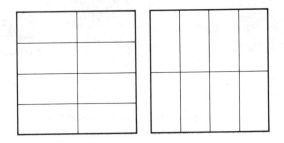

He found these patterns rather boring and was more interested in this arrangement:

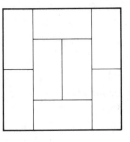

Draw further patterns, making sure each one is symmetrical.

Are there any interesting arrangements of blocks which are not symmetrical?

Do you have reasons for preferring any particular type of pattern?

Are the patterns more interesting for a 6 × 6 square area, which is also in the garden?

The garden does have a 5 × 5 square and a 9 × 9 square. The gardener will allow us to cut off pieces of the blocks this time if we wish. Why did he say that?

Investigate the differences between the odd-order squares (5 and 9).

How is symmetry affected by the use of odd numbers?

The gardener has now discovered that he can buy blocks of a different shape. Here is the complete range of blocks on offer at the Garden Centre:

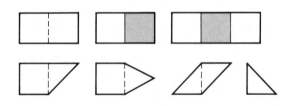

How does this new information affect your suggestions?

Draw some patterns for covering the two even-order squares and the two odd-order squares.

If the garden was your own, which particular pattern would you choose for the blocks in each square? Give your reasons.

Investigation 2

This investigation is carried out on a pin-board using an elastic band. Nine pins are used in a square formation as shown. Your results are best recorded on squared paper.

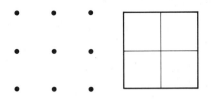

Investigate:

(a) the different ways in which the elastic band can be positioned around the pins to form a triangle

(b) the different quadrilaterals which can be made using the elastic band

(c) the different polygons which can be formed.

These notes will help you to begin the enquiry:

(a) Here is an example showing one triangle on the pin-board and how its shape is recorded on the squared paper:

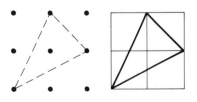

Note that the same shape can be constructed in four ways:

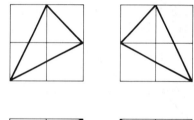

Each time a different triangle is found, a similar line of enquiry must be followed to discover the total number of possibilities. There are many triangles which can be constructed and included in your investigation.

You may like to examine the next drawing and comment on its importance in these experiments:

(b) The next sequence of diagrams demonstrates how a particular quadrilateral can be drawn in four different positions:

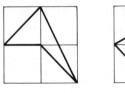

There are many more examples which can be discovered and investigated in this way. The relationship between triangles and quadrilaterals is also worth recording.

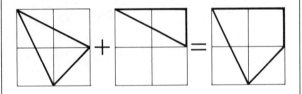

(c) When investigating different polygons you must be aware that irregular as well as regular shapes can be constructed:

You should comment on shapes which possess special properties.

The remainder of these investigations do not include such a comprehensive range of suggestions. Some assistance is offered, but the depth and success of each enquiry will depend upon the extent to which the given facts are questioned and examined.

Investigation 3

Number plates for various purposes are produced by the Special Plate Company. The firm claims that its plates are always manufactured in a sequence. Their catalogue includes this illustration showing plates from a part of one sequence:

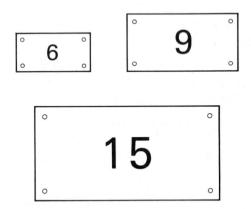

The description of the plates states that they follow each other in a sequence and are drawn full size.

Investigate this particular sequence of plates.

Is there anything significant about the three numbers shown?

Is it likely that number 6 is the smallest number in this sequence?

Are the shapes linked in any way with the numbers?

Is there a doubt about the next number in the sequence?

If there is a doubt, draw some alternatives for the next plate.

On the second page of the catalogue, these three plates appeared:

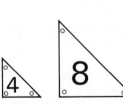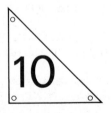

The three plates are part of the second sequence produced by the company. There are ten plates in this particular sequence.

Investigate the plates and their numbers. Work out the first and last plates in this set. Can you be certain that your answers are the only ones possible?

What further information would you like to see in the catalogue?

Investigation 4

Each horizontal and vertical line drawn on the following grid is labelled with a number in the same way that axes are labelled when a graph is to be drawn.

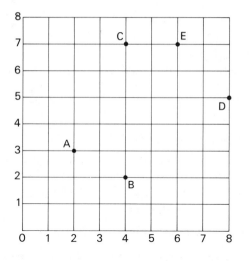

Each of the points A, B, C, D and E is positioned where two lines cross. These points are labelled by their coordinates, e.g. A(2, 3). To travel from one point to another, we must always use the lines of the grid. The shortest route from A to C, for instance, involves moving along a total of 6 lines; from B to D, we must follow 7 lines, etc.

Investigate any connection between shortest routes and the positions occupied by the five points on the grid.

Further investigate the number of squares contained within the shape formed by just

two points. For example, the points B and D lie at the corners of a rectangle containing 12 squares:

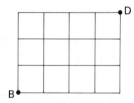

How do your results apply to overlapping rectangles such as those produced by BE and BD?

Is it possible to choose points such that the shortest distance in lines is equal to the number of squares contained within the rectangle formed?

If two arrangements produce the same minimum number of lines, will the rectangles so formed automatically enclose the same number of squares?

Examine circuits involving four or five points.

If no line is to be repeated in a sequence, is there any significance in the maximum number of lines which must be followed to travel between two points?

Investigation 5

Everyone is familiar with the game of noughts and crosses. Using nine squares, two players in turn place a nought or a cross in empty squares. The winner is the one who successfully positions three noughts or crosses in a straight line horizontally, vertically or diagonally. The game shown here, for instance, should be won by the crosses, even though the player using noughts is next to play. Can you see why?

Play the game with an opponent or play both the noughts and the crosses yourself in order to carry out this enquiry.

Investigate:

(a) all possible positions for your opening move if you are the first player

(b) the chances of preventing your opponent from winning if you are the second to play.

Many games end in a draw, so mathematicians have suggested variations and improvements which should make the game more interesting.

Draw a grid of nine squares. Use three cards marked with a nought and three cards each bearing a cross. The three cards are played in turn as in the familiar game of noughts and crosses.

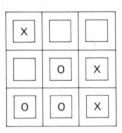

There are now some new rules proposed as follows:

(1) In turn, the cards may slide one square horizontally or vertically until a player produces three in a line.

(2) Diagonal moves are allowed one square at a time.

(3) A player may pick up any of his own cards and place it on an empty square.

(4) The game is played on a grid of 16 squares using four noughts and four crosses. The winner must produce a line of four of the same on this grid instead of three.

Investigate each rule separately and decide whether the game is improved by any or all of them.

Investigation 6

Hundreds of years ago, a puzzle was invented about two jugs. One jug can hold 3 litres of water when full and the other can hold 5 litres when full. There are no measuring marks on the jugs. The puzzle involved measuring out exactly 4 litres using only these two jugs and a tap for filling them with water when necessary.

Here is the solution:

	3-litre	5-litre
Both jugs are empty	0	0
Fill the 5-litre jug	0	5
Fill the 3-litre jug from the 5-litre jug	3	2
Empty the 3-litre jug	0	2
Transfer the 2 litres	2	0
Fill the 5-litre jug	2	5
Top up the 3-litre jug using the 5-litre jug	3	4

We now have exactly 4 litres remaining in the 5-litre jug.

Notice the method:

Fill the 5-litre jug when it is empty.
Pour into the 3-litre jug when possible.
Empty the 3-litre jug when it is full.
Carry out *one* operation at a time.

Investigate these operations further.

Record the results in two columns as shown in the puzzle. The operation which would follow, for instance, is 0–4 and the next stage would be 3–1.

Carry the sequence on until 0–0 (both jugs empty) is reached again. Study the number of stages required, including the 0–0 at each end. Is there anything you notice about the number of operations and the capacity of the two jugs being used?

Repeat the experiment using jugs of capacity 3 litres and 7 litres. Count the number of operations again and study all the stages appearing in the two columns.

See what happens when you use 5-litre and 7-litre jugs.

Will 7-litre and 9-litre jugs produce a similar result? What results are obtained using 3-litre and 9-litre jugs? Why are these results different from the others?

Suggest other pairs of jugs which will produce a different pattern.

Investigation 7

This investigation involves wire strips measuring either 2 cm or 3 cm in length. The wires are each marked in 1 cm steps along their lengths. A strip may be straight or bent at any or all of the centimetre marks.

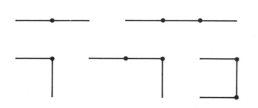

There is, of course, an extra possibility with the 3 cm strip, but it would serve no useful purpose in this investigation.

Using the strips, we can construct rectangles. This diagram shows one measuring 6 cm by 1 cm, producing an area of 6 cm²:

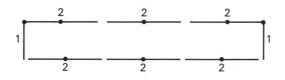

There are other ways of making a rectangle of area 6 cm²:

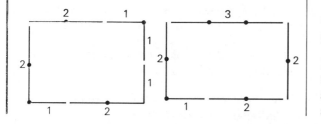

The first effort used 6 separate strips and each of the others used 4 strips. Is it possible to make an area of 6 cm² with less than 4 strips?

Investigate all the whole number areas from 1 cm² to 25 cm².

Is it always possible to make a rectangle (or square) for each area?

What is the smallest possible number of strips in each case?

Are there any similarities or number patterns in the results?

Why do certain areas produce particular rectangles?

Can you make any predictions for areas larger than 25 cm²?

Investigation 8

This is the map of Alphabet district. It shows the roads which link the villages of Axe, Bye, Clay, Dee, Elf and Fay.

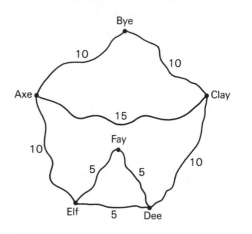

The numbers on the roads represent distances in kilometres.

Each village has a cricket team and a tournament has been organised to take place over a period of 5 days. Each team will play another different team each day.

Before the competition begins, each team is at its own village. They will stay overnight at the villages where they play cricket each day and they need not return to their own villages until after their fifth match, unless travel arrangements and fixture lists demand it.

Investigate fixture lists showing three matches per day, so that each team plays each of the others once only.

Arrangements must be as fair as possible, permitting each team to have two or three games at home. Each team would wish the travelling expenses to be as low as possible. Make notes so that you can be prepared for any criticism you may receive from the team managers.

Investigation 9

A toy shop sells two different types of children's building blocks. Set A contains six flat plastic shapes as shown in the diagram:

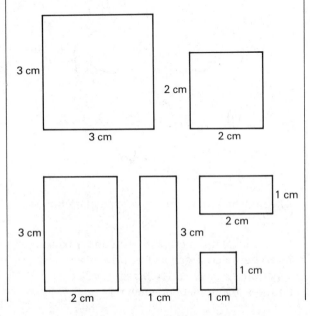

Investigate:

(a) the kind of box suitable for containing set A.

(b) the different rectangles which can be constructed using (i) 2 shapes, (ii) 3 shapes, (iii) 4 shapes, (iv) 5 shapes, if you can only afford to buy one set.

(c) the different methods of constructing squares of various sizes if you can afford to buy four of the sets of blocks.

Set B contains these three solid building bricks:

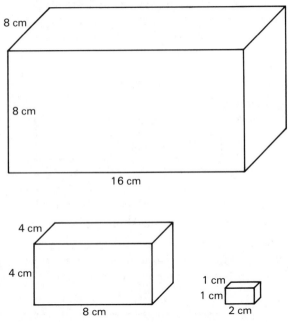

The blocks are used for building to certain heights. Piled on top of one another, for instance, they could give a maximum height of 26 cm.

You may buy these blocks singly instead of buying the whole sets and we shall assume that you can afford as many as you need.

Investigate different combinations for height building up to 36 cm.

Are there different methods needed for building each of the heights?

Investigation 10

If the digits of a number are added together, we obtain the **digital sum**. Here is an example:

Number: 73865 $7 + 3 + 8 +$
$6 + 5 = 29$

If the process is continued further, we shall eventually obtain a single-digit result. This is called the **digital root**:

Number: 73865 $7 + 3 + 8 +$
$6 + 5 = 29$
$2 + 9 = 11$
$1 + 1 = 2$

The digital root of 73865 is therefore 2.

Investigate the digital roots of the following:

(a) odd and even numbers

(b) prime and composite numbers (non-primes)

(c) square numbers and cubes

(d) numbers divisible by the same factor.

Record all points of significance and patterns of behaviour which you discover. Explain any rules which emerge from your investigation.

Further Studies

Further Study 1

We can construct a perennial calendar (one which can be adjusted to produce any date in the year) using four cubes arranged on this stand:

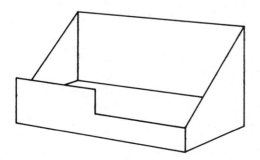

It is possible to construct such a stand using the following two nets. Folds are to be made inwards along each broken line.

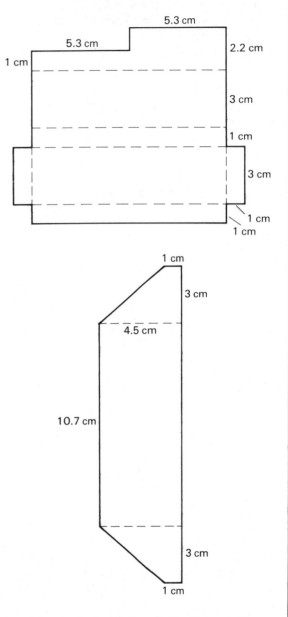

You must now construct the four cubes which fit on the stand. Think carefully about the parts of the cubes which will be visible when they are in position.

Label the cubes so that they show the correct day, month and date throughout the year, e.g. THURSDAY September 01.

Further Study 2

This study is concerned with solids whose faces are formed of equilateral triangles. The following nets will fold up to make a tetrahedron, an octahedron and an icosahedron:

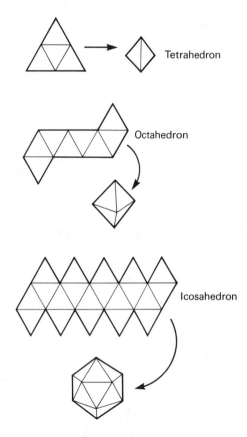

Tetrahedron

Octahedron

Icosahedron

(a) Can a different net of four triangles be used for the tetrahedron?

(b) Find different nets formed of eight triangles which will produce an octahedron.

(c) Is it possible to construct solids which have nets containing 5, 6, 7 or more equilateral triangles?

Further Study 3

Inventing codes to send secret messages is a very sophisticated procedure now that computers can be used to devise complicated systems. Early codes were quite simple and often based on letters of the alphabet. The frequency with which different letters appear in written English was often a clue towards solving many coded messages. Look at this message written in code:

> QEB BKBJV QOFBA QL QXHB LRO
> CLOTXOA MLPFQFLK. TB PEXII
> QOV QL BKQBO QEB QLTK YV QEB
> OLXA COLJ QEB BXPQ XKA TB
> PEXII KBBA PLJB DLLA ZLSBO CLO
> XQ IBXPQ QBK ELROP. MIBXPB
> XOOXKDB QL OBZBFSB QEB
> XASXKZB MXQOLI YV QBK QEFP
> BSBKFKD.

In the message, the letter B appears 30 times, Q appears 21 times, L 17 times and X 15 times. If we know the most frequently used letter in the alphabet, it should help us to decode the message.

In fact, we might investigate the frequency of every letter in the alphabet. This is quite a task for one person and is best carried out by a large group.

Members of the group should record the frequency of letters in different extracts from a book. The recording should be carried out in a specified period of time, say 20 minutes. The diagram shows how a tally chart could be drawn up by each person in the group:

A	⁣卌 III	N	卌 II
B	II	O	卌 IIII
C	IIII	P	II
D	卌	Q	
E	卌 卌 III	R	卌 I
F	III	S	卌 I
G	III	T	卌 卌 I
H	卌	U	IIII
I	卌 II	V	I
J	I	W	IIII
K	II	X	
L	卌	Y	II
M	III	Z	I

One person in the group would need to co-ordinate all the tallies. It should then be possible to guess what the letters B, Q, L and X represent in the code, and eventually the message should be revealed.

Frequency lists can also be attempted using literature from different periods covering other centuries (Chaucer, Shakespeare, etc.). Letter frequency in various foreign languages could also be analysed and compared. The various results could then be displayed in graphical form.

Further Study 4

A house is to be built on the piece of land shown in the diagram. The house will require rectangular foundations measuring 9 metres by 12 metres.

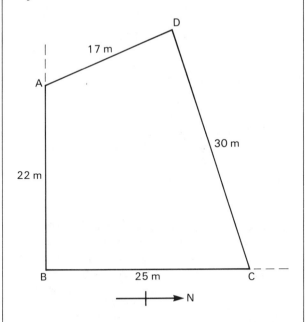

The main road already exists along AB and there is a minor road running along BC. Sides AD and DC are existing fences separating the property from the neighbour's land. The fences must remain or be replaced with equally permanent boundaries. BC runs due North.

Draw a plan of the site to a suitable scale, showing the position you would choose for the house. Include in your plan a garage large enough for a family car, a drive and a separate path to the house. Explain why you have selected a particular position for the house.

On a separate scale drawing, submit your design for the remainder of the ground area. Supply labels or a key to identify the various features of the design.

Further Study 5

Compile an 'Everyone's Dictionary of Useful Information'. The book could be arranged in sections. Here are a few ideas to guide you:

(1) Mathematical terms. The more common terms such as angle, bisector, chord, division, equal, fraction etc. would be included, together with various shapes, signs and properties.

(2) Money. As well as the different units of currency in use throughout the world, there are many obsolete units which you should be able to find such as ducat, pieces of eight, shekel, talent, etc. Also there are many slang names for coins and notes, such as 'buck', 'grand', 'quid', 'tanner', etc. which you might include.

(3) Measures. These could be treated in the same way as the money. Current measures should be recorded together with many older measures with a fascinating history attached to them. Units such as cubit, dram, furlong, grain, minim, sack and stadium are worth investigating.

(4) Codes. Letters, numbers and symbols are often used to record measures,

values and other information. Here are some examples:

GL2 0DZ

086 87 20079

003905   7:28

There are further instances which can be found on packaged food, clothing, furniture, precious metals, books and other products.

(5) Miscellaneous. In our modern world of technology, we regularly encounter new terms and measures, some of them rather obscure. You will need to research some of the following terms to discover their meaning: byte, Celsius, decibel, light year, millennium, 5% inflation, dress size 12, 20% proof, 13 amp, A4 paper, etc. You should be able to find many other terms connected with weather, computers, high finance, various retail trades, clothing and other areas of commerce.

Further Study 6

Semaphore is a way of sending signals using flags:

H   E   L   P

Imagine you have invented a semaphore machine. Here is a drawing of your invention:

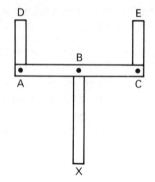

BX is the pole which supports the arm AC. There are pins at the joints A, B and C so that the arm AC can move like a see-saw and the two shorter arms AD and CE can swing upwards, downwards or outwards, or lie inwards on top of arm AC.

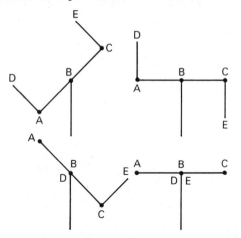

Notice that in the third diagram the arm AD is lying inwards along AC and cannot be seen. In the last example, both arms AD and CE have disappeared because they are both lined up inwards along AC. The see-saw arm may be horizontal or swung at 45° in either direction.

(a) Suggest a code so that your machine will transmit messages.

(b) Improve the machine design so that it can include more symbols.

(c) Explain how your machine could be used at night.

Further Study 7

Produce and display evidence to show that smoking is not a habit to be recommended. You will probably have ideas of your own, but these notes will help you to explore three different aspects in your enquiries.

(a) Work out the length of cigarette which is burnt in a lifetime of smoking by light, average and heavy smokers. Deduce the average length of a cigarette and use it to carry out your calculations for 50 years of smoking.

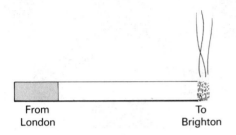

From
London

To
Brighton

Compare your answer with distances on a map.

(b) A smoker would be very surprised if presented with a figure representing the amount he or she spent on smoking over a period of 50 years. You will need to use an average figure for the cost of a packet of 20 cigarettes over the 50 years and once again make calculations for each of the various types of smoker. Work out the cost per day, per week and per year first of all and then obtain a figure for a lifetime of smoking.

£20 000? £40 000? £60 000?

Estimate how much interest could have been collected on that amount of money if it had been invested for 50 years. This can only be an approximate figure and you may need to seek information on interest rates at a local bank or building society.

(c) A graph is always useful for displaying a lot of information. You could produce one to demonstrate the damaging effects of smoking on a person's general state of health.

These statistics were published in a magazine in 1988:

1976:	53 100	1977:	54 900
1978:	56 750	1979:	58 750
1980:	61 000	1981:	63 500
1982:	66 500	1983:	71 000
1984:	76 200	1985:	82 750
1986:	90 750	1987:	100 500

The figures indicate approximately the number of smokers who died from lung cancer, cancer of the bronchus and various related forms of heart disease in the UK from 1976 to 1987.

Use the figures to construct a graph and try to project a figure for the current year, assuming the incidence of smoking does not decrease and the trend continues. Include all your information with some imaginative art work in a poster which might be considered for inclusion in a competition.

Further Study 8

You have completed your compulsory education and have decided to start up a window-cleaning service called Cleanshine. You have the opportunity to clean windows in Grove Street (60 houses), Victoria Square (20 houses), Lime Avenue (40 houses) and

London Road (50 houses). You can assume every house has four windows, each measuring 2 metres by $1\frac{1}{2}$ metres, together with four other panes of glass, each measuring 1 metre by $\frac{1}{2}$ metre.

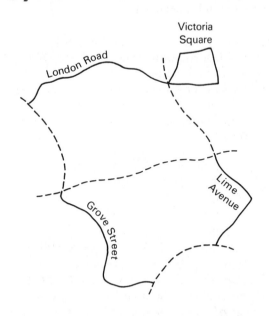

At the moment, you have no equipment or materials. You own a pedal cycle and you have a friend who runs a small van. Grove Street and Lime Avenue are half a mile apart at the lower end, and London Road is half a mile from each of them. Victoria Square is at one end of London Road. There are good access roads in the area.

What items do you need immediately? How much money is required to start up the business? Would it be profitable to invite your friend to join the business? What will you charge each household? How often will you visit each house? Will you employ anyone to help you? How much would you expect to pay an employee? Would your service be subject to VAT? Would you be paying Income Tax? If you were successful, how might you expand the business?

Investigate fully all the different avenues necessary to establish your Cleanshine business.

Further Study 9

Copy the diagram and construct your own 'cross-figure' puzzle. (This is like a crossword, but all the clues are figures instead of words.)

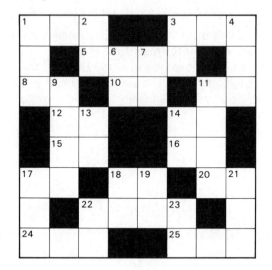

ACROSS clues will be needed for numbers 1, 3, 5, 8, 10, 11, 12, 14, 15, 16, 17, 18, 20, 22, 24 and 25.

DOWN clues will be needed for 1, 2, 3, 4, 6, 7, 9, 11, 13, 14, 17, 18, 19, 21, 22 and 23.

Each answer will be a number and your clues can use simple mathematical processes, or you may invent some more complex problems.

The finished product should include a blank copy of the diagram together with a neatly set out list of Across and Down clues so that someone can attempt the puzzle.

Further Study 10

An old game played with dominoes called 'Fives and Threes' requires some simple but accurate arithmetic. Two players start off with fourteen dominoes each or four players have seven each. To score during the game, you add the spots at each end of the line of dominoes and the player lucky enough to have the double six goes first. Any end total which *divides exactly by 5 or 3* scores points.

The first domino played (double six) has 12 spots. Dividing that by 3 scores 4 points. Watch how this game proceeds:

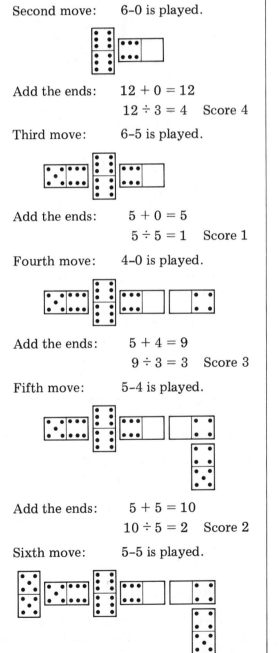

Second move: 6-0 is played.

Add the ends: $12 + 0 = 12$
                         $12 \div 3 = 4$ Score 4

Third move: 6-5 is played.

Add the ends: $5 + 0 = 5$
                         $5 \div 5 = 1$ Score 1

Fourth move: 4-0 is played.

Add the ends: $5 + 4 = 9$
                         $9 \div 3 = 3$ Score 3

Fifth move: 5-4 is played.

Add the ends: $5 + 5 = 10$
                         $10 \div 5 = 2$ Score 2

Sixth move: 5-5 is played.

Add the ends: $10 + 5 = 15$

This is the best score, because

$$15 \div 5 = 3$$

and $15 \div 3 = 5$ Score 8

A domino must be played even if there is no score.

Draw up a table to show all the possible scores in the game.

Are there any other similar games which could be played with dominoes? Could other factors be used? Could a game based on multiplication be invented?

Problems

Problem 1

A road is 99 kilometres long and at every kilometre there is a signpost. Two numbers appear on each sign to indicate how far it is from each end of the road. A sign 23 kilometres from one end, for instance, reads 23-76.

This is one end of the road:

Some signs have four different digits, e.g. 13-86.

Other signs have three different digits, e.g. 1-98.

How many signs contain only two different digits?

The road is to be lengthened to 199 kilometres, so the first sign would then read 0-199. How many of the new set of signs will contain only two different digits?

Problem 2

The numbers 1 to 9 are arranged in the form of a cross as shown:

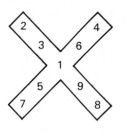

The numbers 2, 3, 1, 9 and 8 along one arm of the cross produce a total of 23 and the numbers 4, 6, 1, 5 and 7 on the other arm also total 23. This makes it a **magic cross.**

Exchanging numbers within one arm of the cross does not alter this relationship. A different formation is produced by switching numbers from one arm to the other.

Draw a similar cross with the number 1 in the middle. There are four other numbers which could also appear in the centre of this cross. What are they?

The next magic cross uses numbers from 0 to 8. Once again, each arm has the same total:

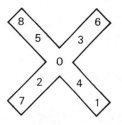

Which other numbers could now be used in the centre to retain the magic property, and what would be the total in each arm?

Now use the even numbers from 2 to 18 to make another magic cross. Can the odd numbers from 1 to 17 be used in the same way?

Problem 3

This is the board on which the game of darts is played:

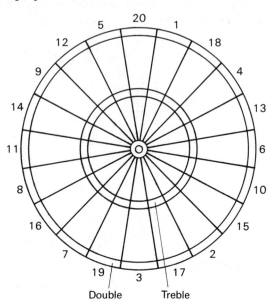

A dart between the outer circles scores double the number in that sector and a dart landing between the inner circles counts treble the number in the sector. In the centre are two small circles: the outer one gives a score of 25 and the centre, or **bull,** scores 50 (double 25).

At each turn, players throw three darts and the winner is the one who reaches a score of 501. The last winning dart must hit a double.

If, for example 111 is needed, this is one way to finish with three darts:

> 60 (treble 20), 27 (treble 9) and
> 24 (double 12)

For a finish of 127 in three darts, a player may score as follows:

> 60 (treble 20), 17 and 50 (bull)

(a) What is the highest total you could make with three darts to finish if the last one must score a double?

(b) In televised darts games, a player can receive thousands of pounds for achieving the 501 with only nine darts (the last must still be a double). Here is one method:

> Treble 20, Treble 20,
> Treble 20   Score: 180
>
> Treble 20, Treble 20,
> Treble 19   Score: 177
>
> Treble 20, Treble 18,
> Double 15   Score: 155
>   Total: 501

How many different ways can you find of scoring a nine-dart finish?

(c) There are quite a few scores less than 60 which cannot be made if only one dart is used. Find the lowest score which is not possible with a single dart.

(d) If there was a game in which the winner has to reach 101, could the first player win with the first two darts and end on a double?

Problem 4

To solve this problem, you must first follow these instructions concerning odd and even numbers:

(a) Choose one odd number and one even number. Multiply the odd number by any other odd number and multiply the even number by any other even number. Add your two answers. Repeat the process with other numbers and note down any pattern which you notice.

(b) Choose any odd number and any even number. Multiply the odd number by any even number and multiply the even number by any odd number. Once again, repeat this procedure and record any patterns which appear.

> Problem: I write down two numbers, one odd and one even. I shall call one my first number and the other my second number. I shall multiply the first by 2 and the second by 3.

The two answers will be added together and you will be told the total. Will you be able to work out whether my first number was odd or even?

Problem 5

The numbers 1 to 10 are placed in the rectangle as shown. The two centre sections are not used.

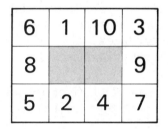

(a) The top row has a total of 20 and the bottom row a total of 18. The numbers down the left-hand side total 19 and those on the right-hand side also total 19.

Change the positions of only two numbers to make four totals of 19 each.

(b) Start a new arrangement of numbers. Place 1 and 2 in opposite corners and arrange the numbers 3 to 10 in the other spaces to give four totals of 18.

(c) Rearrange them all to give four totals of 20 each. (No clues!)

(d) There is one other way of arranging the numbers 1 to 10 in a similar rectangle to give equal totals other than 18, 19 or 20.

Find this last formation.

Problem 6

This problem was devised about two centuries ago.

There are four containers which, when full, will hold 24 litres, 13 litres, 11 litres and 5 litres respectively. The largest vessel is full of wine and the other three are empty.

The problem is to divide the 24 litres of wine equally among three people. Only those four containers can be used and the three equal measures of wine must be in three of those vessels when the division is complete.

Problem 7

You have probably seen a plastic puzzle sold in stores which has fifteen numbers sliding around inside a grid divided into sixteen squares. The two parts to the following puzzle are similar. Your best approach would be to use numbered squares of card which you can move around inside a grid marked out on paper.

(a) In a prison block, there are nine cells and eight are occupied as shown in this diagram:

7	5	6
8	3	2
4		1

The governor would like the prisoners to be arranged in their proper numerical order so that he can place a new prisoner, number 9, in the last cell. This is the arrangement he would prefer:

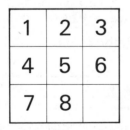

1	2	3
4	5	6
7	8	

A prisoner can only be moved if his cell shares a wall with an empty cell and no two prisoners must be in the same cell at any time. This means, for instance, that only prisoners 1, 3 or 4 can make the first move.

(b) In another prison block, there are six cells containing five prisoners:

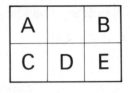

A		B
C	D	E

The same conditions for moving prisoners apply in this case. Prisoners B and E must change places. It does not matter in what order A, C and D finish as long as B and E have swapped over.

There are many ways the swap can be achieved, but the governor would prefer that it was accomplished in seventeen moves.

Problem 8

This problem is also easier to solve if a drawing is made and pieces of card, numbered from 1 to 8, are used.

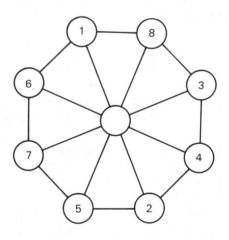

Numbers can move along one of the lines into the space which happens to be empty at any particular time. When the centre circle is empty, any of the eight numbers can move, but if one of the outer spaces is empty, then clearly only three of the numbers will be able to move.

The problem is to arrange the numbers clockwise in correct numerical order from 1 to 8. There are many ways this can be done, but the quickest routine requires only ten moves.

Problem 9

You will need seven playing cards from a pack: the four aces and any three of the kings. (If a proper pack is not available, you can use seven pieces of plain card, four of them marked with an 'A' and three with a 'K'.)

Your problem is to arrange them in a particular order face down so that you can carry out the following instructions:

> Place the top card on the bottom of the pack.
> Lay the next card on the table face upwards.
> Place the third card underneath the pack and lay the fourth card on the table, face upwards.
> Carry on in this fashion, one underneath, one on the table, etc., until all seven are on the table.

The cards must come out on the table in this sequence: A-K-A-K-A-K-A. In what order will you place them at the start to solve the problem?

Problem 10

A puzzle called the Tower of Hanoi was first devised in 1883 and it was probably based on a much older problem called the Tower of Bramah. The problem starts with three discs placed on one peg and two empty

pegs. The discs can only move according to certain rules.

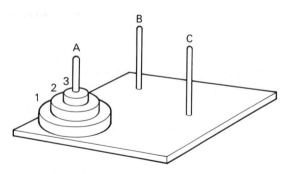

Discs are moved one at a time to another peg. No disc can be placed on a peg on top of a disc smaller than itself. The discs must all finish on peg B or peg C in the order in which they are presently arranged on peg A — largest at the bottom and smallest at the top.

Since this is the easy part of the problem, it will be done for you:

> Move 3 to C, 2 to B, 3 to B, 1 to C, 3 to A, 2 to C, 3 to C. This makes 7 moves altogether.

Your problem:

The same three pegs, A, B and C are used. There are now four discs numbered 1, 2, 3 and 4 from the bottom up on peg A. Using the same rules, transfer the discs to peg B or peg C, so that they are again in their correct order of size. This should take you 15 moves.

Use three crosses marked on paper to represent the pegs and four circles of card graded in size for the discs. Number the discs 1 to 4.

The Tower of Hanoi can be extended further by adding extra discs and you might attempt to transfer five discs in 31 moves.

Answers

Answers to Chapter 1

Exercise 1.1

1. 47
2. 434
3. 403
4. 260
5. 3463
6. 7813
7. 72 000
8. 25 335
9. 700 000
10. 433 000 000
11. 563 340 507
12. 989 000 000
13. Three hundred and fifty-four
14. Six hundred and eight
15. Two hundred and two
16. Three hundred and forty
17. Eight thousand seven hundred and sixty-five
18. Seventy-two thousand
19. Five hundred and sixty-two thousand
20. Nine hundred and five thousand
21. Eighty-seven thousand five hundred and nine
22. Two hundred and four thousand and thirty-seven
23. Six hundred and ninety thousand and fifty-eight
24. Six million eight hundred and ninety-three thousand and twenty-four
25. Six million thirteen thousand seven hundred and four
26. 123, 132, 213, 231, 312, 321
27. (a) 4 units (4)   (b) 4 units (4)
    (c) 4 tens (40)
    (d) four hundred thousand (400 000)
    (e) four million (4 000 000)

Exercise 1.2

1. 17
2. 21
3. 30
4. 30
5. 31
6. 17
7. 23
8. 23
9. 13
10. 22
11. 11
12. 12
13. 38
14. 13
15. 29

Exercise 1.3

1. 40
2. 10
3. 90
4. 90
5. 80
6. 700
7. 700
8. 100
9. 400
10. 900
11. 3000
12. 1000
13. 8000
14. 9000
15. 6000

Exercise 1.4

1. 2334
2. 4504
3. 1830
4. 2772
5. 5481
6. 2 616 525
7. 8 097 820
8. 3581
9. 5645
10. 17 397 707

Exercise 1.5

1. 2
2. 5
3. 6
4. 3
5. 7
6. 5
7. 4
8. 7

Exercise 1.6

1. 486
2. 3568
3. 11 586
4. 230 251
5. 44 616
6. 13 847
7. 1594
8. 599

Exercise 1.7

2. $-7, -5, -3, -1, +5, 6, 7, +9$
3. $-3\,^{\circ}C$ and $-5\,^{\circ}C$
4. $11\,^{\circ}C$
5. $-8\,^{\circ}C$
6. $5, 3, 0, -3, -5$
7. -4
8. 12 seconds

Exercise 1.8

1. 7
2. 3
3. -5
4. -20
5. 8

Exercise 1.9

1. 17
2. -25
3. 0
4. 84
5. 439
6. 1990
7. 12 401

Exercise 1.10

1. 15
2. 42
3. 72
4. 36
5. 140
6. 54
7. 56
8. 15
9. 120
10. 70

Exercise 1.11

1. 3103
2. 3 394 260
3. 2 165 744
4. 66 624
5. 16 119 136
6. 6 757 002
7. 3 242 616
8. 7 648 751

Exercise 1.12

1. 80; 800; 8000
2. 150; 1500; 15 000
3. 800; 8000; 80 000
4. 7230; 72 300; 723 000
5. 8170; 81 700; 817 000
6. 43 210; 432 100; 4 321 000
7. 50 600; 506 000; 5 060 000
8. 30 000; 300 000; 3 000 000
9. 70 500; 705 000; 7 050 000
10. 200 000; 2 000 000; 20 000 000

Exercise 1.13

1. 72
2. 47
3. 15
4. 57
5. 32
6. 652
7. 31
8. 983

Exercise 1.14

1. 7 remainder 1
2. 7 remainder 2
3. 4 remainder 3
4. 1 remainder 5
5. 6 remainder 3
6. 5 remainder 7
7. 3 remainder 2

Exercise 1.15

1. 18
2. 19
3. 28
4. 38
5. 7
6. 2
7. 23
8. 35
9. 7
10. 8

Exercise 1.16

1. 798
2. 3562
3. 0
4. 5
5. 14
6. 0
7. 18
8. 11
9. 0
10. 0
11. 42
12. 6

Exercise 1.17

1. 697
2. 417
3. 2
4. 3061
5. 864
6. 7737
7. 808
8. 924

Exercise 1.18

1. 250, 1250
2. 17, 21
3. 27, 33
4. 22, 11
5. 9, 11
6. 13, 9
7. 162, 486
8. 256, 1024
9. 6, 8, 16
10. 12, 4
11. 192, 3072
12. 23, 35
13. 6, 2
14. 48, 192
15. 11, 26

Exercise 1.19

1. 30 mm
2. 22 mm
3. 816 litres
4. £465
5. 8055 g
6. 6 m
7. 18 mm
8. 6p
9. £463
10. 1780 g
11. 26
12. 210
13. 7000 min = 116 h 40 min
14. 70 litres

Miscellaneous Exercise 1

Section A

1. (a) 486   (b) 2, 6, 54
2. 10 degrees
3. 4
4. £608
5. 110 minutes
6. (a) 15   (b) 272
7. 22, 43, 50

Section B

1. 28 000
2. 180 metres
3. 3711 ohms
4. (a) 40   (b) 19   (c) 25
   (d) 40
5. (a) 212 calories   (b) 198 calories
   (c) 540 calories
6. (a) 22   (b) 2, 7, 16
7. 322

Multi-Choice Questions 1

1. A
2. D
3. B
4. D
5. B
6. C
7. B
8. B

Mental Test 1

1. 160
2. 8
3. 41 035
4. 21
5. 7
6. 36
7. 19
8. 49
9. 36
10. 17

Answers to Chapter 2

Exercise 2.1

1. Odd	2. Even	3. Even
4. Odd	5. Odd	6. Even
7. Even	8. Odd	9. Even
10. Odd		

11. (a) $34, 88, 126$ (b) $15, 55, 63, 91, 139$
12. $16, 18, 20, 22, 24, 26$
13. $17, 31, 79, 127, 254$
14. 56

Exercise 2.2

1. 49	2. 64	3. 169
4. 729	5. 1936	6. 1225
7. 7921	8. 15 129	9. 294 849
10. 126 736	11. 27	12. 729
13. 4913	14. 54 872	15. 205 379
16. 2 299 968	17. 12 812 904	18. 56 623 104
19. 1 728 000	20. 8 000 000	21. 4
22. 6	23. 8	24. 9
25. 12	26. 31	27. 62
28. 89	29. 134	30. 456
31. 18	32. 64	33. 20
34. 90	35. 336	

Exercise 2.3

1. $144, 169$ 2. $25, 26$
3. $45, 55,$ 4. 81
5. any from $15, 34, 57, 63, 81$
6. 15

Exercise 2.4

1. 8	2. 9	3. 9
4. 1	5. 1	6. 3
7. 2	8. 6	9. 5
10. 5		

Exercise 2.5

1. Yes	2. No	3. No
4. No	5. Yes	6. No
7. No	8. Yes	9. Yes
10. Yes	11. No	12. Yes

Exercise 2.6

1. Yes	2. Yes	3. No
4. Yes	5. No	6. No
7. No	8. Yes	9. Yes
10. No	11. Yes	12. Yes
13. No	14. Yes	15. No

Exercise 2.7

1. Yes	2. No	3. No
4. Yes	5. No	6. Yes
7. No	8. Yes	9. No
10. Yes	11. Yes	12. No
13. Yes	14. No	15. Yes

Exercise 2.8

1. No	2. Yes	3. Yes
4. No	5. Yes	6. No
7. Yes	8. No	9. No
10. Yes	11. Yes	12. Yes
13. Yes	14. No	15. Yes

Exercise 2.9

1. (a) Yes	(b) No	(c) Yes
(d) Yes	(e) No	
2. (a) No	(b) No	(c) Yes
(d) Yes	(e) No	
3. (a) No	(b) No	(c) Yes
(d) Yes	(e) Yes	
4. (a) No	(b) Yes	(c) No
(d) Yes	(e) Yes	

5. $3, 9, 21$
6. $1, 3, 6, 8, 9, 12$
7. $1, 3, 9$

Exercise 2.10

1. $1, 2, 3, 4, 6, 9, 12, 18, 36$
2. $1, 2, 4, 7, 8, 14, 28, 56$
3. $1, 2, 4, 5, 10, 20, 25, 50, 100$
4. $1, 2, 4, 7, 12, 21, 42, 84$
5. $1, 2, 3, 4, 5, 6, 10, 12, 15, 20, 30, 60$
6. $1, 3, 5, 7, 15, 21, 35, 105$
7. $1, 2, 5, 7, 10, 14, 35, 70$

Exercise 2.11

1. 6	2. 8	3. 20
4. 18	5. 30	6. 30
7. 30	8. 48	9. 15
10. 30	11. 36	12. 60

Miscellaneous Exercise 2

Section A

1. (a) $2, 4, 6, 8$ (b) $3, 6, 9$
   (c) $2, 3, 5, 7$
2. (a) $55, 60$ (b) 204 (c) 9
3. (a) 61, yes   (b) 5291, no
4. (a) $13, 31$ (b) 5
5. (a) 81 (b) $16, 64, 81$
   (c) $16, 32, 64, 132$

Section B

1. 5, 11
2. (a) 8, 16, 24, 32, 40   (b) 29, 31, 37
3. 36   4. 25   5. 28
6. 60   7. 36   8. 30

Multi-Choice Questions 2

1. A   2. D   3. A
4. C   5. C   6. A
7. D   8. B   9. B
10. D

Mental Test 2

1. 36, 38   2. Yes
3. 8, 10, 12, 14   4. 3
5. 64   6. 35
7. Yes   8. 9, 12, 15, 18
9. 1, 2, 5, 10   10. No
11. No   12. 18, 38, 48
13. 1, 2, 3, 4, 6, 8, 9   14. 7, 31, 53
15. 2 and 5   16. 16

Answers to Chapter 3

Exercise 3.1

1. $\frac{1}{6}$ 2. $\frac{3}{4}$ 3. $\frac{3}{8}$
4. $\frac{3}{5}$ 5. $\frac{5}{7}$

6.

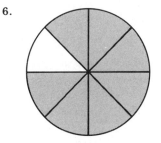

7.

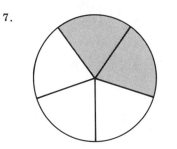

8.

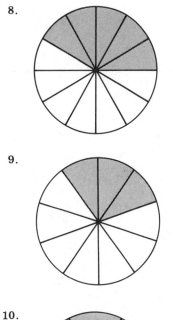

9.

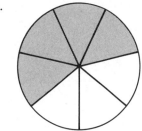

10.

(shaded circle)

Exercise 3.2

1.

2.

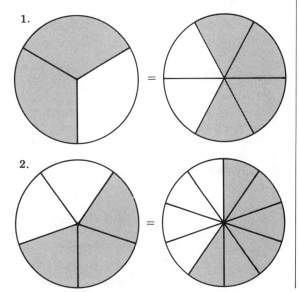

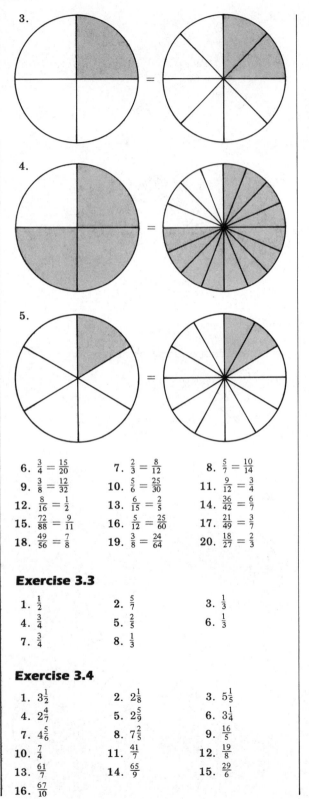

3.

4.

5.

6. $\frac{3}{4} = \frac{15}{20}$ 　　7. $\frac{2}{3} = \frac{8}{12}$ 　　8. $\frac{5}{7} = \frac{10}{14}$

9. $\frac{3}{8} = \frac{12}{32}$ 　　10. $\frac{5}{6} = \frac{25}{30}$ 　　11. $\frac{9}{12} = \frac{3}{4}$

12. $\frac{8}{16} = \frac{1}{2}$ 　　13. $\frac{6}{15} = \frac{2}{5}$ 　　14. $\frac{36}{42} = \frac{6}{7}$

15. $\frac{72}{88} = \frac{9}{11}$ 　　16. $\frac{5}{12} = \frac{25}{60}$ 　　17. $\frac{21}{49} = \frac{3}{7}$

18. $\frac{49}{56} = \frac{7}{8}$ 　　19. $\frac{3}{8} = \frac{24}{64}$ 　　20. $\frac{18}{27} = \frac{2}{3}$

Exercise 3.3

1. $\frac{1}{2}$ 　　2. $\frac{5}{7}$ 　　3. $\frac{1}{3}$

4. $\frac{3}{4}$ 　　5. $\frac{2}{5}$ 　　6. $\frac{1}{3}$

7. $\frac{3}{4}$ 　　8. $\frac{1}{3}$

Exercise 3.4

1. $3\frac{1}{2}$ 　　2. $2\frac{1}{8}$ 　　3. $5\frac{1}{5}$

4. $2\frac{4}{7}$ 　　5. $2\frac{5}{9}$ 　　6. $3\frac{1}{4}$

7. $4\frac{5}{6}$ 　　8. $7\frac{2}{5}$ 　　9. $\frac{16}{5}$

10. $\frac{7}{4}$ 　　11. $\frac{41}{7}$ 　　12. $\frac{19}{8}$

13. $\frac{61}{7}$ 　　14. $\frac{65}{9}$ 　　15. $\frac{29}{6}$

16. $\frac{67}{10}$

Exercise 3.5

1. $\frac{11}{20}, \frac{3}{5}, \frac{7}{10}$ 　　2. $\frac{2}{3}, \frac{7}{10}, \frac{4}{5}$ 　　3. $\frac{1}{2}, \frac{2}{3}, \frac{5}{6}$

4. $\frac{7}{12}, \frac{3}{4}, \frac{5}{6}$ 　　5. $\frac{17}{32}, \frac{9}{16}, \frac{5}{8}, \frac{3}{4}$ 　　6. $\frac{3}{8}, \frac{5}{9}, \frac{4}{7}, \frac{3}{5}$

7. $\frac{1}{5}, \frac{1}{4}, \frac{1}{3}, \frac{1}{2}$ 　　8. $\frac{4}{7}, \frac{3}{5}, \frac{5}{8}, \frac{3}{4}$

Exercise 3.6

1. $\frac{8}{15}$ 　　2. $\frac{7}{8}$ 　　3. $1\frac{1}{4}$

4. $\frac{5}{6}$ 　　5. $1\frac{3}{10}$ 　　6. $1\frac{2}{9}$

7. $1\frac{7}{15}$ 　　8. $1\frac{1}{28}$ 　　9. $7\frac{5}{6}$

10. $6\frac{3}{8}$ 　　11. $8\frac{19}{24}$ 　　12. $10\frac{3}{20}$

13. $3\frac{4}{5}$ 　　14. $6\frac{11}{12}$ 　　15. $10\frac{2}{3}$

16. $\frac{13}{24}$ 　　17. $\frac{1}{15}$ 　　18. $\frac{1}{2}$

19. $\frac{1}{2}$ 　　20. $\frac{7}{20}$ 　　21. $1\frac{1}{12}$

22. $2\frac{1}{16}$ 　　23. $\frac{3}{8}$ 　　24. $\frac{25}{33}$

25. $\frac{19}{20}$

Exercise 3.7

1. $\frac{8}{15}$ 　　2. $\frac{1}{6}$ 　　3. $\frac{9}{20}$

4. $\frac{6}{35}$ 　　5. $\frac{3}{10}$ 　　6. $\frac{8}{9}$

7. $2\frac{2}{5}$ 　　8. $2\frac{5}{8}$ 　　9. $4\frac{9}{10}$

10. $5\frac{5}{6}$

Exercise 3.8

1. $\frac{2}{3}$ 　　2. $\frac{1}{4}$ 　　3. $\frac{1}{6}$

4. $\frac{2}{9}$ 　　5. $1\frac{1}{3}$ 　　6. 4

7. $\frac{7}{16}$ 　　8. 8 　　9. $\frac{7}{16}$

10. 60 　　11. 12 　　12. 100

13. 40 　　14. 32 　　15. 30

Exercise 3.9

1. $\frac{8}{9}$ 　　2. $\frac{24}{25}$ 　　3. $\frac{2}{3}$

4. $1\frac{1}{3}$ 　　5. $\frac{3}{5}$ 　　6. 8

7. $1\frac{1}{2}$ 　　8. 50 　　9. 40

10. $\frac{25}{26}$ 　　11. $\frac{1}{6}$ 　　12. $\frac{1}{16}$

Exercise 3.10

1. $1\frac{1}{4}, 1\frac{1}{2}$ 　　2. $\frac{8}{5}, \frac{10}{5}$ 　　3. $1, 2$

4. $18, 54$ 　　5. $\frac{4}{27}, \frac{4}{81}$ 　　6. $\frac{1}{16}, \frac{1}{32}$

7. $1\frac{2}{5}, 1\frac{4}{5}$ 　　8. $2\frac{1}{2}, 1$

Exercise 3.11

1. $\frac{2}{3}$ 　　2. $\frac{3}{5}$ 　　3. $\frac{5}{7}$

4. $\frac{2}{3}$ 5. 13 6. 6

7. 4 8. $\frac{5}{4}$ 9. $\frac{13}{3}$

10. $\frac{4}{7}$

Exercise 3.12

1. 135 p 2. 45 min 3. $\frac{11}{12}, \frac{1}{12}$
4. $\frac{7}{15}$ 5. 40 min 6. 181 ℓ
7. (a) 4 hours (b) 12 hours (c) 18 hours
8. (a) $\frac{2}{5}$ (b) $\frac{2}{3}$
9. 57 min 10. $\frac{1}{4}$ ℓ 11. 320
12. (a) 480 (b) 270

Miscellaneous Exercise 3

Section A

1. $\frac{1}{3}$ 2. $\frac{5}{12}$ 3. $\frac{39}{40}$
4. $\frac{9}{10}$ 5. $1\frac{17}{24}$ 6. $\frac{2}{3}$
7. 1 hour 8. $\frac{4}{6}$

Section B

1. $\frac{1}{4}$ 2. $\frac{3}{4}$ 3. $5\frac{13}{20}$
4. $2\frac{3}{4}$ 5. $1\frac{1}{12}$ 6. 412
7. $\frac{1}{8}, \frac{1}{16}$
8. (a) $\frac{1}{2}$ (b) $\frac{1}{6}$ (c) $\frac{1}{15}$

Multi-Choice Questions 3

1. C 2. A 3. C
4. C 5. D 6. D
7. C 8. D

Mental Test 3

1. 6 2. $\frac{1}{4}$ 3. $1\frac{1}{4}$
4. $\frac{7}{2}$ 5. No 6. $\frac{3}{4}$
7. $\frac{1}{3}$ 8. 10 9. 2
10. 6

Answers to Chapter 4

Exercise 4.1

1. (a) 7 hundredths (b) 7 thousandths
   (c) 7 tens (70)
2. (a) 5 hundreds (500) (b) 5 hundredths
   (c) 5 thousandths
3. (a) 8 thousandths (b) 8 tenths
   (c) 8 hundreds (800)
4. $\frac{3}{10}$ 5. $3\frac{7}{10}$ 6. $20\frac{8}{100}$

7. $25\frac{27}{100}$ 8. $\frac{308}{1000}$ 9. $60\frac{9}{1000}$
10. $\frac{4}{1000}$ 11. 7992
12. (a) 5 (b) 6 (c) 8
13. (a) 0.9 (b) 0.273 (c) 0.45
    (d) 0.058 (e) 0.03
14. (a) $\frac{8}{10}$ (b) $\frac{57}{100}$ (c) $\frac{603}{1000}$
    (d) $\frac{9}{1000}$ (e) $\frac{53}{1000}$
15. (a) 5 tenths (b) 8 hundreds (c) 7 tens

Exercise 4.2

1.-8.

9. 0.9 10. 4.8 11. 2.0
12. 0.3 13. 6.3 14. 2.0
15. 1.5 16. −0.5 17. −1.2
18. 0.1

Exercise 4.3

1. (a) Yes (b) No, it is a negative fraction
   (c) Yes
2. (a) Yes (b) No (c) Yes
   (d) No
3. (a) 7 (b) 60 (c) 500
   (d) 400
4. (a) 0.103, 1.03, 10.3, 103
   (b) 0.504, 5.04, 50.4, 504
   (c) −2.5, −0.25, 0.025, 0.25, 25
   (d) 0.0151, 0.115, 0.15, 0.151
   (e) −52.05, −5.052, 5.025, 5.205, 5.502
5. (a) 48.9 (b) 4 (c) 0.5
   (d) 500 (e) 108 070 (f) 0.065
   (g) 0.9 (h) 14 000 000

Exercise 4.4

1. 2.714 2. 17.36 3. 895.619
4. 0.625 5. 88.335 6. 421.562
7. 1.064 8. 15.713 9. 1.12
10. 18.553 11. 57.524 12. 89.025
13. 123.205 14. 20.9 15. 57.29
16. 61.63 17. 77.05 18. 0.247
19. 92.72 20. 21.55

Exercise 4.5

1. (a) 3.5   (b) 35   (c) 350
2. (a) 59.83   (b) 598.3   (c) 5983
3. (a) 0.38   (b) 3.8   (c) 38
4. (a) 982.345   (b) 9823.45
    (c) 98 234.5
5. (a) 81.624   (b) 816.24   (c) 8162.4
6. (a) 0.46   (b) 4.6   (c) 46
7. (a) 0.0058   (b) 0.058   (c) 0.58
8. (a) 0.09   (b) 0.9   (c) 9
9. (a) 18.9   (b) 1.89   (c) 0.189
10. (a) 1.813   (b) 0.1813
    (c) 0.018 13
11. (a) 52.731   (b) 5.2731
    (c) 0.527 31
12. (a) 0.003   (b) 0.0003
    (c) 0.000 03
13. (a) 0.0325   (b) 0.003 25
    (c) 0.000 325
14. (a) 0.000 28   (b) 0.000 028
    (c) 0.000 002 8
15. (a) 0.562   (b) 0.0562
    (c) 0.005 62

Exercise 4.6

1. 2   2. 1.8   3. 1.5
4. 0.012   5. 0.01   6. 0.0006
7. 0.000 03   8. 0.012   9. 1.23
10. 1.24   11. 0.0106   12. 0.6
13. 0083   14. 0.6   15. 60
16. 0.3   17. 0.1   18. 15

Exercise 4.7

1. 34.5   2. 30.1   3. 101.4
4. 130.5   5. 407.85   6. 11 680.2
7. 0.32   8. 7.67   9. 12.7464
10. 11.0019   11. 409.64   12. 0.126
13. 11.56   14. 3.914   15. 2.093 43
16. 14.364   17. 0.0234   18. 0.2445
19. 1.254 26   20. 3.905 25

Exercise 4.8

1. 0.9   2. 0.5   3. 2.3
4. 3.2   5. 0.8   6. 2
7. 40   8. 0.3   9. 3000
10. 2   11. 70   12. 12
13. 0.05   14. 60   15. 1300

Exercise 4.9

1. 17.5   2. 9.7   3. 11.33
4. 178.9   5. 0.091   6. 10.6
7. 510   8. 0.16   9. 80
10. 112.4   11. 4.08   12. 0.009
13. 0.14   14. 0.17   15. 8.64

Exercise 4.10

1. 0.75   2. 0.2   3. 0.875
4. 0.8125   5. 0.05   6. 0.12
7. 0.45   8. 0.8   9. 3.04
10. 1.625   11. 7.5   12. 6.25
13. 4.125   14. 6.2   15. 9.3125

Exercise 4.11

1. $\frac{3}{10}$ 2. $\frac{11}{20}$ 3. $\frac{19}{50}$
4. $\frac{13}{25}$ 5. $\frac{4}{25}$ 6. $\frac{7}{40}$
7. $\frac{1}{8}$ 8. $\frac{5}{8}$ 9. $7\frac{9}{25}$
10. $2\frac{3}{50}$ 11. 0.18   12. 3.138
13. 0.16   14. 6.56   15. 1.585

Exercise 4.12

1. 4.676 tonnes   2. 18.3 litres
3. 13.49 mm   4. 12.75 km
5. 744.7 mm   6. £39.38
7. 86.95 m; 13 rolls   8. 1038.34 ohms

Exercise 4.13

1. (a) 19.37   (b) 19.4
2. (a) 0.007 52   (b) 0.008   (c) 0.01
3. (a) 4.970   (b) 4.97
4. (a) 153.262   (b) 153.26   (c) 153.3
5. (a) 34.157   (b) 34.16   (c) 34.2

Exercise 4.14

1. 308.89   2. 1167.0   3. 32.85
4. 1.19   5. 2.9

Exercise 4.15

1. $0.2 \times 30 = 6$; 6.81
2. $5 + 1 + 27 = 33$; 32.954
3. $80 + 40 - 10 - 30 = 80$; 79.075
4. $1 \times 0.1 \times 2 = 0.2$; 0.18
5. $90 \div 30 = 3$; 2.92
6. $0.09 \div 0.03 = 3$; 2.706
7. $\dfrac{3 \times 8}{4} = 6$; 6.46
8. $\dfrac{30 \times 30}{10 \times 3} = 30$; 29.17

Miscellaneous Exercise 4

Section A

1. (a) 14.4   (b) 0.4   (c) 1.44
    (d) 4
2. (a) 332.1   (b) 2.64   (c) 1.612

3. (a) 0.625   (b) 0.4375
4. (a) $\frac{3}{4}$ (b) 0.75
5. 1328.7 pence
6. (a) 73 800   (b) 0.738
7. 23; 17.4
8. 1.43
9. (a) 6 thousandths
   (b) $40 - 0.04 = 39.96$
10. 9500
11. 0.3, 0.11, 0.03
12. 177.194
13. (a) 0.875   (b) 0.156 25
14. (a) $\frac{7}{10}$ (b) $\frac{3}{25}$

Section B

1. (a) 1275   (b) 216.09   (c) 6.7
   (d) 158.6
2. 9
3. 54.9
4. (a) 73.14   (b) 0.67
5. (a) 4 lengths   (b) 0.7 metre
6. 700
7. 0.4573; 0.0198
8. 346.71 kg

Multi-Choice Questions 4

1. B   2. D   3. D
4. C   5. B   6. C
7. C

Mental Test 4

1. $\frac{5}{100}$ 2. 2.7   3. 0.31
4. 0.009   5. 5.017   6. 7
7. 6.5   8. 23.00
9. 2.03, 20.3, 203   10. 5
11. 2.31   12. 4.5   13. 3.6
14. 2.4   15. 0.34   16. 0.8
17. £270   18. 900   19. 7.4
20. 0.77

Answers to Chapter 5

Exercise 5.1

1. (a) 30 mm   (b) 280 mm   (c) 1340 mm
   (d) 5000 mm   (e) 63 000 mm   (f) 4600 mm
2. (a) 6 cm   (b) 24 cm   (c) 6.8 cm
   (d) 400 cm   (e) 5600 cm   (f) 374 cm
3. (a) 5 m   (b) 7 m   (c) 8.9 m
   (d) 5.643 m   (e) 5000 m   (f) 6420 m
4. (a) 7 km   (b) 6.34 km   (c) 8.325 km

Exercise 5.2

1. (a) 2 ft   (b) 5 ft   (c) 12 ft
   (d) 15 ft   (e) 240 ft
2. (a) 2 yd   (b) 24 yd   (c) 300 yd
   (d) 3520 yd   (e) 49 280 yd
3. (a) 60 in   (b) 240 in   (c) 408 in
   (d) 108 in   (e) 540 in
4. (a) 3 miles   (b) 9 miles   (c) 20 miles

Exercise 5.3

1. (a) 8000 g   (b) 19 000   (c) 15 000 g
   (d) 12 g   (e) 27 g
2. (a) 5 kg   (b) 18 kg   (c) 3000 kg
   (d) 18 000 kg
3. (a) 7000 mg   (b) 24 000 mg   (c) 500 mg
4. (a) 8 t   (b) 427 t   (c) 0.6 t

Exercise 5.4

1. (a) 4 oz   (b) 8 oz   (c) 72 oz
   (d) 80 oz   (e) 128 oz
2. (a) 0.75 lb   (b) 3 lb   (c) 0.5 lb
   (d) 6 lb   (e) 560 lb   (f) 6720 lb
3. (a) 100 cwt   (b) 160 cwt   (c) 5 cwt
4. (a) 3 tons   (b) 3 tons   (c) 8 tons

Exercise 5.5

1. (a) 7 cℓ   (b) 56 cℓ   (c) 600 cℓ
2. (a) 50 mℓ   (b) 260 mℓ   (c) 8000 mℓ
3. (a) 6 ℓ   (b) 0.5 ℓ   (c) 3 ℓ
4. (a) 60 fl oz   (b) 10 fl oz   (c) 320 fl oz
5. (a) 2 pt   (b) 6 pt   (c) 48 pt
6. (a) 2 gal   (b) 8 gal

Exercise 5.6

1. 144 lb   2. 150 cm   3. 3 ft
4. 200 yd   5. 8 kg   6. 112 km/h
7. 150 mℓ   8. 12 ℓ   9. 21 pt
10. 6 gal

Exercise 5.7

1. 6.7 m   2. 19 cm   3. 23 m
4. 150   5. 1.85 ℓ   6. 12.065 m
7. 282.15 m   8. 31   9. 7
10. 1500

Exercise 5.8

1. £15.97   2. £4.42   3. £100.64
4. £2.39   5. £9.11   6. £701.69
7. £1098.81   8. £9.80   9. £5.72
10. £2.49

Miscellaneous Exercise 5

Section A

1. (a) 0.155 m   (b) 1.36 ℓ   (c) 3.5 kg
2. £8.16
3. 18 lb
4. (a) 3000 mm   (b) 3400 mℓ
5. 6
6. 30 cm
7. 18 lb
8. £231.24

Section B

1. £97.92
2. £59.85
3. £1.32
4. (a) 12 340 g   (b) 1.234 m   (c) 1.234 ℓ
5. 14
6. 6.75 lb
7. 870 g
8. 40
9. 100 miles
10. 48 km/h

Multi-Choice Questions 5

1. D	2. C	3. D
4. C	5. C	6. A
7. A	8. C	9. C
10. C	11. B	12. C
13. B	14. C	

Mental Test 5

1. 8000 mm	2. 50 mm	3. 8.3 m
4. 3.4 m	5. 80 km	6. 93 000 m
7. 6 ft	8. 3 yd	9. 24 ft
10. 6 km	11. 4 in	12. 1.8 kg
13. 9000 g	14. 23 g	15. 6000 kg
16. 48 oz	17. 120 cwt	18. 7 oz
19. 300 cℓ	20. 9000 mℓ	21. 8 ℓ
22. 5 ℓ	23. 56 pt	24. 80 fl oz
25. 3 pt	26. £4.00	27. £32.50
28. £12	29. 18 p	30. 1.5 p
31. £3.00	32. 20 p	33. £2.50
34. £1.50	35. £30	

Answers to Chapter 6

Exercise 6.1

1. (a) 800 g butter; 640 g caster sugar; 2 eggs;
        1600 g self-raising flour
   (b) 1200 g   (c) 3
2. 120 kg sand; 320 kg water
3. 40 kg
4. (a) 15 kg   (b) 20 kg
5. 25 cm

6. (a) 4   (b) 25 g
   (c) 2 onions; 4 carrots; 500 g beef; 680 g soup;
        4 teaspoons of curry powder;
        100 g of raisins

Exercise 6.2

1. 1:4	2. 1:2	3. 1:3
4. 6:7	5. 5:6	6. 5:6
7. 2:3	8. 5:4	9. 3:2
10. 7:8	11. 2:3	12. 7:6

Exercise 6.3

1. 400:1	2. 1:4	3. 1:5
4. 1:2	5. 3:100	6. 2:1
7. 10:3	8. 100:3	9. 32:1
10. 5:1		

Exercise 6.4

1. $\frac{7}{20}$	2. $\frac{1}{2}$	3. $\frac{1}{2}$
4. $\frac{4}{5}$	5. $\frac{2}{3}$	6. $\frac{4}{5}$
7. $\frac{1}{3}$	8. $\frac{7}{8}$	

Exercise 6.5

1. £250, £150   2. 224 kg, 96 kg
3. 192 m, 48 m
4. 336 mm, 1176 mm, 1848 mm
5. £280   6. 150 kg
7. £176

Exercise 6.6

1. 10 units   2. 84 km
3. (a) £2.70   (b) 45 p
4. 7 p
5. (a) 25 miles   (b) 75 miles
6. (a) £5   (b) £30
7. (a) 2 p   (b) 4 p
8. (a) £2.50   (b) £22.50

Exercise 6.7

1. 140 p   2. 72 p   3. 168 p
4. £22   5. 52 litres   6. £112
7. (a) 120   (b) 30
8. (a) £9   (b) £1.50

Exercise 6.8

1. 50 g for £1.00   2. 850 g for £2.04
3. 250 g for 28 p   4. 750 g for £1.12
5. 200 g for £2.50

Exercise 6.9

1. 36 days	**2.** 5 days	**3.** 5
4. 10 min	**5.** 5 hours	**6.** 8 hours
7. 24	**8.** 50 days	

Exercise 6.10

1. 44 km	**2.** 20 min	**3.** 80 min
4. (a) 40 miles	(b) 200 miles	
5. £1.32	**6.** 10 hours	**7.** 40 min
8. (a) £1	(b) £8	
9. £160	**10.** $2\frac{1}{2}$ weeks	

Exercise 6.11

1. (a) 12 litres per minute   (b) 10 min
    (c) 96 litres
2. (a) 40 miles per gallon   (b) 200 miles
    (c) 6 gallons
3. 8100 kg
4. (a) 2 metres per second   (b) 300 m
    (c) 40 s
5. (a) 60 lb   (b) 4 square feet

Exercise 6.12

1. 234.6	**2.** 192.5	**3.** 36 750
4. 2250	**5.** £6.19	**6.** £32.52
7. 217	**8.** 2700	**9.** £90.03
10. 136.8		

Exercise 6.13

1. (a) 2 km   (b) 3.5 km
2. 1.6 km
3. 25 cm
4. (a) (i) 1.5 km   (ii) 20 km   (iii) 75 km
    (b) (i) 2 cm   (ii) 5 cm   (iii) 0.5 cm
5. (a) 1:200   (b) 6 m
6. (a) 1:50 000   (b) 5.4 km
7. 405 km

Miscellaneous Exercise 6

Section A

1. 14 kg	**2.** 1.44 in	**3.** $\frac{2}{3}$
4. 1:20	**5.** £350, £150	**6.** 96p
7. 500 ml for 48p		**8.** 20 days
9. 600 m	**10.** 4 cm	

Section B

1. £1.08	**2.** $31\frac{1}{2}$ km	**3.** $4\frac{1}{2}$ km
4. (a) 15	(b) £43.50	
5. 7:8	**6.** 4:1	**7.** $\frac{1}{8}$
8. No		**9.** 160 kg, 320 kg
10. £400, £300, £200		**11.** 75 cm
12. $66\frac{2}{3}$ cm		

Multi-Choice Questions 6

1. B	**2.** A	**3.** D
4. B	**5.** B	**6.** C
7. C	**8.** C	

Mental Test 6

1. $\frac{1}{2}$	**2.** $\frac{8}{1}$	**3.** £2.80
4. $\frac{1}{2}$	**5.** £2	**6.** 250 km
7. 20 days	**8.** $\frac{1}{4}$	**9.** 20 litres
10. 30 miles per gallon		**11.** 2000
12. £15		

Answers to Chapter 7

Exercise 7.1

1. 90%	**2.** 40%	**3.** 35%
4. 38%	**5.** 62.5%	**6.** 60%
7. 87%	**8.** 3%	**9.** 56.2%
10. 91.7%	**11.** $\frac{2}{25}$	**12.** $\frac{3}{20}$
13. $\frac{2}{5}$	**14.** $\frac{19}{20}$	**15.** $\frac{8}{25}$
16. 0.44	**17.** 0.09	**18.** 0.083
19. 0.952	**20.** 0.333	

	Fraction	Decimal	Percentage
21.	$\frac{11}{20}$	0.55	55
22.	$\frac{17}{50}$	0.34	34
23.	$\frac{7}{8}$	0.875	87.5
24.	$\frac{19}{25}$	0.76	76
25.	$\frac{2}{25}$	0.08	8
26.	$\frac{3}{8}$	0.375	37.5
27.	$\frac{3}{20}$	0.15	15
28.	$\frac{27}{100}$	0.27	27
29.	$\frac{9}{20}$	0.45	45
30.	$\frac{31}{50}$	0.62	62

Exercise 7.2

1. £4	**2.** £24	**3.** £36
4. £18	**5.** £15	**6.** £0.45
7. £2.40	**8.** £7.29	**9.** £18.62
10. £79.36		

Exercise 7.3

1. £150	**2.** £189	**3.** £2.20
4. £664	**5.** £280.80	

Exercise 7.4

1. $0.48, \frac{1}{2}, 55\%$ 2. $23\%, 0.24, \frac{1}{4}$

3. $35\%, 0.53, \frac{3}{5}$ 4. $0.33, \frac{1}{3}, 34\%$

5. $66\%, \frac{2}{3}, 0.67$ 6. $34\%, 0.43, \frac{3}{4}$

7. $0.8, \frac{1}{8}, 8\%$ 8. $0.62, \frac{5}{8}, 63\%$

Miscellaneous Exercise 7

Section A

1. $\frac{1}{5}, 0.22, 25\%$

2. (a) 65%   (b) 60%

3. (a) $\frac{7}{20}$ (b) $\frac{12}{25}$

4. £74.80   5. £96   6. less than
7. greater than

Section B

1. (a) $\frac{6}{25}$ (b) $\frac{3}{50}$

2. (a) 0.64   (b) 0.07

3. $a = \frac{17}{20}$, $b = 85\%$, $c = 0.35$, $d = 35\%$,

   $e = \frac{3}{25}$, $f = 0.12$

4. (a) £1.80   (b) £26.60
5. (a) £12   (b) £68
6. (a) 12   (b) 60%
7. $\frac{5}{6}, 85\%, 0.875$

Multi-Choice Questions 7

1. B   2. B   3. D
4. C   5. C   6. D
7. B

Mental Test 7

1. 80%   2. $\frac{19}{100}$ 3. 0.38

4. 89%   5. £24   6. £3.50

7. 61.2%   8. $\frac{1}{4}$ 9. No

10. No

Answers to Chapter 8

Exercise 8.1

1. £160   2. £5
3. (a) £70   (b) £87.50   (c) £105
4. £4   5. £156.80

Exercise 8.2

1. (a) £6.25   (b) £7.50   (c) £10
2. (a) £4.80   (b) £6.40
3. £280

4. (a) £4   (b) £5   (c) £40
   (d) £200
5. (a) £105   (b) £4.50   (c) £22.50
   (d) £127.50

Exercise 8.3

1. £59.00   2. £62.50   3. £30.50
4. £80   5. £37.50

Exercise 8.4

1. (a) £4   (b) £7   (c) £16
2. £450   3. £165   4. £115
5. £200

Exercise 8.5

1. £443   2. £500   3. £700
4. £508   5. £620

Exercise 8.6

1. £1200
2. (a) £4675   (b) £1402.50
3. (a) £23 000   (b) £6305
4. £1080   5. £37.50   6. £122.65
7. £1155

Miscellaneous Exercise 8

Section A

1. £116   2. £750   3. £4750
4. £5   5. £3.75   6. £240
7. £90   8. 12%

Section B

1. £180   2. £3.53   3. £50.40
4. £600
5. (a) £203.75   (b) £396.25
6. (a) £6350   (b) £1905
7. £10.50
8. (a) £180   (b) £240

Multi-Choice Questions 8

1. C   2. B   3. C
4. C   5. D   6. B
7. C   8. B

Mental Test 8

1. £5   2. £120   3. £6
4. £8   5. £150   6. £600
7. £130   8. £6000   9. £1500
10. £400

Answers to Chapter 9

Exercise 9.1

1. £25
2. £80
3. £400
4. £960
5. £3000
6. £100
7. £20
8. £120

Exercise 9.2

1. £330.75
2. £1331.00
3. £6612.50
4. £463.20
5. £1587.00
6. £1553.00
7. £5706.00

Exercise 9.3

1. £810
2. £12 282.50
3. £15 360
4. £281.25

Miscellaneous Exercise 9

Section A

1. £500
2. £318
3. £121
4. £2560
5. £23.50 in favour of the Savings Bond

Section B

1. £80
2. £20
3. (a) £24   (b) £48
4. (a) £20   (b) £42   (c) £242
5. £500

Mental Test 9

1. £10
2. £100
3. £60
4. £100
5. £2000
6. £10
7. £5
8. £108

Answers to Chapter 10

Exercise 10.1

1. £20
2. 30 p
3. 50%
4. £1
5. £500
6. 20%
7. (a) £10   (b) $33\frac{1}{3}$%
8. 10%

Exercise 10.2

1. (a) £40   (b) £160
2. (a) £18   (b) £102
3. £36
4. £240
5. £58.20

Exercise 10.3

1. £45
2. £276
3. £184
4. £69
5. £24

Exercise 10.4

1. (a) £160   (b) £240   (c) £525
   (d) £360   (e) £650
2. £80 000
3. £4860 000
4. £46 000
5. 20 p in the pound

Exercise 10.5

1. £45
2. £20
3. £84
4. £144
5. £2100
6. £2080
7. £1000
8. (a) £600   (b) 6 p
9. (a) £8329   (b) £27 222
10. £16.16
11. (a) £6000   (b) £5241

Miscellaneous Exercise 10

Section A

1. £140
2. £4.50
3. £230
4. (a) £50   (b) 25%
5. £114
6. £260
7. £4.17
8. (a) £1460   (b) 16 p per mile

Section B

1. (a) £80   (b) $66\frac{2}{3}$%
2. (a) £150   (b) 50%
3. £60
4. (a) £7.50   (b) £57.50
5. £500
6. £90
7. £84

Multi-Choice Questions 10

1. C
2. B
3. B
4. D
5. C
6. A
7. A

Mental Test 10

1. 10 p
2. £30
3. £1000
4. 20%
5. £20
6. 10%
7. £65
8. £150

Answers to Chapter 11

Exercise 11.1

	House price	Deposit at 10%	Mortgage reqd.
1.	£20 000	£2000	£18 000
2.	£25 000	£2500	£22 500
3.	£30 000	£3000	£27 000
4.	£40 000	£4000	£36 000
5.	£50 000	£5000	£45 000

Amount of mortgage	Monthly repayments per £1000 borrowed	Monthly repayments to the building society
6. £10 000	£12.00	£120.00
7. £20 000	£10.00	£200.00
8. £25 000	£11.50	£287.50
9. £40 000	£10.80	£432.00
10. £50 000	£11.70	£585.00

Weekly rent	Annual rates	Rates per week	Rent + rates per week
11. £20.00	£156.00	£3.00	£23.00
12. £30.00	£208.00	£4.00	£34.00
13. £17.50	£109.20	£2.10	£19.60
14. £45.00	£364.00	£7.00	£52.00
15. £63.00	£483.60	£9.30	£72.30

Exercise 11.2

1. £600   2. £120
3. (a) £600   (b) £200
4. (a) £580   (b) £180
5. (a) £160   (b) £640   (c) £1060
    (d) £260

Exercise 11.3

1. £60
2. (a) £200   (b) £1200   (c) £100
3. (a) £103.50   (b) £12 420
4. (a) £780   (b) £180
5. (a) £2700   (b) £700

Exercise 11.4

1.

Charge per therm	Number of therms used	Cost of gas used in pence	in pounds
30p	100	3000	30
40p	80	3200	32
45p	120	5400	54
35.3p	200	7060	70.60
37.5p	300	11 250	112.50

2. £45
3. (a) 352   (b) 352   (c) £140.80
4. (a)

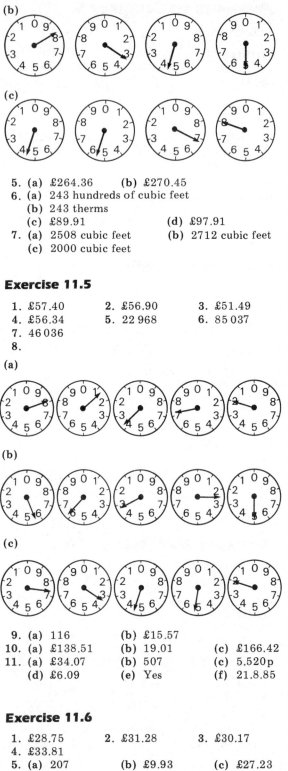

(b)

(c)

5. (a) £264.36   (b) £270.45
6. (a) 243 hundreds of cubic feet
   (b) 243 therms
   (c) £89.91   (d) £97.91
7. (a) 2508 cubic feet   (b) 2712 cubic feet
   (c) 2000 cubic feet

Exercise 11.5

1. £57.40   2. £56.90   3. £51.49
4. £56.34   5. 22 968   6. 85 037
7. 46 036
8.

(a)

(b)

(c)

9. (a) 116   (b) £15.57
10. (a) £138.51   (b) 19.01   (c) £166.42
11. (a) £34.07   (b) 507   (c) 5.520p
    (d) £6.09   (e) Yes   (f) 21.8.85

Exercise 11.6

1. £28.75   2. £31.28   3. £30.17
4. £33.81
5. (a) 207   (b) £9.93   (c) £27.23
    (d) £4.08   (e) £31.31

Exercise 11.7

1. £15   2. £40.46   3. £10
4. (a) Majorca, 31st Jan   (b) £59
   (c) £70
   (d) £189 per person; £69 per person
5. (a) £717.50   (b) £107.62   (c) £825.12

Miscellaneous Exercise 11

Section A

1. £95   2. £960
3. (a) £3000   (b) £17 000
4. (a) £1100   (b) £1200   (c) £200
5. (a) £400   (b) £200
6.

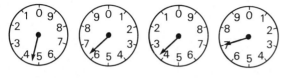

7. 5236
8. (a) £119   (b) £459
9. (a) 524   (b) £39.44
10. £40.25

Section B

1. £6
2. (a) £45   (b) £345   (c) £39
   (d) £387
3. (a) 68571   (b) 69331

(c)

4. Quarterly rental charge   = £9.75
   Dialled units: 600 at 4.7 p
   per unit   = £28.20
   Total (exclusive of VAT) = £37.95
   VAT at 15%   ≑ £5.69
   Total payable   = £43.64
5. (a) 1238 hundreds of cubic feet
   (b) 225 hundreds of cubic feet
   (c) 225 therms   (d) £87.77

Multi-Choice Questions 11

1. D   2. D   3. B
4. C   5. B   6. C
7. C

Mental Test 11

1. £20   2. £22   3. £40
4. £100   5. £2000   6. £80
7. £200   8. £80   9. £35
10. £36   11. £99   12. 24/1
13. £371   14. Algarve; 2/2
15. Same travel expenses
16. Palmeras Pl. Apt., Tenerife

Answers to Chapter 12

Exercise 12.1

1. 195   2. 330   3. 7
4. 14 400   5. 39   6. £750
7. (a) 20   (b) 32   (c) 44
8. £6   9. £2.80
10. Cartoons $\frac{1}{2}$ hour; plays $1\frac{1}{4}$ hours; music $\frac{3}{4}$ hour
11. 1784, 1940

Exercise 12.2

1. 5 h 4 min   2. 5 h 7 min
3. 15 h 48 min   4. 10 h 30 min
5. 15 h 6 min   6. 0920 hours
7. 3.10 p.m.   8. 1150 hours

Exercise 12.3

1. 1030, 28 min
2. 1100 from Hester; 1430 from Hester
3. 58 min   4. 32 min   5. 25 min
6. 11.07   7. 50 min   8. 1013
9. 1222   10. 1517; 2 h 21 min

Exercise 12.4

1. 89   2. Tuesday
3. 5   4. 11th Feb
5. Monday 8th Sept   6. 30
7. Friday

Exercise 12.5

1. 4 mile/h   2. 4 km/h   3. 20 m/s
4. 8 ft/s   5. 60 km/h   6. 25 mile/h
7. 9 km/h   8. 60 mile/h   9. 10 miles
10. 24 miles   11. 100 km   12. 120 m
13. 60 ft   14. 280 km   15. 75 miles
16. 4 h   17. 4 h   18. 2 s
19. 2 s   20. 10 h   21. 8 h
22. 2.5 h
23. (a) 1325 hours   (b) 30 mile/h
24. (a) 3 h   (b) 48 mile/h
25. (a) 60 miles   (b) 120 miles
   (c) 36 mile/h

26. (a) 150 miles   **(b)** 270 miles
    (c) 5 hours   **(d)** 54 mile/h
27. (a) 210 miles   **(b)** 6 hours
    (c) 35 mile/h
28. (a) 2800 km   **(b)** 560 km/h

Miscellaneous Exercise 12

Section A

1. 7200   **2.** 1440   **3.** 2 h 13 min
4. (a) 1217   **(b)** 49 min   **(c)** 1227
5. Monday, 15th Sept
6. 55 mile/h
7. (a) 120 miles   **(b)** 2 hours   **(c)** 5 hours
    (d) 42 mile/h

Section B

1. (a) 2015 hours   **(b)** 30 min
2. 1815 hours
3. (a) $2\frac{1}{2}$ hours   **(b)** 50 km/h
4. (a) $1\frac{1}{2}$ hours   **(b)** 8 mile/h
5. (a) 1235 hours   **(b)** 50 mile/h

Multi-Choice Questions 12

1. B   **2.** A   **3.** B
4. B   **5.** D   **6.** D
7. B

Mental Test 12

1. 2 min   **2.** 300 s   **3.** 4 h
4. 180 min   **5.** 120 h   **6.** 7 h
7. 12 h 30 min   **8.** 50 mile/h   **9.** 150 miles
10. 8 h
11. (a) 10 min   **(b)** 25 min   **(c)** 30 min
12. 1000 from Wroughton; 35 min

Answers to Chapter 13

Exercise 13.1

1. $7y$ **2.** $5b - 7$ **3.** $5p + 3q$
4. xyz **5.** $8rs$ **6.** $\dfrac{2m}{n}$
7. $9a - 3b$ **8.** $\dfrac{pq}{r}$

Exercise 13.2

1. $3m$ **2.** $2x$ **3.** $4t$
4. $3r$ **5.** $5p$ **6.** $6q$
7. $3w$ **8.** $5y$

Exercise 13.3

1. 8   **2.** 5   **3.** 22
4. 15   **5.** 13   **6.** 4
7. 5   **8.** 3   **9.** 20
10. 23

Exercise 13.4

1. $8x$ **2.** $10y$ **3.** $9c$
4. $16a$ **5.** $8d$ **6.** $6y$
7. x **8.** $3q$ **9.** $5t$
10. $10m$ **11.** x **12.** $11q$
13. $3a + 4b - 5c$ **14.** $5x - 5y$
15. $2x + 3y$ **16.** $12y + 3$
17. $p + q$ **18.** $5a + 3b + 1$
19. $6a + 5b + -8c$ **20.** $x + y + z$

Exercise 13.5

1. b^2 **2.** c^3 **3.** n^4
4. p^3 **5.** q^6 **6.** r^5
7. 8   **8.** 9   **9.** 16
10. 13   **11.** 16   **12.** 35
13. -16 **14.** 21   **15.** 33

Exercise 13.6

1. 18   **2.** 12   **3.** 36
4. 32   **5.** 144   **6.** 320
7. 216   **8.** 384

Exercise 13.7

1. xy **2.** $2pq$ **3.** $6py$
4. $24pq$ **5.** $6abc$ **6.** $6xyz$
7. $15pn$ **8.** $30pnm$

Exercise 13.8

1. $3a + 3b$ **2.** $2x + 6y$ **3.** $10x - 5y$
4. $6x - 8y$ **5.** $3m + 6n$ **6.** $10x - 15y$
7. $19p - 12q$ **8.** $13a + 9b$ **9.** $11x + 9y$
10. $18p - 24q$ **11.** 22   **12.** 12
13. 8   **14.** 22   **15.** 5
16. 60   **17.** 10   **18.** 36
19. 26   **20.** 54

Exercise 13.9

1. $2x$ **2.** 3   **3.** 2
4. $2b$ **5.** $7a$ **6.** $2y$
7. $4c$ **8.** $2y$

Exercise 13.10

1. (a) 36   (b) $12n$
2. (a) 4800 miles   (b) $1200t$ miles
   (c) 40 miles   (d) $\dfrac{1200m}{60}$
3. $(1 + w + m)\,\text{kg}$
4. (a) 2   (b) $\dfrac{k}{x + y}$
5. (a) 40   (b) $\dfrac{8k}{x}$
6. $5y$ pence
7. $(5x + y + z)$ hours

Miscellaneous Exercise 13

Section A

1. $n + 10$ 2. $x^2 + 5$ 3. $3(x + y)$
4. (a) 1   (b) 56   (c) 56
5. (a) 21   (b) 42
6. (a) $3x + 2$ (b) $5x + 2y$ (c) $12xy$
   (d) $10x^2$ (e) $4y$
7. (a) $9x - y$ (b) $13p - 9q$

Section B

1. 28   2. $\frac{1}{4}$ 3. 74
4. 30   5. $54\,^\circ\text{F}$ 6. 270
7. 2

Multi-Choice Questions 13

1. B   2. A   3. B
4. C   5. D   6. A
7. C   8. B

Mental Test 13

1. $8b$ 2. $\dfrac{a}{b}$ 3. 6
4. 25   5. 8   6. $6a^2$
7. $8x$ 8. $3x + 6y$ 9. 3
10. $5x$

Answers to Chapter 14

Exercise 14.1

1. $x = 5$ 2. $x = 7$ 3. $x = 3$
4. $x = 2$ 5. $x = 8$ 6. $x = 3$
7. $x = 8$ 8. $x = 11$ 9. $x = 9$
10. $x = 14$

Exercise 14.2

1. $x = 6$ 2. $x = 4$ 3. $x = 3$
4. $x = 4$ 5. $x = 12$ 6. $x = 9$

7. $x = 10$ 8. $x = 9$ 9. $x = 2$
10. $x = 2$ 11. $x = 2$ 12. $x = 6$
13. $x = 8$ 14. $x = 15$ 15. $x = 21$
16. $x = 10$

Exercise 14.3

1. $x = 5$ 2. $x = 2$ 3. $x = 1$
4. $x = 2$ 5. $x = 4$ 6. $x = 5$
7. $x = 3$ 8. $x = 3$

Exercise 14.4

1. $x = 4$ 2. $x = 7$ 3. $x = 7$
4. $x = 15$ 5. $x = 6$ 6. $x = 3$
7. $x = 6$ 8. $x = 4$ 9. $x = 5$
10. $x = 4$ 11. $x = 3$ 12. $x = 3$
13. $x = 2$ 14. $x = 8$ 15. $x = 4$

Exercise 14.5

1. 195 min   2. 195 min   3. 26 cm
4. 20 km/h   5. $70\,^\circ\text{F}$ 6. £160
7. (a) £6500   (b) £1625
8. £650

Exercise 14.6

1. $2 < 5$ 2. $10\,\text{mm} = 1\,\text{cm}$
3. $£1 > 80\text{p}$ 4. $9 > 1$ 5. $-3 > -7$
6. $2 < 4$ 7. $\frac{1}{2} > \frac{1}{3}$ 8. $\frac{2}{3} > \frac{4}{9}$
9. $0.25 < \frac{1}{3}$ 10. $0.3 = 30\%$

Exercise 14.7

1.

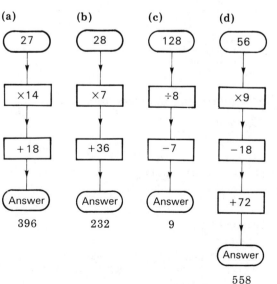

(a) 27 ×14 +18 Answer 396
(b) 28 ×7 +36 Answer 232
(c) 128 ÷8 −7 Answer 9
(d) 56 ×9 −18 +72 Answer

558

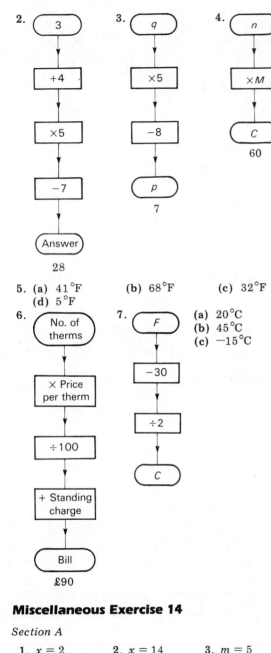

2. 3 → +4 → ×5 → −7 → Answer: 28

3. q → ×5 → −8 → p: 7

4. n → ×M → C: 60

5. (a) 41°F   (b) 68°F   (c) 32°F
(d) 5°F

6. No. of therms → × Price per therm → ÷100 → + Standing charge → Bill: £90

7. F → −30 → ÷2 → C
(a) 20°C
(b) 45°C
(c) −15°C

Miscellaneous Exercise 14

Section A

1. $x = 2$ 2. $x = 14$ 3. $m = 5$
4. $x = 5$ 5. £16
6. (a) $\frac{1}{2}$ is greater than $\frac{1}{4}$ (b) $100\,cm = 1\,m$
   (c) 50p is less than £1
7. $x = 4$ 8. 150 min

Section B

1. 40
2. (a) 80 min   (b) 117 min
3. (a) $5^2 - 3^2 = 4^2$ (b) $9^2 - 3^2 > 3^2$
   (c) $0.5^3 < 0.5^2$

4. 165   5. 8
6. (a) $x = 4$ (b) $x = 21$ (c) $x = 2$
7. 16

Multi-Choice Questions 14

1. A   2. D   3. D
4. C   5. B   6. B
7. C

Mental Test 14

1. $a = b + c$ 2. $p = xy$ 3. $x = 2$
4. 100 min   5. 45   6. 15
7. $<$ 8. $x = 8$

Answers to Chapter 15

Exercise 15.1

1. 240°   2. 135°   3. 72°
4. 324°   5. 54°   6. 46.5°
7. 57°   8. 170.3°   9. 36.3°
10. 8.2°

Exercise 15.2

1. A acute, B reflex, C acute, D obtuse,
   E reflex, F acute, G obtuse
2. (a) 60°   (b) 22.5°   (c) 180°
   (d) 135°   (e) 54°   (f) 31.5°
3. (a) 45°   (b) 120°   (c) 72°   (d) 60°
4. (a) 50°   (b) 133°   (c) 52°
   (d) 125°   (e) 270°
5. A 60°, acute B 145°, obtuse
   C 130°, obtuse D 240°, reflex
   E 82°, acute F 48°, acute
6.

Acute	Obtuse	Reflex
27°	165°	195°
75°	173°	220°
64°	126°	245°
82°	153°	280°
30°	173°	325°
15°	110°	340°
	98°	220°

7. (a) 33°   (b) 56°   (c) 153°   (d) B = 82°

Exercise 15.3

1. 103°
2. $x = 101°$, $y = 101°$
3. $m = 40°$, $n = 40°$, $p = 130°$, $q = 310°$
4. $y = 117°$
5. $x = 110°$, $y = 70°$, $z = 110°$
6. $x = 72°$, $y = 134°$
7. $a = 85°$
8. $u = 62°$, $v = 72°$, $w = 53°$, $x = 55°$,
   $y = 118°$, $z = 108°$

Multi-Choice Questions 15

1. C	2. C	3. A
4. B	5. B	6. C
7. C	8. C	9. D
10. D		

Mental Test 15

1. $\frac{1}{360}$ 2. $60°$ 3. $108°$
4. Acute 5. $30°$ 6. $50°$
7. (a) $140°$ (b) $130°$ (c) $270°$
8. $A = 50°, B = 130°$
9. $b = 110°$, $c = 110°$, $d = 70°$, $e = 70°$,
   $f = 110°$, $g = 70°$, $h = 110°$
10. $80°$
11. (a) $75°$ (b) $140°$ (c) $320°$

Answers to Chapter 16

Exercise 16.1

1.

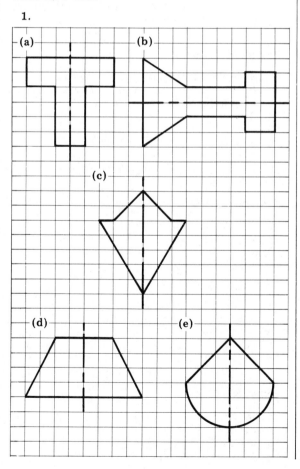

2.

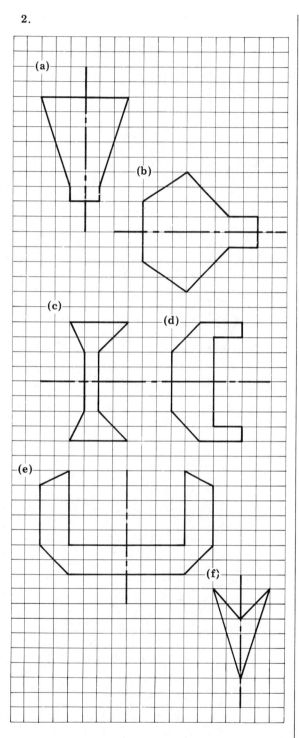

3.

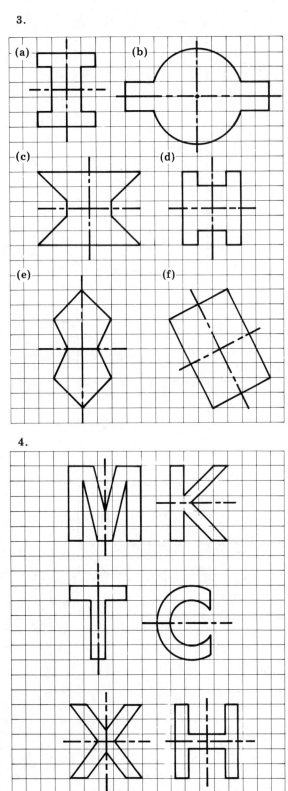

4.

Exercise 16.2

1. (a) 2	(b) 2	(c) Yes
2. (a) 5	(b) 5	(c) Yes
3. (a) 2	(b) 2	(c) Yes
4. (a) 8	(b) 8	(c) Yes
5. (a) 5	(b) 5	(c) No
6. (a) 2	(b) 2	(c) Yes
7. (a) 1	(b) 1	(c) No
8. (a) 3	(b) 3	(c) No

Answers to Chapter 17

Exercise 17.1

1. G, I, K   2. D, H, J
3. A, B, E, F, G, I, K   4. C, F, L
5. C, L

Exercise 17.2

1. A $85°$, B $61°$, C $20°$, D $62°$, E $13°$, F $90°$
2. (a) $x = 59°$, $y = 121°$ (b) $x = 85°$, $y = 43°$
   (c) $x = 11°$, $y = 47°$ (d) $x = 79°$, $y = 84°$
   (e) $x = 111°$, $y = 29°$

Exercise 17.3

1. $B = 71°$, AB = 5.9 cm, BC = 10.2 cm
2. $B = 19°$, AB = 32.6 cm, BC = 28 cm
3. $B = 25°$, $C = 39°$, BC = 8.5 cm
4. $A = 80°$, $C = 50°$, AC = 7.0 cm
5. $A = 56°$, $B = 83°$, $C = 41°$
6. $A = 52.4°$, $B = 67.6°$, AB = 6.6 cm
7. $A = 71°$, AB = 7.3 cm, AC = 8.9 cm
8. $A = 67°$, $C = 58°$, AC = 7.4 cm

Exercise 17.4

1. $y = 50°$, $x = 80°$ 2. $x = y = 70°$
3. $x = 70°$, $y = 20°$ 4. $x = y = 62°$
5. $x = y = 60°$ 6. $x = 30°$, $y = 60°$

Exercise 17.5

1. $x = 93°$ 2. $x = 72°$ 3. $x = 105°$
4. $w = 80°$, $x = 100°$, $y = 80°$
5. (a) Trapezium   (b) Rhombus   (c) Triangle
6. $w = 90°$, $x = 40°$, $y = 80°$
7. $a = 45°$, $b = 90°$

Exercise 17.6

1. (a) 4	(b) 3	(c) 4
(d) 6	(e) 3	(f) 4
(g) 5	(h) 4	(i) 4
(j) 8	(k) 3	

2. (a) $120°$ (b) $135°$ (c) $60°$
3. (a) ABFE   (b) BCGKJE
4. (a) semicircle   (b) sector
5. (a) 5   (b) 4   (c) 3
   (d) 8
6. $108°$
7. $198°$
8.

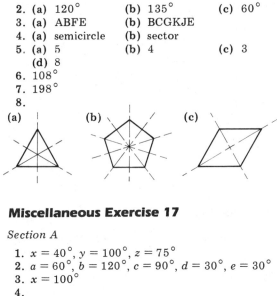

(a)   (b)   (c)

Miscellaneous Exercise 17

Section A

1. $x = 40°, y = 100°, z = 75°$
2. $a = 60°, b = 120°, c = 90°, d = 30°, e = 30°$
3. $x = 100°$
4.

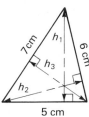

7 cm h_1 6 cm h_3 h_2 5 cm

     $h_1 = 2.9$ cm, $h_2 = 2.45$ cm, $h_3 = 2.1$ cm
5. (a) BCJF
   (b) AEGF, FGIH, BCGJ
   (c) ACJF, BEJF, EGIJ

Section B

1. 40 ft
2. $w = 118°, x = 155°, y = 25°$
3. $a = 70°, b = 110°$
4. $x = 78°, y = 120°$
5. (a) $x = 65°$ (b) $y = 25°$ (c) BAC $= 50°$
6. $x = 140°$
7. ACB $= 110°$, BAC $= 20°$, ABC $= 50°$
   The sum of the angles is $180°$.

Multi-Choice Questions 17

1. A   2. C   3. C
4. B   5. B   6. B
7. A

Mental Test 17

1. 6   2. rhombus, square
3. 2   4. sector
5. equilateral   6. $80°$
7. 4 cm   8. chord

Answers to Chapter 18

Exercise 18.1

1. (a) 13 cm   (b) 12 cm   (c) 21 cm
   (d) 21 cm   (e) 17 cm
2. 18 cm   3. 24 m   4. 19 in
5. (a) 12 ft   (b) 16 m   (c) 32 in
   (d) 24 yd   (e) 20 cm
6. 18 cm   7. 28 in   8. 32 m
9. 16 cm   10. 20 ft
11. (a) 24 cm   (b) 12 cm   (c) 8 cm
    (d) 16 cm   (e) 18 cm   (f) 14 cm
12. 18.84 cm   13. 6.28 ft   14. 66 in
15. 44 ft   16. 31.4 cm   17. 50.24 m
18. 352 cm   19. 176 in   20. 15.7 m
21. 44 ft   22. 220 cm   23. 75.36 cm
24. (a) 26 cm   (b) 24 cm   (c) 20 cm

Exercise 18.2

1. (a) 8 cm^2 (b) 15 cm^2 (c) 32 cm^2
   (d) 18 cm^2
2. 220 mm^2 3. 280 ft^2 4. 43.16 cm^2
5. 133 in^2 6. 62.32 ft^2 7. 468 yd^2
8. 4.81 m^2 9. 9600 in^2 10. 28.42 yd^2
11. 64 in^2 12. 900 cm^2 13. B and C
14. DEFG and HIJK
15. B and C
16. (a) 10 cm^2 (b) 12 cm^2
    (c) 18 cm^2 (d) 21 cm^2
17. 384
18. $15\,000$ cm^2 $(1.5$ m$^2)$
19. 1.04 m^2
20. £153

Exercise 18.3

1. 29 cm^2 2. 103 m^2 3. 56 in^2
4. 52 mm^2 5. 72 mm^2 6. 68 cm^2

Exercise 18.4

1. 14 cm^2 2. 12 cm^2 3. 18 ft^2
4. 20 m^2 5. 18 in^2 6. 24 cm^2
7. 66 in^2 8. 236 yd^2
9. (a) 56 m^2 (b) 30 m^2 (c) 26 m^2

Exercise 18.5

Section A

1. 30 cm^2 2. 10 in^2 3. 3 m^2
4. 9 ft^2 5. 17.5 cm^2 6. 20 cm^2
7. 27 cm^2 8. 24.64 cm^2 9. 10 in^2
10. 27.84 cm^2 11. 28 cm^2 12. 66 in^2
13. 16.91 cm^2 14. 4500 mm^2 15. Y

Miscellaneous Exercise 18

1. (a) 16 cm   (b) 13.11 cm^2
2. (a) 28 m   (b) 88 m
3. (a) 18 cm   (b) 18 cm^2
   (c) 4 cm, 5 cm
4. R and S
5. Area = 24 cm^2, Perimeter = 26 cm

Section B

1. 40 ft^2 2. 96 yd^2 3. 220 cm
4. (a) 24 cm^2,   (b) 38 cm^2 (c) 38 cm^2
5. Area = 71 cm^2; perimeter = 40 cm

Multi-Choice Questions 18

1. D   2. D   3. B
4. B   5. D

Mental Test 18

1. 20 cm   2. 24 cm^2 3. 36 cm
4. 81 cm^2 5. 18 cm^2 6. 12 cm
7. 25 cm   8. 17 cm

Answers to Chapter 19

Exercise 19.1

1. (a) 6   (b) 12   (c) 8
2. (a) 8   (b) 18   (c) 12
3. (a) 4   (b) 6   (c) 4
4. (a) 6   (b) 12   (c) 8
5. (a) 5   (b) 8   (c) 5
6. (a) 5   (b) 8   (c) 5

Exercise 19.2

1.
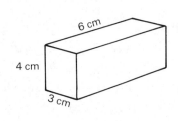

2.   3.
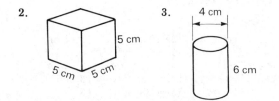

4.   5.
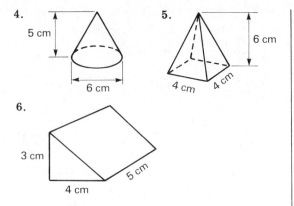

6.
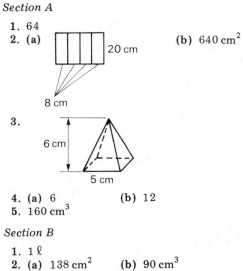

Exercise 19.3

6. (a) Cylinder   (b) Square pyramid
   (c) Rectangular cuboid
7. C and D   8. 88 cm^2
9. 108 in^2 10. 214 m^2

Exercise 19.4

1. 440 cm^3 2. 360 cm^3 3. 936 in^3
4. 300 ft^3 5. 306 cm^3

Exercise 19.5

1. 160 grams   2. 0.1625 ℓ   3. 17.5 mℓ
4. 16 m^2
5. (a) 1296 in^3 (b) 27

Miscellaneous Exercise 19

Section A

1. 64
2. (a)   (b) 640 cm^2

3.

4. (a) 6   (b) 12
5. 160 cm^3

Section B

1. 1 ℓ
2. (a) 138 cm^2 (b) 90 cm^3
3. (a) Cube   (b) Regular tetrahedron
4. 192 cm^3
5. Circumference = 22 cm

Multi-Choice Questions 19

1. D	2. D	3. D
4. D	5. D	6. B
7. D	8. D	

Mental Test 19

1. 8 cm^3
2. 30 cm^3
3. Hexagonal prism
4. 60 cm^3
5. 8
6. 6
7. 27 m^3
8. 10

Answers to Chapter 20

Exercise 20.1

1. 22 (to nearest mile)
2. 230 km
3. 1 : 200
4. 20
5. 40 m × 35 m
6. (a) 14 km (b) (i) 7.3 cm (ii) 35 km
7. (a) 29 miles (b) 36 miles (c) 35 miles
8. (a) 15 (b) 165 km

Exercise 20.2

1. (a) 15 m by 12.5 m (b) 187.5 m^2
2. (a) 40 m × 16 m (b) 640 m^2

3.
Name of room	Length (m)	Width (m)	Area (m²)
Lounge	5	5	25
Dining room	5	3	15
Kitchen	5	3	15
Hall	5	2	10
Bathroom	4	3	12

4.
Name of room	Length (m)	Width (m)	Area (m²)
Lounge	7	6	42
Bedroom	5	4	20
Hall	3	2	6
Kitchen	4	2	8
Bathroom	4	3	12

5. (a) 3 (b) 3
   (c) 25.5 m, 13.5 m (d) 344.25 m^2

Exercise 20.3

5. (a) $48°$ (b) $75°$ (c) $57°$
   (d) $132°$ (e) $296°$ (f) $280°$
   (g) $153°$ (h) $285°$
6. (a) $50°$ (b) $20°$ (c) $80°$
   (d) $30°$ (e) $60°$ (f) $40°$

Exercise 20.4

1. (a) 3 cm (b) 4.8 cm (c) 5.6 cm
   (d) 3.7 cm
2. CD = 3 cm, EF = 6 cm, GH = 4.5 cm
   KL = 8 cm, MN = 2 cm, PQ = 10 cm
6. 6.2 cm, 2.5 cm
7. BD = 7.25 cm, AC = 9.84 cm

Exercise 20.5

6. N30°E 7. N80°E 8. S40°E
9. S25°W 10. N45°W

Exercise 20.6

6. $035°$ 7. $080°$ 8. $135°$
9. $225°$ 10. $330°$

Exercise 20.7

1. (a) $323°$ (b) $190°$
2. $335°$
3. $248°$
4. (a) $300°$ (b) $75°$
5. (a) Oak tree (b) $030°$
   (c) About $300°$

Miscellaneous Exercise 20

Section A

1. (a) $225°$ (b) $315°$
2. (a) 12.5 km (b) $068°$
   (c) 13.75 km (d) $113°$
4. 3 km

5.
Name of room	Length (m)	Width (m)
Lounge	7	6
Dining room	7	4
Bedroom 1	7	4
Bedroom 2	4	3
Hall	11	2
Bathroom	3	3
W.C.	2	2
Kitchen	5	3

Section B

2. (a) 130 miles (b) $62°$ (c) 25 miles
3. BW = 3 cm, AX = 8 cm
4. 173 ft
5. (b) (i) $122.5°$ (ii) 10.7 miles

Multi-Choice Questions 20

1. C	2. C	3. A
4. C	5. D	6. C
7. C	8. B	

Mental Test 20

1. (a) 1 m   (b) 4 m
   (c) 0.5 m
2. (a) 150 km   (b) 10 km
3. (a) 2 cm   (b) 6 cm
4. (a) 1.2 cm   (b) 1.5 cm
   (c) 1.8 cm   (d) 2.7 cm
   (a) 12 km   (b) 15 km
   (c) 18 km   (d) 27 km
5. (a) 135°   (b) 045°
6. (a) 60°   (b) 120°
   (c) 210°
7. 220°
8. (a) 120°   (b) 315°
9. 6 cm
10. 60°

Answers to Chapter 21

Exercise 21.1

1. (a) High 12.12; low 6.12, 18.38
   (b) High 6.41; low 0.37, 12.50
2. (a) 1450   (b) 0830
   (c) 45 min   (d) 1356
   (e) 1150, 50 min
3. (a) 111   (b) 112
   (c) 115
4. (a) Montreal, Milan, Rome, Athens, Bangkok
   (b) London, Rome, Nairobi
   (c) Rome, Teheran, Bombay, Singapore, Sydney
   (d) Chicago, Montreal, Milan, Rome, Athens, Bangkok, Hong Kong, Tokyo
5. (a) City centre, Longlevens, Churchdown; No. 48
   (b) City centre, Quedgeley, Hardwicke; No. 53
   (c) City centre, Hucclecote, Brockworth, Witcombe; No. 50
6. (a) (i) West Auckland, Darlington, Houghton; West Auckland, Bishop Auckland, Shildon, Darlington, Houghton; West Auckland, Bishop Auckland, Rushyford, Coatham, Houghton
       (ii) West Auckland, Darlington, Houghton
       (iii) West Auckland, Bishop Auckland, Rushyford, Coatham, Darlington, Houghton
   (b) 7 miles
   (c) Rushyford, Long Newton, Stockton; 9 miles; Rushyford, Coatham, Houghton, Long Newton, Stockton; 11 miles; difference = 2 miles
7. (a) 12.40, 51   (b) 6.36, 7.34, 58 min
   (c) 51   (d) 55 min, 9 min
   (e) Yes

Exercise 21.2

1. (a) 4.15 pm, X2
   (b) 6.48 arrives 7.36
       8.05 arrives 8.53
       3.15 arrives 4.10
   (c) 12 min
2. (a) 28.3°C   (b) 45.9°C   (c) 131°F
   (d) 155.7°F
3. (a) 488 miles (b) 130 miles   (c) 63 miles
4. (a) 14.3%   (b) 1:14
5. (a) 53p   (b) £1.28   (c) £3.38
6. (a) 28p   (b) 46p   (c) 66p
7. (a) 24.13 km   (b) 30 mile/h
   (c) 109.42 km   (d) 914.12 km/h
8. (a) 37.8°C   (b) 176°F   (c) 17.8°C
   (d) 118°F
9. (a) 1.5 kg   (b) 258 kg   (c) 4.0 lb
10. (a) 1.54 kg/cm^2 (b) 30 lb/in^2
    (c) 43 lb/in^2

Exercise 21.3

1. (a) 40   (b) 11

2.

0.6 cm	160	Other
1.5 cm	420	Private car
5 cm		
2.9 cm	780	Bus

3.

3.1 cm	£95	Food & drink
1.3 cm	£42	Housing
0.9 cm	£29	Transport
1.1 cm	£33	Clothing
1.6 cm	£51	Other

8 cm

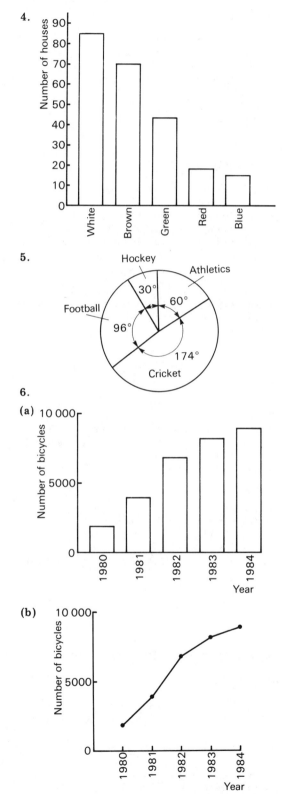

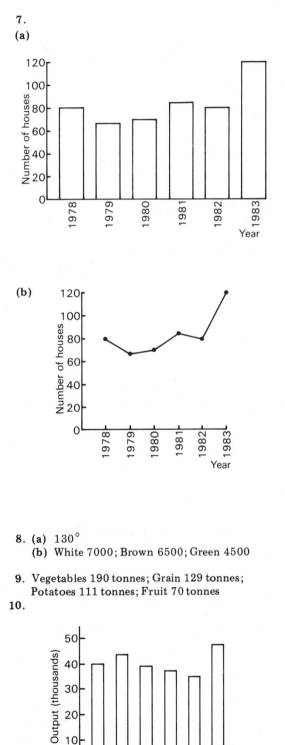

8. (a) 130°
   (b) White 7000; Brown 6500; Green 4500

9. Vegetables 190 tonnes; Grain 129 tonnes;
   Potatoes 111 tonnes; Fruit 70 tonnes

10.

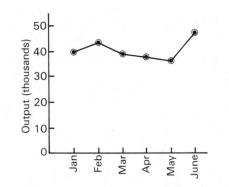

11. Food £22.22, rent £12.50, other £15.28

Multi-Choice Questions 21

1. A	2. B	3. B
4. B	5. C	6. B
7. A	8. B	9. D

Mental Test 21

1. (a) 110 miles   (b) 240 miles
   (c) 110 miles   (d) 72 miles
2. (a) −40°F   (b) 68°C   (c) 38°C
3. (a) 1600   (b) 880   (c) 460
4. 90, 150
5. (a) £60 000   (b) £200 000
6. (a) 1978   (b) 50 000   (c) 35 000
   (d) 200 000
7. 72°

Answers to Chapter 22

Exercise 22.1

1. $A(1, 1)$ $B(2, 3)$ $C(3, 4)$ $D(4, 5)$ $E(5, 2)$ $F(5, 1)$
2. $R(2, 25)$ $S(4, 10)$ $T(8, 20)$ $U(6, 5)$ $V(10, 15)$
4. (a) $AB = 4$; $BC = 2$ (b) $AC = 4.5$
   (c) 8 square units
5. (a) Trapezium   (b) 3 units

Exercise 22.2

1. (a) 19 m   (b) 5 s
2. (a) 5 years   (b) 122 cm
3. (a) 32 years   (b) 140 cm
4. (a) 49 cm^2 (b) 5 cm
6. (a) Year 4   (b) Year 2
7. (a) £5   (b) Standing charge
   (c) £21   (d) 2 p
8. (a) Inconsistent scale   (b) No scale

Exercise 22.3

1. (a) 80 marks   (b) £35
2. (a) £10   (b) 40 p   (c) £110
   (d) 200 therms
3. (c) (i) 200 mm   (ii) 16 inches
4. (b) (i) 64°F   (ii) 27 °C
5. (c) (i) 75 francs   (ii) £45
6. (c) (i) £39   (ii) 400

Exercise 22.4

1. (a) 20 miles   (b) 3 hours
   (c) 10 mile/h
2. (a) 200 miles   (b) $\frac{1}{2}$ hour
   (c) 50 mile/h   (d) 50 mile/h
3. (a) 120 km   (b) 40 km/h   (c) $\frac{3}{4}$ hour
   (d) 96 km/h
4. (a) $1\frac{1}{2}$ hours   (b) 60 miles
   (c) $1\frac{1}{2}$ hours   (d) 60 mile/h
5. (a) 132 km   (b) 60 km/h   (c) 30 km
   (d) 6 km/h   (e) 22 km/h

Miscellaneous Exercise 22

Section A

1. (a) D3   (b) E9   (c) A8
2. (a) $W(1, 2)$ (c) $T(4, 1)$
3. (a) 6 ounces   (b) 110 grams
4. (a) 1224 hours   (b) 3 hours 24 min
   (c) 1 hour   (d) 65 miles
5. (a) 19.5 amperes   (b) 20 mm

Section B

1. 3.2 min
2. (a) 110 miles   (b) $\frac{3}{4}$ hour
   (c) 40 mile/h
3. (a) 2.8 units   (b) 2 square units
5. (a) 24 ft   (b) 25 ft   (c) 11 ft

Answers to Chapter 23

Exercise 23.1

1. (a) $\begin{pmatrix} 4 \\ 8 \end{pmatrix}$ (b) $\begin{pmatrix} 8 \\ 4 \end{pmatrix}$ (c) $\begin{pmatrix} 0 \\ 7 \end{pmatrix}$

2. $\begin{pmatrix} 15 \\ 15 \end{pmatrix}$

3. $A'(6, 8)$, $B'(8, 8)$, $C'(8, 7)$, $D'(6, 7)$
4. (b) $A'(4, 8)$, $B'(8, 7)$

5.

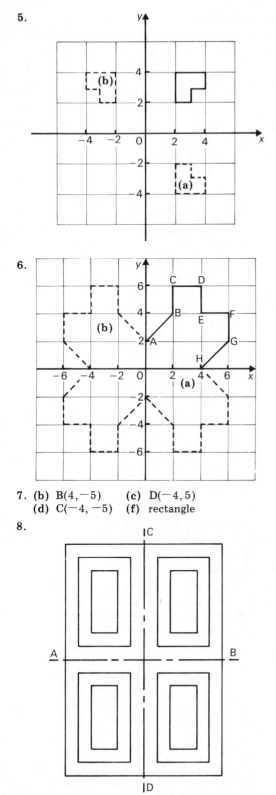

6.

7. (b) B(4, −5) **(c)** D(−4, 5)
   (d) C(−4, −5) **(f)** rectangle

8.

9.

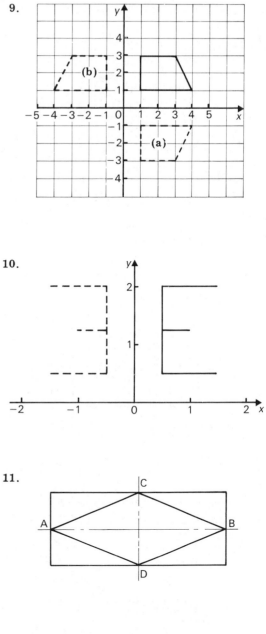

10.

11.

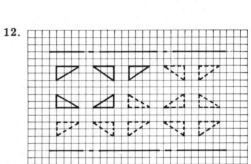

12.

13. (c) P′(−3, 4)   **(b)** P″(−4, −3)

14.

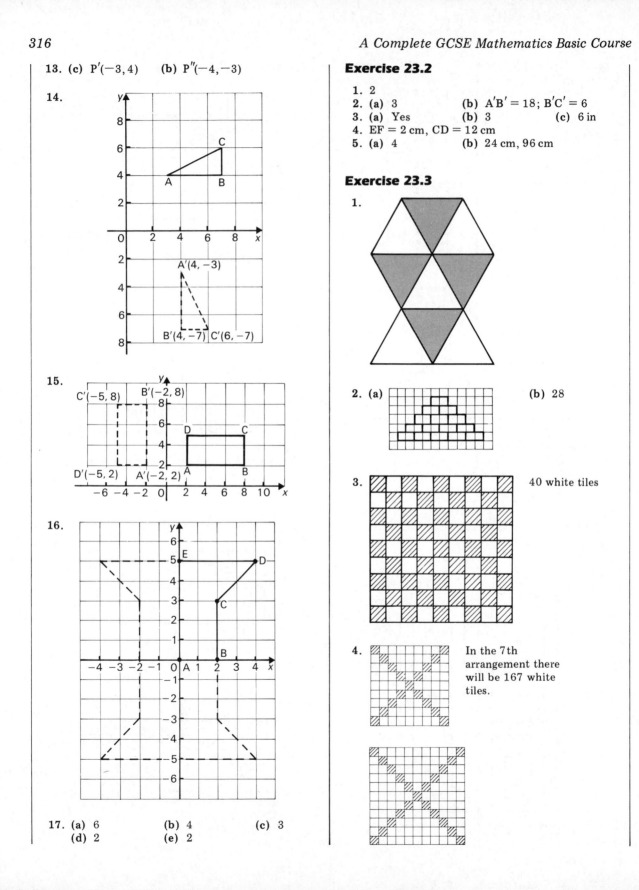

15.

16.

17. (a) 6   **(b)** 4   **(c)** 3
(d) 2   **(e)** 2

Exercise 23.2

1. 2
2. (a) 3   **(b)** A′B′ = 18; B′C′ = 6
3. (a) Yes   **(b)** 3   **(c)** 6 in
4. EF = 2 cm, CD = 12 cm
5. (a) 4   **(b)** 24 cm, 96 cm

Exercise 23.3

1.

2. (a)   **(b)** 28

3.   40 white tiles

4.   In the 7th
            arrangement there
            will be 167 white
            tiles.

5. (a)

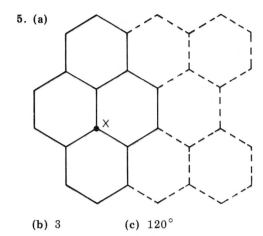

(b) 3   **(c)** 120°

Miscellaneous Exercise 23

Section A

1.

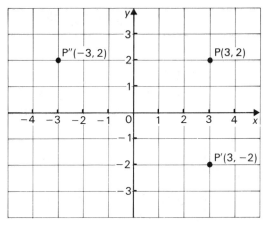

2.

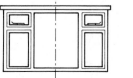
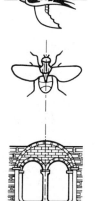

3.

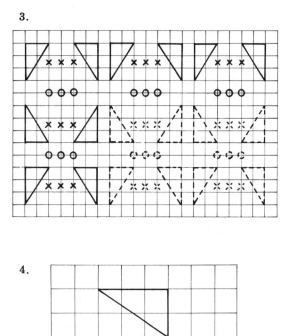

4.

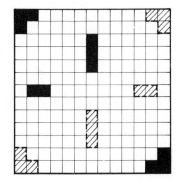

5. (a) 3   **(b)** 24 in   **(c)** 4 in²
(d) 36 in²

Section B

1.

2.

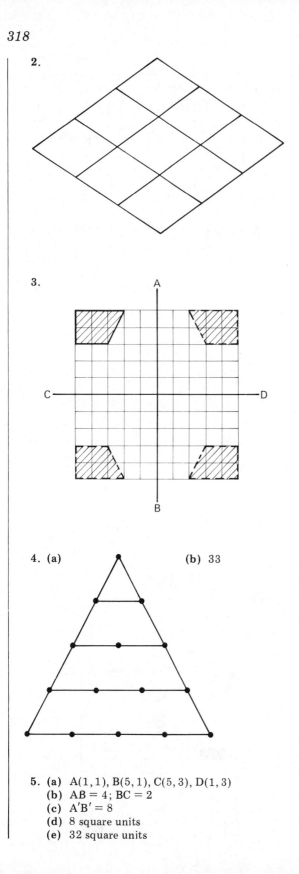

3.

4. (a)   **(b)** 33

5. (a) A(1, 1), B(5, 1), C(5, 3), D(1, 3)
   (b) AB = 4; BC = 2
   (c) A'B' = 8
   (d) 8 square units
   (e) 32 square units

Answers to Chapter 24

Exercise 24.1

1.

Time (seconds)	Frequency
24	1
25	2
26	4
27	3
28	1
29	2
30	1
31	3
32	3

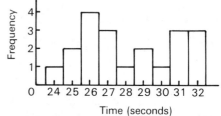

2. (a)

No. of goals	0	1	2	3	4	5
Frequency	7	13	5	1	2	2

(b)

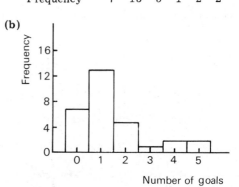

3.

Height (cm)	Frequency
150	2
151	3
152	4
153	5
154	6
155	7
156	8
157	6
158	5
159	4
160	3
161	1

4.

Age (years)	14	15	16	17	18
Frequency	10	9	5	4	2

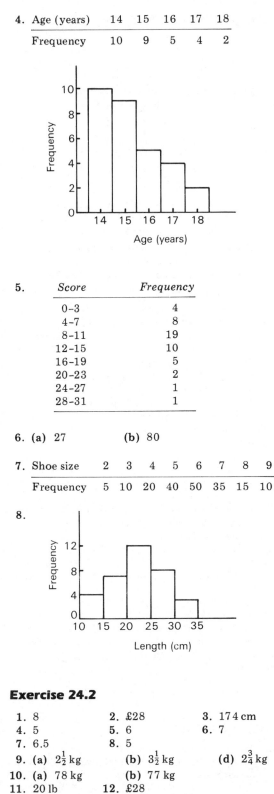

5.

Score	Frequency
0–3	4
4–7	8
8–11	19
12–15	10
16–19	5
20–23	2
24–27	1
28–31	1

6. (a) 27   **(b)** 80

7.

Shoe size	2	3	4	5	6	7	8	9
Frequency	5	10	20	40	50	35	15	10

8.

Exercise 24.2

1. 8   **2.** £28   **3.** 174 cm
4. 5   **5.** 6   **6.** 7
7. 6.5   **8.** 5
9. (a) $2\frac{1}{2}$ kg   **(b)** $3\frac{1}{2}$ kg   **(d)** $2\frac{3}{4}$ kg
10. (a) 78 kg   **(b)** 77 kg
11. 20 lb   **12.** £28

Exercise 24.3

1. (a) $\frac{1}{6}$ **(b)** $\frac{1}{2}$ **(c)** $\frac{1}{2}$
2. (a) $\frac{1}{8}$ **(b)** $\frac{1}{4}$
3. (a) $\frac{1}{52}$ **(b)** $\frac{1}{13}$ **(c)** $\frac{1}{2}$
    (d) $\frac{1}{4}$
4. (a) $\frac{3}{10}$ **(b)** $\frac{1}{2}$ **(c)** $\frac{1}{5}$
    (d) $\frac{4}{5}$
5. 0.4   **6.** $\frac{3}{7}$ **7.** $\frac{1}{3}$
8. 0.6   **9.** $\frac{1}{100}$ **10.** $\frac{3}{50}$

Miscellaneous Exercise 24

Section A

1. (a)

Score	3	4	5	6	7	8	9
Frequency	1	3	4	6	7	5	4

(b)

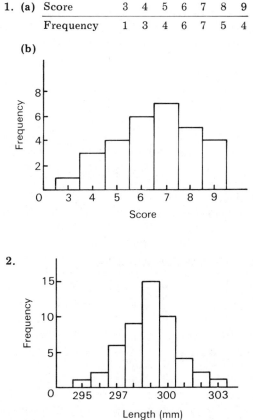

2.

3.

Length (mm)	Frequency
150	10
151	25
152	35
153	45
154	55
155	60
156	70
157	55
158	40
159	30
160	15
161	5

4. (a) $10\,^{\circ}$C   (b) $4\,^{\circ}$C
5. 2134   6. 0.7   7. Size 4

Section B

1. 61
2. (a) $\frac{1}{3}$ (b) $\frac{7}{12}$
3. (a)

Length (mm)	Frequency
120–129	4
130–139	7
140–149	14
150–159	8
160–169	5
170–179	2

(b)

Frequency vs Length (mm) histogram with bars: 120–130: 4, 130–140: 7, 140–150: 14, 150–160: 8, 160–170: 5, 170–180: 2.

4. 0.86
5. (a) 137 cm   (b) 140 cm

Multi-Choice Questions 24

1. B   2. C   3. C
4. C   5. A   6. B
7. C   8. C

Mental Test 24

1. 7   2. £15   3. 7
4. 2   5. 2   6. $\frac{3}{10}$
7. $\frac{1}{3}$ 8. 7 mm

Answers to Chapter 25

Investigations

2. 68 triangles; 58 quadrilaterals.
   Many 7-sided polygons are possible:

3. There could be plates before Number 6. The next plate may be 24 or 27. The numbers relate to the perimeters of the plates.

   The first triangular plate is probably 2, but may be 3.2. The last plate may be 28 or 94. If fractions are included, it may be 156.25.

   The three series are:

   2, 4, 8, 10, 14, 16, 20, 22, 26, 28
   2, 4, 8, 10, 20, 22, 34, 46, 92, 94
   3.2, 4, 8, 10, 20, 25, 50, 62.5, 125, 156.25
   There is a further possibility in which the first is 3.2 and the tenth is 38.593 75.
   Each number is double the sum of the triangle's base and height.
   The most practical solutions have the tenth numbers of 28 and 94.

4. Assuming coordinates $(a, b), (c, d)$:
   Distance: $(c - a) + (d - b)$
   Squares: $(c - a)(d - b)$
   The same number of lines do not necessarily produce the same number of squares, e.g. A to E and A to D.

6. The number of operations is double the sum of the two capacities, except when there is a common factor.

7. Strips for areas 1 to 25: 2, 2, 3, 3, 4, 4, 6, 4, 4, 5, 8, 5, 10, 6, 6, 6, 12, 6, 14, 6, 7, 9, 16, 7, 7.
   Prime number areas should be studied and also later multiples.

9. A 5×5 box will hold all the blocks.
   Heights of 15, 19 and 23 are impossible unless extra blocks are bought. For 36 cm and many other heights, there are several methods. The binary system is the key.

10. Patterns for odd and even numbers are interesting. Rules emerge for squares and cubes. Divisibility rules also appear for 3, 6 and 9.

Further Studies

3. Order of letter frequencies:
   ETOANIRSHDLCWUMFYGPBVKXQJZ
   Message reads: THE ENEMY TRIED TO TAKE
                    OUR FORWARD POSITION.

WE SHALL TRY TO ENTER
THE TOWN BY THE ROAD
FROM THE EAST AND WE
SHALL NEED SOME GOOD
COVER FOR AT LEAST TEN
HOURS. PLEASE ARRANGE
TO RECEIVE THE ADVANCE
PATROL BY TEN THIS
EVENING.

Problems

1. The key lies in digits with a sum of 9. There are twenty signs. When lengthened to 199, eight signs have only two different digits.

2. Four with 1 in the middle, three with 3 in the middle, four with 5 in the middle, three with 7 in the middle, four with 9 in the middle. Numbers 0 to 8: centres of 0, 2, 4, 6 and 8. Even numbers: centres of 2, 6, 10, 14 and 18.

3. (a) Treble 20, treble 20, bull (50): total 170.
   (b) There are many methods. Here is one more possibility:
       Treble 20, treble 20, treble 20,
       Treble 20, treble 20, treble 20,
       Treble 20, treble 17, double 15.
   (c) The lowest is 23.
   (d) Treble 17 and bull (50).

4. If the answer is even, the first number must be odd. If the answer is odd, the first number must be even.

5. (a)

5	1	10	3
8			9
6	2	4	7

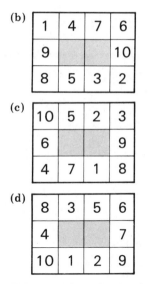

(b)

1	4	7	6
9			10
8	5	3	2

(c)

10	5	2	3
6			9
4	7	1	8

(d)

8	3	5	6
4			7
10	1	2	9

6. Fill the 11 from 24; fill the 5 from the remaining 13; pour the 5 into the 13; fill the 13 from the 11; fill the 5 from the 13; pour the 5 into the 11.

7. (a) Move the numbers in this sequence:
       1, 2, 6, 5, 3, 1, 2, 6, 5, 3, 1, 2, 4, 8, 7,
       1, 2, 4, 8, 7, 4, 5, 6
   (b) Move the letters in this sequence:
       B, E, D, B, A, C, B, D, E, A, D, B, C, D,
       A, E, B

8. Move in this order: 8, 1, 8, 2, 4, 3, 2, 7, 6, 7

9. Order of the cards: K A K K A A A

10. 4 to B, 3 to C, 4 to C, 2 to B, 4 to A, 3 to B, 4 to B, 1 to C, 4 to C, 3 to A, 4 to A, 2 to C, 4 to B, 3 to C, 4 to C.

    General solution: If pegs are assumed to be in cyclic order, the odd numbers rotate one way, even numbers the other. On every alternate move, the smallest disc is transferred to another peg.

Index